PUBLIC WATERS

Green River Lakes, Wyoming headwaters of the Colorado River.
Courtesy of Rita Donham / Wyoming Aero Photo.

PUBLIC WATERS

LESSONS FROM WYOMING FOR THE AMERICAN WEST

ANNE MACKINNON

University of New Mexico Press | Albuquerque

© 2021 by the University of New Mexico Press
All rights reserved. Published 2021
Printed in the United States of America

ISBN 978-0-8263-6241-4 (paper)
ISBN 978-0-8263-6242-1 (electronic)

Library of Congress Cataloging-in-Publication Data is on file with the Library of Congress.

Cover photograph courtesy of Rita Donham / Wyoming Aero Photo.
Maps by Rachel Savage, Casper, WY
Designed by Felicia Cedillos
Composed in Adobe Garamond Pro 10.25/14.25

For Cy and Wig

CONTENTS

ACKNOWLEDGMENTS

I have been lucky to interview everyone who served as Wyoming state engineer since 1963: Floyd Bishop, George Christopherson, Jeff Fassett, Dick Stockton, Pat Tyrrell, and Greg Lanning; and almost every Wind River tribal water engineer: Kate Vandemoer, Wold Mesghinna, Gary Collins and Mitch Cottenoir. I've been welcomed to meetings by members of their respective boards, including Earl Michael, Bill Jones, Randy Tullis, Brian Pugsley, Mike Whitaker, Carmine Loguidice, Dave Schroeder, Craig Cooper, Loren Smith, John Teichert, Jade Henderson, and Kevin Payne on the Wyoming Board of Control; and John Stoll, Sandra C'Bearing, Dick Baldes, Scott Ratliff, Leslie Shakespeare, Merl Glick, Jeremy Washakie, Ron Givens, Kenneth Trosper, Pat Lawson, Howard Brown, and Garrett Goggles on the Tribal Water Board. Several have taken time to talk or tour ditches with me, plus review book drafts. Jeff Fassett, former Wyoming Water Development director Mike Purcell, and former Bureau of Reclamation Wyoming manager John Lawson spent countless hours explaining water to me.

Thanks to academic colleagues Stephen White and Dan Meltzer of Harvard University, and early funding from the Mark DeWolfe Howe Fund at Harvard Law School; to Kristi Hansen, Ginger Paige, Harold Bergman, and Nicole Korfanta at the University of Wyoming; in Germany, to Konrad Hagedorn of Humboldt University, Berlin, Insa Theesfeld of Martin Luther University at Halle, Andreas Thiel, Uta Schuchman; and to Elinor and Vincent Ostrom at Indiana University. And to colleagues at the *Casper Star-Tribune*, WyoFile, and WyoHistory.org who taught me a lot about Wyoming, especially Dan Neal, Joan Barron, Paul Krza, Katharine Collins, Tom Rea, and Nadia White.

Thanks to those who, in addition to the state and tribal officials above, were always ready to talk water: Dan Budd, Kate Fox, Larry Wolfe, Bern Hinckley,

John Shields, Steve Wolff, Tom Annear, Chris Brown, Baptiste Weed, Berthenia Crocker, Geoff O'Gara, Ernie Niemi, Jodee Pring, Barry Lawrence, Jane Caton, Sue Lowry, Marion Yoder, Albert Sommers, Randy Bolgiano, Ron Vore, Jodee Pring, Jason Baldes, Kim Cannon, Dave Palmerlee, Keith Burron, Arch McClintock, John Jackson, Jennifer Gimbel, Jon Wade, Larry McDonnell, Ramsey Kropf, Dennis Cook, Jim Jacobs, and Pete Ramirez. Thanks also to those who, in addition to several people above, read book drafts and offered helpful comments: Ted Ballard, Jim Boddy, Mike Cassity, Carol Rose, and Emlen Hall. And to the staff at the University of New Mexico Press: Clark Whitehorn, Sonia Dickey, Katherine White, and James Ayers. Librarians at the Wyoming State Library; the American Heritage Center at the University of Wyoming; Cindy Smith and Suzi Taylor at the Wyoming State Archives; Laurie Lye at Casper College Library; Lida Volin at Natrona County Library; the county libraries in Buffalo and Wheatland; the Homesteader Museum in Powell; the Buffalo Bill Museum in Cody; and the archives of the Midvale and Shoshone Irrigation Districts were unfailingly helpful. Many thanks to my indefatigable citations and format editor, Josh Moro of Caldera Communications.

Finally, friends and family: MacKinnons Cecil, Steve, and Peter and their families; Petre Osmanliev, Bernie Barlow, Sarah Gorin, George Jones, Connie Wilbert, Reed Zars, Megan Hayes, Tom Arrison, Marion Yoder, Ann Rochelle, Jane Ifland, Will Robinson, Mary Katherman, Eleanor Bliss, Betsy Dodd, Nancy Ranney, and Betsy Weiss. And finally, I am ever grateful to my husband, Jon Huss, and son, Ted Huss, who staunchly supported my every foray into water matters.

All errors are mine.

INTRODUCTION

Believing fully in the doctrine that public waters should remain
a public property, and that to grant private perpetual rights is to sacrifice
the welfare of future generations.

—ELWOOD MEAD, WYOMING STATE ENGINEER, 1890–1899[1]

Heading to Yellowstone past Buffalo Bill Reservoir. Courtesy of Park County Archives.

UNCERTAIN WATER SUPPLY IS one of the most unsettling consequences of climate change and rapid population growth, the legacies of the twentieth century. Finding a way to govern the use of water so that this liquid resource can continue to sustain us and the world we live in is one of the major challenges of our time.

This book tells the story of how water has been used, shared, and sometimes used until it was gone, under a system of rights to water in one state in the western US: the high, cold, and dry state of Wyoming, where mountains gather the snowpack that feeds rivers crossing the dry lands below. It is the story of the rules people have made about using water, in a secluded place where the story is easy to trace.

Water is not to be taken for granted. It moves among us all, coursing through an endless cycle yet possible to deplete at any point on land where people might try to use it. And it can be hard to preclude people from using it. So one person's water use affects the water use of others. It creates interdependence among people. It creates interdependence between people and the natural world. How water is used affects the well-being of whole societies and landscapes. What we deal with, therefore, are fundamentally public waters. For the welfare of future generations, and of our own, those waters demand governance.[2]

What has become of public waters? Over a century ago, the arid states of the US West, fearing monopoly private control over all access to water, adopted the idea that water resources should be public property. Then, to attract people and development, these states worked to put the use of public waters into private hands. State governments gave first comers, at no charge, a first right to use water, even when very little water was available; individuals could keep that right—perpetually—if they regularly put the water to use. Not surprisingly, water users and many courts came to see a water right as simply a matter of private property.[3]

Yet there remain odd twists to those private rights to use water. Not only can a water right be lost for non-use, but it can also be difficult to change to a new use or to use in a different place. Such changes typically shrink the amount of water that can be used. Private water rights are an "incomplete" property right, say the law treatises. That is because of the powers over water that remain in public hands—such as the crucial power to restrict changes to a new use or a new place. There is public oversight, with veto power, to ensure that the interdependency and overall social welfare issues inherent in water use are respected.[4]

The idea of public property in water thus manages to glimmer through western water law. By giving state governments license to guide their water systems into new paths, that idea may help western states to refresh their water governance. There are examples of that license in action now, as state water management agencies work with entities of all kinds—other states, water right holders, and groups of people who don't have water rights—to find ways that water can serve all their needs.

The twists and turns of water law have usually been traced in prominent western states—Colorado, California, or Arizona. There, however, the development of the water governance system has been overlain by the growth of substantial and often corporate agriculture, a largely urban population, major hydropower generation and related industries, and considerable investment in massive infrastructure to turn wild western rivers into steady water suppliers.[5]

Wyoming, by contrast, is home to neither corporate agriculture nor genuine urban centers; its dominant industry has been energy, largely shipping raw materials out of state, plus running a few hydro- and coal-fired power plants and mineral-processing plants that use water locally. There are a few big federal reservoirs producing hydropower and irrigation water, but they serve only a small number of people on irrigated lands. The population is hardly urban.[6]

Wyoming is therefore ideal territory for documenting how western water law developed independent of the pressures of urban or industrial growth. Wyoming shows how water law is likely to be animated, everywhere, by local forces easily overlooked in the shadow of urban growth and massive water projects. Local experience with local terrain creates ideas about water that still play a role in water management.

In Wyoming, 130 years ago, there was an engineer who tried to set down a rational system, intending to improve upon California and Colorado water law. Committed to "public waters," he sought to achieve a model system that brought new life and power to the idea of water as public property. But an ideal water code cannot simply be dropped down on a western landscape. The terrain asks for a system that responds to it. It requires water governance that reflects longtime interaction among people, land, and water, incorporates a learning process, and embodies the capacity to adapt to change upon change.[7]

Wyoming water law therefore morphed considerably over time, as the people using it encountered the high, cold, and dry landscape in which they sought to live. People experimented in shaping a water-law system that sustained them in that place. At first, they simply wanted their own secure, individual rights to water; they then felt their way to an understanding of the mutual interdependence that water use creates. They experimented and built on their experience of a harsh landscape. Water governance became the work of the community, with some powers over water in private hands and other powers held by the state government, as representative of the public—the relative extent of the powers of the private users or the public shifting back and forth over time. In the web of rules for water that the community devised to make it possible to live where they did, the concept of public property in water glints through.

By the end of its first hundred years, however, Wyoming's system for governing water had become somewhat rigid. Local society and politics—and accordingly, the water-law system—responded stingily to two demands: the water needs of native people asserting their sovereignty, and the attention that other residents sought for the ecological needs of rivers throughout the state. The effect of climate change, particularly on major interstate rivers, where impacts can ripple back up to the headwater states like Wyoming, has since added to the issues that the state water-law system must address. Now the challenge for Wyoming is refreshing its water governance system. There is reason to believe it can do so, since water governance has been integral and responsive to people and place there for so long—and because the concept of public waters is still alive.

What has happened in Wyoming with water is a distilled version of what has been at play at the local level in other, bigger, more complicated places in the

US West. Those other states share with Wyoming the evolution of local water governance in response to people and place; they share its potential for revitalized water management that draws on local experience.

Wyoming covers many square miles but hosts few people, and almost every moment of its water management history has been meticulously documented. The story is not too long to tell or to remember. It is worth a look.

THE SETTING

Red Canyon. Courtesy Merrill J. Mattes Papers,
American Heritage Center, University of Wyoming.

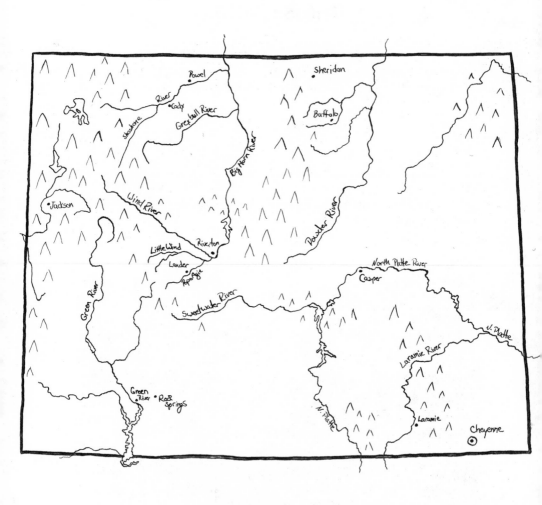

WYOMING IS A HIGH, cold, dry land, encompassing nearly ninety-eight thousand square miles—more than twice the size of Ohio or Pennsylvania. Agricultural lands occupy the state's lowest elevations, but even those average some three to four thousand feet above sea level. More than one-third of the state has an elevation over seven thousand feet, and there are about fifty mountain peaks thirteen thousand feet or higher. Even in the lowest places, there are typically only about 125 frost-free days per year. In some parts of the state, people say the last snow is on June 30 and the first is on July 1. They're only half-joking. Growing tomatoes in the backyard requires considerable thought and care.

Rain or snow amounts to about twelve to sixteen inches a year in the eastern part of the state (contrasting with forty inches or more in Ohio or Pennsylvania). Precipitation is lower still in some western Wyoming basins, but much higher in the mountains. Trees grow only on the mountains, right along the rivers, or in places where people have spent serious time nurturing them. The rest is sagebrush plains and some grasslands in the eastern part of the state.

The population has always been small. Ancestors of some native people lived there for thousands of years, and some native groups came there centuries before the 1870s, when white settlers and their armies launched the last assaults to take control of the land. In 1890, when territorial leaders made a bid for Wyoming to be recognized as an independent state in the US federal system, Wyoming had sixty thousand new settlers. That same year, Ohio had about 3.5 million people and Pennsylvania 5 million. By 2017, Wyoming's population managed to grow to just ten times its 1890 size. Those people are in small towns and cities far from one another, and on ranches and farms. The largest city in the state today has a population just a little more than the 1890 population for the whole state—63,600 people. Statewide, native people are under 3 percent of the population.[1]

Federal troops forced the native population onto a reservation in the wake of emigrant wagon trains and gold seekers. The settlers who came next went "against the grain" of a rapidly urbanizing nation, as one eminent historian of the state has put it.[2] They tried to create family farms and ranches in a difficult landscape, starting in the late nineteenth century and continuing well into the twentieth. One cow in Wyoming was said to require forty acres of grazing land; a herd of a hundred would require four thousand acres, or more than six square miles. Federal settlement laws took a long time to adjust to that reality. A combination of topography and federal policies has meant that people settled and acquired private ownership of lands either in the state's eastern grasslands, or close to streams, or on a few private or (later) federal irrigation projects. Much of the rangelands, some of the grasslands, and nearly all the forested mountains are now permanently in federal hands, often leased out for grazing. Some 57 percent of Wyoming is federal, tribal, or state-owned land, while 43 percent is privately owned.

Mineral production—particularly for oil, and later, natural gas—took over from agriculture as a prime economic driver and employer as long ago as the 1920s; major coal production kicked in during the 1970s. Much of the mineral is federally owned and managed, but the state can tax the production. Even into the twenty-first century, the state has had no income tax because revenues from coal, oil, and gas production have paid much of the bill of a conservatively fashioned government budget. When cheap energy fuels national prosperity, that typically means a bust in Wyoming. Since the 1980s, Wyoming has been the nation's largest coal producer, fueling power plants thousands of miles away. The national turn away from coal has put Wyoming in a quandary about its future. Tourism is the second largest industry, but nowhere near as big as energy production has been.[3]

The landscape is dominated by agriculture and by federal reserves of grazing, forest, and wilderness lands, punctuated by swaths of surface coal mines, oil, and gas development in pockets or major patches and some tentacles of suburban and second-home sprawl. Many ranch and farm operations (and certainly the most vocal) are still family run.

Water in Wyoming comes largely from snowfall on the mountains; of some 18.2 million acre-feet of water flowing in Wyoming streams each year, nearly 16.2 million acre-feet, mostly from snowpack, originate in the state. With its high-elevation lands and low population, Wyoming can't consume much of that water. Some 15.4 million acre-feet flow out of the state—Wyoming is headwaters to the major US river systems of the Mississippi, the Columbia, and the Colorado. As in other western states, most of the water Wyoming people do use goes to agriculture. People have scoured and leveled land for irrigation, yet just over 3 percent of the land is irrigated. More land is classified "agricultural," but only because it can be used for livestock grazing. On the irrigated land, the most common goal is to raise hay to sell or to feed livestock in winter. There are a few places that can raise sugar beets, oats, or even corn. Three-quarters of the state's crop production, however, is some form of hay—in 2019, worth nearly $430 million.[4]

Because the water comes from snowpack, it is naturally delivered with the rush of snowmelt in the warming days of May and June. Since they first arrived, irrigators (plus a few private investors and eventually the federal and state governments) have been building reservoirs to catch that rush and make it possible to extend the water flow as much as possible through summer. In many places, irrigation by a gentle flood system from ditches in the fields is preferred to the more technically efficient and expensive sprinkler systems, which can't always pay for themselves in the short seasons and with limited crops of high-elevation lands. The flood system has its peculiar benefits because flooded soils retain the water for a while and slowly let it back into a creek for the irrigator downstream to use—seen by some as a means of water storage for the end of the growing season. Aside from water flowing in the streams, some farms and towns use underground water, though statewide, groundwater use is dwarfed by surface water use. For agriculture, underground water is typically used in the occasional places where the crop is worth the cost of sprinklers tied to a well tapping groundwater that is handily nearby.[5]

More frequent drought, massive mountain forest fires, and smoky summer skies have appeared in some recent years, brought on by climate change. Climate change in turn has been fueled, in part, by the decades during which the nation demanded coal for power—often, Wyoming coal. Today's challenges make it appropriate to look back, and then forward, in this book. Until lately, Wyoming summers were the most beautiful anywhere. Sleeveless days, cool nights, blue skies, the occasional brief and dramatic thunderstorm. All that was followed by the dazzling winter. Until the erratic climate of the last few years, gray days were rare, and cold sunny blue followed the frequent snowstorms. Winter seems the season the state is made for, with a pure color scheme of brown, white, and every shade of blue in the mountains and the shadows. The cities—most of them just towns, really—are small. People know each other, participate in a very homemade government, and find that they can sometimes genuinely shape what happens in their place. If Wyoming has spoken to you, it can be very hard to live anywhere else.

1. A MODEST PROPOSAL

Public Ownership for the Public Good

I became the voice of John crying in the wilderness
for a more adequate public control.

—ELWOOD MEAD, WYOMING STATE ENGINEER, 1890–1899[1]

Place: Clear Creek and French Creek,
    northeast Wyoming's Big Horn Mountains.
Time: The 1890s.

Buffalo, Wyoming, and Clear Creek. J. E. Stimson Collection,
Wyoming State Archives.

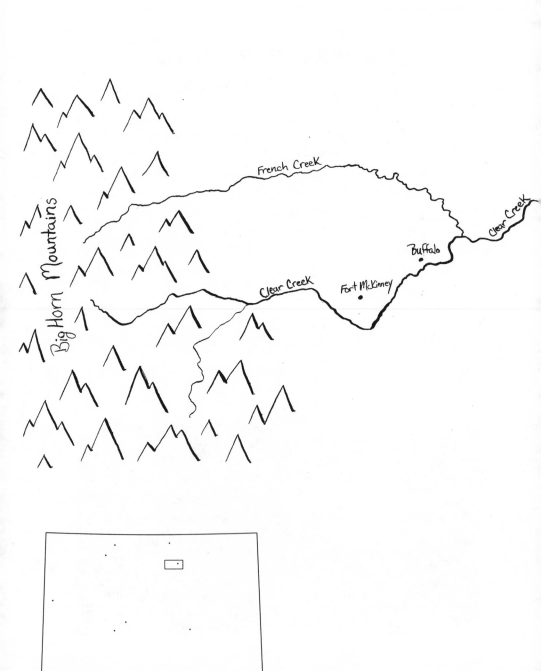

ON A HOT JULY day in 1890 under a clear blue sky, people turned out to celebrate the transformation of their territory into a state.

"With all the Pomp and Ceremony of a Mighty State—A Grand Parade—Eloquent Addresses!—Firing of Cannon!—Elaborate Fireworks!—Reception and Ball!" crowed a newspaper in Cheyenne, capital of the new state.[2] Statehood meant a lot: no more top officials appointed from Washington rather than elected by local people, and hopes for an influx of new people and money to Wyoming's sparsely populated lands.[3]

The parade filled the streets. Marching bands, national guardsmen, and the new state flag flanked by "girl guards" stepped out with ranks of Civil War veterans, two good-looking milk cows to tout a milkman's business, and a float from an ice-cream confectioner featuring "a fat boy dressed up gaily" with a sign testifying that he ate that ice-cream.[4]

Speeches focused on the recognition of voting rights for women—something Wyoming had already done twenty years earlier as a territory and which now, as a state, it was the first in the nation to do. Despite the territorial tradition, there was not universal support for women voting. One of the honored speakers on statehood day, Theresa Jenkins, went door-to-door in Cheyenne the year before statehood, when she was nine months pregnant, to rally women to head to the new capital building and fight a proposal by a few men in the state constitutional convention who wanted to take votes for women out of the constitution and isolate it for voters to decide separately.[5] When she spoke at the capitol on the celebration day the next July, Jenkins graciously omitted any reference to the opposition: "Happy are our hearts today, and our lips but sound a faint echo of the gratitude within our bosoms . . . We have been placed upon the very summit of freedom and the broad plain of universal equality," she said.[6]

Francis E. Warren, the territorial governor—like many of the leaders of the territory, a Civil War veteran—would soon be elected state governor, a post he quickly abandoned to become a US senator. Warren struck the same note in his speech:

Here, in the open air, near the crest of the continent, Wyoming, forming the keystone of the arch of states extending from ocean to ocean, celebrates an event significant in the extreme, new in the history of our country, and without precedent in the world; that is to say, a state, in adopting its constitution, extends free and equal suffrage to its citizens regardless of sex.[7]

There was one other unusual portion of the constitution adopted for the new state of Wyoming, which otherwise largely copied what had been done by other recently admitted western states. The other notable portion in the new constitution dealt with a quite different issue: water.[8]

Water was a resource that was scarce and therefore valuable. It was in demand and not always easy to access, particularly for the agriculture believed to be important to Wyoming's future. The state constitution proposed to sort out and stabilize water use.

The poet of statehood day, after celebrating women's right to vote, led her listeners to appreciate the further promise of the new constitution for a new dawn:

> If we look within the future, our prophetic eyes can see
> Glorious views unfold before us, of joy, wealth, prosperity,
> We can see the sons of Science, Music, Poetry and Art
> Coming to our grand dominion, in our growth to take a part.
> We can see the iron monster, rushing fiercely to and fro,
> We can see the sky o'erspread with smoke from furnaces below.
> . . . See the plains, now dry and barren, where the sage or cactus grows,
> Desert plains, no longer barren, then shall "blossom like the rose."
> Thirsty lands, no longer thirsty, filled with moisture wisely stored,
> Bounteous to the happy farmer, noble harvests will afford.[9]

Two of the young "sons of Science" of whom she spoke had in fact recently come to Wyoming. They were already at work on that celebration day, acting out their ideals of public service. One was Edward Gillette, a Yale-educated railroad surveyor. The other was Elwood Mead, a Purdue-trained engineer. Like the poet, both saw science and planning as the tools to help turn the raw settlements of a territory into a community.

Gillette, at thirty-six, arrived in Wyoming the summer before, heading the first railroad survey into northeastern Wyoming—Powder River country. He laid out routes to ship out cattle and coal. He was earnest and thorough, and the railroad he worked for put his name on a town that ninety years later was the capital of the US coal industry.[10]

Gillette soon saw the impact of his work firsthand in the town of Sheridan, nestled in rolling lands where the Big Horn Mountains meet the Powder River basin right near the Wyoming-Montana line. Fledgling farmers and small stockmen in that area were mired in a swamp of debt. Once Gillette's rail line connected them to markets a couple of years after statehood, though, they quickly made their way to solvency, even prosperity. Gillette liked adventure; he later went off to survey the rail route to the Klondike gold fields. But he had fallen in love with a girl from Sheridan. It was there that he settled down.[11]

The other young Son of Science, Elwood Mead, was thirty in the summer of 1890. He had arrived in Wyoming Territory two years earlier and had spent months driving a wagon across it, gauging creeks and rivers, examining county record books, and mentally drafting water statutes. He had crafted a whole new scheme for water management and had managed to get it written into the new state's constitution. Now he had to put that scheme into action.[12]

Having grown up on an Indiana farm by the Ohio River, with plenty of rain and floods, Mead landed his first job teaching agricultural engineering in Fort Collins, Colorado. In Colorado, Mead saw firsthand what water in a dry country could mean for good and for ill. On the Colorado plains, as in Wyoming and much of the West, precipitation was perhaps one-third of what it was where Mead grew up. There could be raging floods in the spring but dry rivers in the fall. Mead worked in the summers for the top water official of Colorado, gauging irrigation ditches along the South Platte River, which was home to assorted irrigation colonies and ditch companies competing for water. When new ideas for managing that competition failed in the Colorado legislature, Mead took up an offer to move north and become Wyoming's first territorial engineer, in 1888. When he arrived, he started drafting new laws, to avoid the water problems he'd seen in Colorado.[13]

Within a few years, Mead and Gillette joined forces, and together they put Mead's new ideas about water to work in Buffalo, Wyoming, along the Big Horns south of Sheridan.

———

Centuries before Mead and Gillette, the land they saw anointed as a state had been familiar ground for a variety of peoples. By the 1830s, as the United States gazed increasingly westward, a people who became known as Eastern Shoshone were in the southwest portion of what became Wyoming. Northern Arapaho were in the southeast and central areas, Crow in the north and central areas, and Sioux in the northeast. As hunting peoples, they moved back and forth across the landscape with the game and the seasons, using both land and water lightly (unlike the people for whom Mead did his planning). Each dealt with incursions from European-origin people differently. The Eastern Shoshone, who joined the "rendezvous" fur trade meetings in the 1820s and 1830s, held to a friendship policy, even as increasing numbers of newcomers in the 1850s and 1860s beat a path to the Pacific Coast across their lands and killed or drove out game along the way. The Northern Arapaho, Crow, and Sioux tended to clash with those trespassers on their hunting grounds.[14]

Though the US government initially signed treaties recognizing vast areas as exclusive tribal lands, increasing overland wagon trains and construction of the transcontinental railroad led the federal government to force new deals on the tribes. In 1868, the Northern Arapaho and their neighbors, to protect key lands for their own use, agreed to sign with the United States a treaty that was soon shredded. Also in 1868, the Eastern Shoshone accepted treaty establishment of a reservation of over three million acres embracing the Wind River valley in western Wyoming. That was a drastic drop from the 45 million acres (across what are now several states) the government, only five years before, had promised for the exclusive use of the Shoshone and their related tribes.[15]

The idea of boundaries, however, was alien to people who had known none and had moved when they liked—wherever the opportunity for food and shelter, and the lack of opposition, had allowed. Not until the 1870s did scarce game and the US Army ultimately force native people like the Shoshone to settle on the reservations, the only places they could receive promised federal government support.[16]

When Wyoming became a territory in 1869, new legislators immediately attacked the treaties of 1868. The new territorial governor said the treaties had locked away from settlement important parts of "our strong box and our garden spots." Of the reservation on the Wind River, the legislators at first demanded that it be eliminated, and that the people on it be moved elsewhere. Two years later, with that goal not achieved, the territorial legislators insisted that Washington drastically cut back the size of the reservation. Their pressure began a nearly forty-year process that ultimately succeeded. Meanwhile, the treaty signed by the Northern Arapaho and their neighbors in 1868 soon became meaningless, drowned in bloodshed as European-origin settlers claimed gold or land in treaty territory. In 1876, just a little north of the boundary of Wyoming Territory, the Sioux and Cheyenne wiped out George Armstrong Custer's forces—but were ultimately forced out of their most prized lands in northeast Wyoming. The Northern Arapaho, in 1878, were forced by the US Army onto the Eastern Shoshone's Wind River Reservation. When a key US Army general sympathetic to the Arapaho died, Northern Arapaho hopes of getting their own reservation died too. The Shoshone and Northern Arapaho had long been enemies, but the "temporary" arrangement for both tribes to live on the Wind River Reservation became permanent.[17]

―――――

Soon after, the town of Buffalo bustled rapidly into existence. In 1879, the US Army built Fort McKinney on a good stream called Clear Creek in northeast Wyoming, among fertile foothills on the edge of the Big Horn mountains. The fort was intended to solidify the army's presence east of the Big Horns. As soon as the fort was established, its commander and provisions master set the pace for claiming land and water. They filed papers on public land just outside the fort boundaries so they could grow hay to sell back to the fort. Traders, suppliers, hotelkeepers, and brewers materialized (with the first hotel and bank sheltered in a tent), and together they built a town. Nearly all the shop owners in town filed on land, and more importantly, on water. With both in hand, they raised cattle, hay, garden produce, and barley for the humming breweries. Even when the army shut the fort in 1895 and the soldier-customers disappeared, the town found the strength to stay. In the one hundred years since, a remarkable number of the names from the early 1880s have continued to populate Buffalo, its neighboring ranches, and its elected offices. The hotel that emerged from a tent is still operating there, too—with weekly music jams where locals play bluegrass.[18]

Getting access to water was no small part of the enterprise of those first settlers in Buffalo. In 1884, for instance, some went up into the mountains and with teams of horses cut a diversion over the intervening divide to take water out of Clear Creek and put it into the headwaters of the smaller French Creek, which ultimately joins Clear Creek a little east of Buffalo. They managed to get their diversion installed just days before the big Wyoming Land and Cattle Company, south of town, built a diversion to tap Clear Creek for its ranch operations.[19]

French Creek is not fed by the high mountain country that feeds Clear Creek. Left on its own, French Creek would run dry in early summer. So French Creek valley became and remains a green and pleasant place, not so much because of French Creek itself, but because of the water brought to it from Clear Creek.

Meanwhile, the town of Buffalo grew—and it, too, relied on Clear Creek for water. The intricate patterns of what water went where, in the ditches dug by busy settlers with mule teams, were a nightmare to sort out. Young Mead soon cited them as western classics of dizzyingly complex irrigation systems.[20]

For well over a century, Buffalo and French Creek have kept going, supported by their tangle of irrigation ditches. The local economy has been agriculture, but also scenery—since the 1880s, "eastern money" has come in to buy and keep some of the beauty of the Big Horns foothills (and dude ranches sprang up before World War I, bringing in visiting money that some local cowboys managed to marry into). By the mid-1990s, only one ranch on French Creek was making a living from agriculture; the other owners ranged from a retired major oil company CEO, to a New York banker, to the county attorney making a living in town. In 2020, Buffalo is best known as the real-life inspiration for the town featured in the Netflix Western series *Longmire*. Buffalo has grown a little, and with it have bloomed sales pitches for "ranchette" subdivisions. Squabbles over land use planning follow. The question is whether, and how, to manage the beauty of the open spaces that has been an important resource for the town since it began.[21]

Notably, it is newcomers to the area—not the people with big money, but the ones dreaming of ranchettes—who have been the main opponents to land-use planning. They came "out West" for freedom from the rules they knew back in, say, New Jersey, and they are determined to make their idea of freedom stick here. Wyoming's "Cowboy State" logo can attract such people.

Yet while the standard polemic in Wyoming may be anti-government, the standard experience has long been joining together in government. In a big place with few people, government is an intimate affair. It really is by, for, and of the people. People are members of their local boards and agencies; they have often known their legislators and their governor personally. The lone rancher trying to make a living on French Creek in the 1990s, on the place his father built in the 1940s, was elected to the county commission and pushed for land-use planning.[22]

People in Buffalo—and in Wyoming—have wrestled before with how and whether to plan for and manage natural resources. Though it's not always recognized, they, and others like them in the West, have some special history to draw upon. It is the history of water management that starts with bright young Elwood Mead. He helped Wyoming people come up with a unique way to handle the water resource they all depend upon. French Creek, as it happens, was a key testing ground for Wyoming's initial experiment in water management. Buffalo and its neighboring ranch valleys, so attractive to newcomers today, attest to the strengths of that scheme for managing water that has grown and changed to meet assorted challenges over a century.

———

In early 1888, Mead, the twenty-eight-year-old engineering professor, received a letter from his former boss, E. S. Nettleton, who had been Colorado State engineer. The two had become close friends. Nettleton wrote that he had been talking to a Wyoming legislator interested in water, helped draft a bill to flesh out a job overseeing Wyoming water, and recommended Mead for the job.[23]

Mead traveled the forty miles north to Wyoming's capital city, Cheyenne. He discussed the job with some legislators and Thomas Moonlight, the territorial governor. Moonlight, once a losing candidate for governor of Kansas, had been appointed to Wyoming by the Democratic administration then in Washington. He was blustery, chronically at odds with Wyoming's Republican territorial legislature, and certainly not politically adroit.[24]

A few hours after their interview, Moonlight ran into Mead on the street in Cheyenne. The governor stopped him, as Mead recalled years later: "He said, 'You've been on my conscience ever since I first saw you this morning. I didn't know you were so young. If I had known this I would never have offered you the place, and the reason is that if you come here I am sure you will fail.'"

"The Governor had quarreled with the Legislature," Mead went on, "and had an unfavorable opinion of the influence which dominated public life in Wyoming. He ended his talk with me by saying, 'I do hope and pray to God that you will reconsider and not accept this place.'"[25]

Mead went back to Fort Collins, where over the next few days, others painted a more favorable picture of Wyoming politics.

Then Mead met, in person, "the influence which dominated public life in Wyoming"—Francis E. Warren, later a headliner at the statehood celebration. A Massachusetts farm boy who went off to the Civil War, Warren had arrived in Wyoming in 1868 at age twenty-three to work in a dry-goods store. He soon took over the store and ultimately plunged into almost everything he could think of—ranching, starting an electric lighting company, and investing in blocks of downtown Cheyenne real estate. In 1888, Warren, at forty-three, had lately been the territory's staunch Republican-appointed governor, and soon would be again; he despised Moonlight, whom Democratic president Grover Cleveland had appointed to replace him.[26]

Warren was in fact Wyoming Territory's only mostly home-grown governor, and he fought for "home rule" and statehood. His greatest talent was politics, and he managed to unite Wyoming's warring petty political factions to bring in the federal largesse that alone could make up for gaping holes in a place likely to see only slow development. After statehood, he spent more than thirty years doing just that as US senator.[27]

Warren made his way by his ability to judge people—one of his many ventures was assessing business prospects for the predecessors of Dun and Bradstreet—and to use his blunt charm, or money or threats, to get key people to do what he wanted. His voluminous letters between Cheyenne and points east, stored today at the University of Wyoming, make him appear to have been a formidable mix of Theodore Roosevelt (a friend) and Lyndon B. Johnson, without their interest in or impact on national policy. Warren asked Mead to take the territorial water job. (As he told Mead years later, however, Warren saw Mead as "still wearing pinafores" when he first came to Cheyenne.)[28]

Mead, meanwhile, wanted the job, regardless of Moonlight's advice. Mead's baby face belied his determination to put into action his ideas about water in the West. In Colorado, he had seen bitter struggles between would-be farmers and ranchers, and between farmers and would-be irrigation companies—fights over who had rights to how much water, and battles over speculation in water and attempted monopoly of water supplies.[29]

Wyoming Territory, with settlement, government, and water law all in a fledgling condition, presented—far more than did Colorado—a clean slate on which Mead thought he could chart new policy. He believed that western states could avoid destructive conflict by using public control to manage resources for the greater public good. Warren and his associates also saw public order as critical to buttressing their business prospects and investments. Stockmen needed a reliable system of water rights to protect the small clutch of water claims on which their stock increasingly depended. Further, because their open-range stock industry had begun to founder, they were looking toward new kinds of development for a territory where there was not much other economic activity. Warren and other stockmen had in fact already invested in at least one new irrigation venture, which might benefit from an orderly water rights system controlled in Wyoming.[30]

So Mead came to Wyoming backed by Wyoming patrons, both he and they believing that the time had come for firm state government intervention in water rights matters. Mead's commitment to government involvement in water went deeper, however, than his stockmen backers probably understood at the time. Mead felt that those who drafted new water law in the West were shaping legal principles controlling not only water use, but also "the social and economic fabric under which unnumbered millions of people must dwell." For him, that imposed a near sacred duty.[31]

Mead saw resource law, based on public ownership, as a tool that people in new territory could use to foster communities. He had a specific kind of community in mind—a community self-sufficient not only in economic but also in social terms, providing itself with the educational and cultural resources to make rural America a vital and satisfying place to live. These communities would then also become the birthplace to the kind of citizens a democracy needed.[32]

Though his belief in the connection of resource law to democratic societies had a rosy tinge, Mead disparaged the extreme agrarian romanticism of leading irrigation boosters in the 1890s, who imagined the West as a garden spot that would bloom if just a little money and water were added. He partook more in the natural resource conservation thinking that was beginning to come of age in the late-nineteenth century, with the dawn of the Progressive Era and eventually the era of its conservation standard-bearers, Theodore Roosevelt and Gifford Pinchot. Mead believed in government's giving a preeminent role to science, to scientists, and to engineers in managing resources for the greater good. He thought that public resources should be put to work through private use, but they should never lose their public character and their ultimate subjection to public control. The goal was to ensure that public resources would continue to serve changing social needs.[33]

After a few years of work in Wyoming, Mead's thinking matured into a belief that a public grant of private rights to use water should be decades-long, not perpetual, and that it should have rental fees attached—in order to remind users that the public owns the water. He attested to "believing fully in the doctrine that public waters should remain a public property, and that to grant private perpetual rights is to sacrifice the welfare of future generations." After Wyoming, he spent a few years in Washington and then went off to Australia to work where people supported the approach to water he advocated.[34]

In the United States, Mead had not found it easy to put such beliefs into effective action. From Australia in 1908, he fumed, in a letter back to Wyoming:

> The difficulty encountered in maintaining public ownership of water in the United States grows out of the fact that for fifty years we have been a spend-thrift nation, and public opinion has favored the prodigal disposal of public resources.
>
> The consequence is that it is ceasing to be a land of opportunity, and is becoming instead a land where the predatory and powerful rich have most of the resources and privileges, while the great body of the middle and working classes have narrower opportunities than in many older and poorer countries.[35]

Mead was driven by his social and political passion, his sense of moral duty. He did his best all his long life to translate vision into practice in several resource fields. In the 1890s in Wyoming, he not only wrote new water law but proposed a variety of changes to land law. Into the 1930s, (when his seventy-year-old, still-baby-face gave him a benevolent look in the eyes of the young New Dealers he mentored), he agitated for rural planning and credit systems in order to wipe out farm failures in the West and sharecropping in the South—all to be spearheaded by the agency he came to head at the close of his career, the US Bureau of Reclamation.[36]

———

To understand what Mead did in Wyoming, in the first big step of his career, requires a dive into nineteenth-century legal theory and practice.

Water matters in Colorado and California, the western states in the spotlight at the time, were dominated by private scrambling, wrangling, and speculation in water. That pattern had begun to make headway in Wyoming. To defeat it, Mead posited a rational system, animated by a single idea. That idea was *active* public ownership of water through state supervision of private rights to use water.

Public ownership of water was old legal language, borrowed from the Romans and from English common law. It was a phrase worn out and largely emptied of its meaning, though typically adopted by western states and by Wyoming as a territory. But, as water rights really operated in Wyoming in 1889, public ownership was "simply a fiction," Mead told the people of the territory after his first year as territorial engineer.[37] Under Mead's plan, public ownership of water in Wyoming would be a serious matter—there would be state allocation of water, and state-imposed limits on what a private water right meant. That was in stark contrast to water law then prevalent in the West, called "prior appropriation," a product of forty years of local practices, hodgepodge legislation, and a growing pile of court decisions.

Prior appropriation was developed on the western frontier. The concept behind it was that firstcomers could take water from a stream, move it elsewhere—sometimes quite far away—and have a protected right to use as much water as they needed for a productive use.[38] Previous British and US law had said something quite different. The original rule in Britain and in the eastern US was that a person had to own land on a streambank to use water from that waterway *and* had to use the water on that land. It was called a "riparian" right to water, from the Latin word *ripa*, meaning streambank. Under riparian law, any and every streambank owner along the length of the stream had a right to use the natural flow of the stream on the adjacent land, while the stream flowed on with good quality and quantity of water for the next users. It made sense in areas with plenty of rainfall, many streams, and water use that typically amounted to only domestic use, livestock use, and water-powered mills—places like the Ohio farm where Mead grew up. There was little danger there of creeks, or the Ohio River, being significantly drawn down by users. The streams could be shared. With industrial and urban growth in such places in the nineteenth century, further guiding principles in riparian law developed. Where growth caused conflict over water, the courts determined that each water use should be "reasonable," so that others might have opportunity for their own (reasonable) uses. There could, of course, be a variety of considerations defining "reasonable use"—the purpose of the use, the economic and social value, the potential for harming others. Occasionally, courts might decide that riparian law could allow use away from the stream, and they might protect big investments in a use against a later comer. Courts took on a balancing act aiming at fairness to all. That could, of course, make the extent of a water right uncertain as times changed.[39]

People heading West found a situation very different from the one they had known in the East. They wanted to be able to take what water there was and move it wherever they could use it—and they wanted clear and certain rights to that water. As the United States acquired the western region through conquest and sale, most land was considered owned by the federal government and known as "the public domain." People often got to that land, however, before the federal land offices got there to survey and dispose of it by law. Those early arrivals had to set their own rules for using resources. Farming and ranching in an arid land can lead to community sharing of shortages and surpluses, as demonstrated by traditions of the former Spanish colonies in the Southwest. On the public domain in California and Colorado, however, the newcomers wanted gold, not farms. There were struggles in the mining camps over different approaches, but water sharing tended not to become standard.[40]

The typical rule was that the best water rights would go to firstcomers. If supplies should run short, those who had claimed and used the water first would get all their water before later comers got theirs—no balancing act required. That appeared to be a way to support economic development. As farming-minded settlers succeeded miners, the same rule tied into the nineteenth-century American idea that the labor of settlement should be rewarded with ownership of resources, in order to create an independent, self-sufficient society. That belief had led Congress, starting in the 1830s, to give squatters on public lands the ability (called a "preemption" right) to buy land if they had settled there before the land went up for public auction. The hope of creating territories of independent small settlers blossomed during the Civil War into the Homestead Act, when Congress had no slave-state members to block it. The Homestead Act allowed settlers to get title to, usually, 160 acres of land—considered enough to support a family—after living on the land for a period of years and building improvements (plus a small filing fee). A good-faith effort to settle the land was thus rewarded.[41]

If good faith effort should make it possible to acquire rights to land on the public domain, why shouldn't that apply to water on the public domain? Western courts ultimately agreed. They ruled that taking water out of a stream and putting that scarce and precious resource to some productive use would be a "beneficial use" that should be rewarded with a legal right to the water, even if the water was completely consumed. The only restriction would be against wasting water. Diverting water from a stream and using it "beneficially" and regularly, in the season appropriate to the use, became the standard way to get rights to water in most of the West. Just putting water into a ditch amounted to staking a claim that could in time become a right protected by law. Courts used the word "appropriation," which means taking something for one's own. Lawyers sometimes equated it to hunting game on the frontier: "He who wanted water took it." "Appropriating" the water first gave a person a "prior" right—thus, "prior appropriation." A prior appropriation water right was a definite right to a certain amount of water. The right could not, of course, ensure that water would always be there in the many streams that naturally went dry after their mountain snowpack sources ran low. But holding a prior appropriation right meant that if water was flowing, the person with the prior right got his water before people holding later rights got any.[42]

Prior appropriation became common custom. It started primarily among some gold mining communities in California and Colorado and was accepted by courts in those states and later applied to irrigation (California created troubles by recognizing both riparian and prior appropriation rights). Courts determined that holding a prior appropriation right meant a person had a right that could be protected like private property. He kept the right to take, transport, and use the water. He could lose his rights by failing to use the water, for five years or so. But if he kept using the water, it was his. His to use, and his to sell, if he pleased.[43] The hunting analogy that was used when describing the taking of water was soon "stretched," one lawyer noted later: "He who took water without so much as anyone's by-your-leave acquired not only that which actually he took but also a right perpetually to take; hence the difficulties."[44]

In many places around the West, there were fights over exactly *how much* water had been put into use and *when*, to establish the basis of a prior appropriation right to water. The earliest claims, after all, had the highest value as the best rights. As economies in western states changed and developed, proposals for new uses of water grew, and disputes intensified. Some saw opportunities to monopolize water. The fights were sometimes on the riverbank, with fists or guns rather than lawyers. Sometimes, the struggles were in court; sometimes, they could be settled by an exchange of cash. Sometimes, the battles were in the legislatures, where one side might call for wholesale condemnation of the other side's water rights. Courts, not knowledgeable in streamflow or water use, tended to award rights based on ditch size. A man could dig a ditch much larger than his immediate use and hope either to expand his use or sell the excess water at exorbitant prices. Meanwhile, investors put together canal companies to build extensive and expensive ditch systems, believing they must control the water, charging water "royalties" in addition to fees to cover canal construction and profit, and threatening to withhold the water if farmers didn't pay the royalty.[45]

Looking back years later, Mead called the amalgam of mining camp customs, court decisions, and early state statutes in California and Colorado "the ruck of the arid States of America, whose water laws belong to the lower Silurian period." Reading all he could find on water and water management, he came upon publications by John Wesley Powell and William Hammond Hall— respectively, the US Geological Survey chief who had explored the Colorado River, and the first state engineer of California, who later worked for Powell, joining Mead's friend Nettleton there. Powell and Hall inspired in Mead a faith in the power of planning and supervision of resource use to make stable societies possible in the arid West.[46]

Mead subscribed to the general belief that water had to be put into private hands to build economies and societies in the arid frontier states. How to do that while recognizing the special qualities of water was the challenge.

California's engineer Hall had called water "that kind of property which no one can own" yet many could lay hands on to use however they liked. Hall argued that in the early years of any new civilization, "the necessity for guarding the common prosperity of all people is not felt." Now, Hall wrote, the West must move ahead and recognize that necessity.[47]

The solution Hall proposed was imposing state supervision on the reigning chaos in water rights. He and Nettleton were out of office in their states, though they remained active in policy debates. Mead was in position to propel a new system in a new state. Nettleton and Hall wrote to Mead and laid out their objections to "prior appropriation" in water law. As Hall put it to Mead, "It has been the curse of irrigation from time immemorial, that water has been treated like it was a beast—to be shot down and dragged out by the first brute that came in sight of it."[48]

Mead shared that antipathy to pure—unregulated—prior appropriation. As he later wrote, "It is wrong:"

It assumes that the establishment of titles to the snows on the mountains and the rains falling on the public land and the water collected in the lakes and rivers, on the use of which the development of the State must in a great measure depend, is a private matter. It ignores public interests in a resource upon which the enduring prosperity of communities must rest.[49]

With his plan for imposing state supervision, however, Mead admitted "prior appropriation" into the new system for Wyoming. A good deal of his new water-law code would adopt basic prior appropriation ideas. Water rights were to be based on use of water, not ownership of land next to water, and the water could be diverted to other lands. A would-be user had to give notice of her intent to use water. Priority access to water in times of low flows would go to the person who had made clear, before her neighbors did, that she planned to use that water and had followed up by actually putting the water to a beneficial use within a reasonable time and continuing to use it regularly.[50]

Prior appropriation in Mead's system was different, though, because it was subject to severe state control. Giving new life to the tired old language of public ownership of water, Mead made the concept of "public waters" the answer to the riot of private rights in water elsewhere in the West. The public, represented by the state, had key rights in water. The state could decide who got a private right to water and define exactly what that private right to water consisted of, in volume, timing, and use. It could determine whether a person had lost that right to water by non-use.[51]

Significantly, Mead's system also put special controls on private rights in water for irrigation, the highest-volume use, and the one he and others expected to be the most important to future prosperity. Mead planned to require that a water right could not be sold away from the land it irrigated. As Hall and Powell had suggested, Mead said, irrigated regions would flourish when "the land appropriates the water."[52]

Mead had taken to the hustings in 1887 to advocate for that principle at a meeting of Colorado farmers. Citing Hall and Powell, he told the farmers that water scarcity could be a blessing; it could foster a high level of civilization because in living with scarcity, people learn "the value of co-operation to a common end." Yet in Colorado, prior appropriation law made him fear that "the foundation of monopolies a hundred-fold more exacting than Irish landlordism is being laid." The farmer, not the canal companies, put the water into use and, Mead said, made "the barren plain into productive fields, [and] . . . covered this State with thriving and happy homes." It was the farmer who should have secure rights to water. That meant water rights tied to the land.[53]

Mead urged the farmers to back bills in the Colorado legislature to accomplish that, to limit canal companies to charging fees to cover canal construction and maintenance, and to create a state board to supervise and authorize water use. The key bills failed. A year later, Mead was in Cheyenne, ready to put his ideas into action there. Speakers at the eventual constitutional convention backed Mead in spotlighting the threat of water monopoly and decrying pure prior appropriation, advocating state water ownership and supervision to keep prior appropriation in bounds.[54]

In the water law he wrote for Wyoming after adoption of the new state constitution, Mead proudly blocked sales of water rights apart from the land it served. It was a substantial change from standard prior appropriation law, and he made it integral to his new process of state supervision. He believed that it underlined what state ownership meant: while the state would grant some rights to water to individual users, it retained rights he believed would serve the general welfare on behalf of the public. The society was best served by tying water rights to land so the people who farmed the land held the rights to use the water. Looking back a dozen years later, Mead said that in adopting the statutes he had drafted, the Wyoming legislature had "in effect abandoned the doctrine of appropriation, although retaining the word in their statutes."[55]

Prohibiting sales of water rights away from the land "is opposed to the decisions of the Courts of other arid States," Mead told readers of his early reports as state engineer, because Wyoming had adopted "entirely different laws" from other states. Under Wyoming's laws, a water right must include the exact legal description of the land, with an amount of water usually determined by a state standard for how much water was needed per acre. That created a right to use water only for a specific beneficial purpose, Mead told readers, and that right could not be sold separate from the land.[56] He continued:

> There is no question but what absolute ownership would be more valuable to the individual securing control of the stream than the right to use the water for beneficial purposes, but we have never believed that the purpose of the State in assuming control of the water supply and protecting appropriators in its use was for the purpose of conferring a valuable property right on individuals to the exclusion of the rights of the public.[57]

Wyoming state engineers remained equally proud of Mead's effort to block sales of water rights apart from the land. Fifty years later, the Wyoming State Engineer's Office reprinted in its regular report to the state the speech Mead had made to the farmers in Colorado in 1887. For all its passion of a time long past, that speech remained the best answer to "why water is tied to the land in Wyoming," the state engineer of 1942 wrote.[58]

For administration and supervision of water rights by the state, Mead set up two key features. First, the only private rights to use water would be rights examined and permitted by the state. Nobody could obtain a private right to use water just by staking a claim to some water and building a ditch. For a new right, a state permit was required before any work was done; an application had to be filed to give notice of intent to use water, the application would be reviewed, and sometimes it would be returned for improvements or denied. Water claims that predated statehood, meanwhile, would be examined and would be recognized as a water right only for the amounts of water actually in use. All water rights approved by the state were to be strictly defined by the use to which the water was applied. Every water right would include restrictions as to how much water could be used and where. Irrigation rights must include an exact legal description of the lands to be watered.[59]

Second, water rights would be the domain of an expert state staff, not courts and judges. Only experts could make state ownership of water active in the way Mead intended. They would be empowered to measure, delineate, and establish water rights. They would be the Progressive Era ideal—a cadre of experts dedicated to public service.[60]

The combination of those key features, Mead believed, would ensure that his system would always serve the needs of the society, however those needs might change. Time tended to bring new and competing demands for water, as seen in California and Colorado. Change might be inevitable, but Mead believed economic transition in the West had been made particularly difficult by legal acceptance of private property rights to water, "rights" established by people on their own without public supervision. An active doctrine of state ownership granting and defining private rights to use water could instead promote stable societies where the emergent law of private property in water failed.[61]

The expert staff would actively measure and record existing water rights and issue new water permits where water was available and water use proposals were feasible. The permits would have their own deadlines for getting water put into use. The state engineer would determine what was a beneficial use deserving of protection. New beneficial uses might develop over time, and the state engineer could recognize them. The engineer also set a "duty of water," the amount of water to be accepted as necessary to irrigate one acre in Wyoming. "Excess decrees" that were based by the courts on mere ditch size would disappear. No waste of water would be countenanced. Use of water on lands other than those originally intended and permitted would not be allowed. Diversion was not required, though it was assumed in the water right application form. Only rights to water kept in use would be protected. Non-use would lead to loss of the right.[62]

The experts running such a system would be "practical men" who knew the rivers, the land, and the problems of irrigation. The water superintendents would not wait for water disputes to arise. To create order among claims dating from territorial days, they would review and measure water supply and use in an entire watershed in a "stream adjudication." They were armed with the power of the state to create, for the first time in Wyoming, a list of water rights recognized and protected on each stream for the purpose of maintaining order in water-short times. They could require every claimant to submit to the state process of determining their rights. In low-water years, officials would divide up water according to the lists of rights with their priority dates and keep the peace on the streams. With water right lists updated over time with new rights, water use could continue in orderly fashion as the society dependent on water grew and changed.[63]

The superintendents working from practical experience would not be hobbled by traditional legal training in property rights. They could adopt the concept of state ownership of water and its limits on private rights in a way the courts probably could not. And water users would not have to hire lawyers to deal with water rights questions; they could explain their cases themselves to the water supervisors.[64]

Mead adopted the system that Powell had unsuccessfully recommended to Congress ten years earlier—management based on watersheds rather than political dividing lines. There would be four superintendents, each based in one of the state's four major river basins (roughly, their four divisions comprise the basins of the Platte River, the Powder River, the Wind-Big Horn River, and the Green, Snake, Salt, and Bear Rivers). Mead's focus was on surface water; groundwater received attention only sixty years later and eventually became a significant use only in limited areas.[65]

Fundamental state ownership and controls would, in Mead's view, ensure that people who could put water to use would be secure in their water rights. There could also, however, be a peaceful rollover of water rights between old and new water uses and users—between old and new economies. Water unneeded for current actual use, or water left unused when an old irrigation scheme or an old industry failed, would be available to the next person who could successfully put it to use. A new project could get a new permit to use water on lands where an old project failed. Mead projected that eventually, instead of issuing "perpetual rights" to water, the state would lease water rights, with a rental fee, to individuals for several decades and have a chance to review the use when the lease term ended. But for that favorite idea of his, he conceded that in 1890s Wyoming, "it is probably too early to seriously consider its adoption." Mead also expected growth in towns and cities, providing in constitutional language that they could acquire irrigated lands and put the water to urban use. Water law so designed could offer western states a flexible stability. Future growth in the West, Mead said, would depend on establishment of a straightforward system like the one he planned for Wyoming to determine, limit, and protect rights to use water.[66]

———

Mead arrived in Wyoming with all these ideas, but first he had to deal with the mess at hand.

Wyoming water law at the time reflected a bare twenty years of Wyoming's life as a territory—underpopulated and undeveloped. Water use had demanded little serious attention until just a few years earlier. The resources of interest were the grasses of what seemed endless ranges and some coal that could be mined to fuel the railroads carrying people across the "Great American Desert" to the fertile lands of California and Oregon.

As people who had lived and hunted in the land that became Wyoming were killed or forced onto reservations, the buffalo herds in Wyoming were slaughtered, both as a military objective and as a commercial venture for hides. The grazing land was empty. It was called "open range"—federal land open for anyone's use. The great advantage of these ranges was that typically, despite sometimes fierce winter weather, there were bouts of warmer winds, known as chinooks. Those winds blew enough snow away that cattle could find food and fend for themselves all winter. And the grass growing in such tough terrain was high in protein and very nutritious. The buffalo had thrived on it, and now so could the cattle and sheep.[67]

As economists might analyze it now, both the water in the streams and the great grazing ranges of Wyoming were "common pool" resources: hard to limit to only a few users yet easily reduced or even destroyed if users piled on. Such resources can be governed well and last for centuries by several means: by public management, private management, or by a group of people who work together to control overall access as well as each other's uses of the resource they hold together as common property. There are examples of all three kinds of successful management and many in-between combination versions around the world and through history. Grazing lands are among the natural resources that can be managed well under a system of common property.[68]

In case of the great expanses of Wyoming, there was some attempt at joint management of grazing herds. The men and women who brought cattle and sheep into these ranges were at first relatively few, and they created a way to manage their herds among themselves. Their joint management efforts were the famous "roundups" where cowboys from different outfits joined forces to gather and sort cattle after the long winters. But more people could and did come on the ranges—the prospect of free grass was hard to resist, for everyone from Civil War veterans to Scottish investors—and the roundups became almost the only moment of joint management. The open range became a land not of joint management but of what economists call "open access." Nominally owned and run by the government in the name of the public, the range was in fact not managed at all and so was left open to any comer, with no limits on use. As a Wyoming cattleman bemoaned in 1902, after some thirty years working on the open range, "What is everybody's is nobody's."[69] Such places—with rich resources unmanaged by any system—are indeed almost always, and everywhere, an invitation to disaster.[70]

Before the Civil War, the lands of a future Wyoming were crossed by Europeans first in the fur trade and then on emigrant and gold rush trails—for people going elsewhere. A few chose to try to settle in river and creek valleys. With the completion of the transcontinental railroad after the war ended, more settlement was possible, and towns built along the railway grew, as did a few small farms and livestock operations nearby. Making use of the preemption laws, those few settlers ventured into small valleys, growing gardens, getting water to their livestock, moving water here and there. The first territorial water law, in 1875, declared that landowners along a stream had a (riparian) right to draw water from it—but so did others farther away. It then focused on the problems of how someone at a distance from a stream could convey that water to his land—condemning, if necessary, the route for the ditch he needed to build across a neighbor's place. Further, in water-short time, the law said, a local board should be appointed to divvy up supplies fairly.[71]

As the 1870s wore on into the 1880s, a bubble of "range cattle industry" boom briefly overtook the new territory. Cattle companies fueled by distant investors temporarily dominated a place of formerly slow-growth, small-ranch, and homestead settlement. Congress in faraway Washington cast about for ways to encourage settler ownership of the public lands, following up the homestead idea with other measures like the 1877 Desert Land Act intended to be more suitable for high, dry areas like Wyoming. Significantly, that act specifically authorized taking water from a stream and moving it elsewhere, a death blow to riparian water law in the West. The new law gave people the right to claim and receive title to more acreage than the homestead law did—if they brought water to it. No one was required to live on the land while trying to get water to it, and that encouraged the use of "dummy entry-men" who would sign up for the land without ever seeing it and convey the title to a backer. Across the West, many a stockman—like Mead's future patron Francis Warren—used dummy entry-men and had a web of useless ditches dug if necessary, just to claim land under federal land law. Their claims spread in patterns simply snaking along the key stretches of streams where they could water livestock. The company Warren and other top stock growers had formed also experimented in investing in irrigation projects that might turn out to be the real thing.[72]

In 1886–1887, the invited disaster arrived. On the ranges overstocked by eager companies, a brutal winter followed a summer of drought. The winter made itself famous far across the northern plains. Charlie Russell, a cowboy in Montana, made his celebrated drawing of a wasted steer in the snow surrounded by coyotes, "Waiting for a Chinook, the last of 5,000." In much of Wyoming, substantial sections of herds and fortunes were lost. Banks failed. For western Wyoming, the terrible winter came two years later—and became known as the "Equalizer Winter" for decimating the herds of small and large cattlemen alike (Warren, in a partnership on the Green River, was one of the latter).[73]

Those winters signaled (though not all could see it) the end of the open-range cattle industry. Stockmen reacted with often desperate efforts to exercise control of the ranges, with illegal fences on public land and violence toward ex-cowboys or new settlers who started up small herds of their own. After 1888, the smaller ranches and farms, whose numbers had kept growing around and among the temporary burst of livestock barons, kept steadily building. The era of domination by the big operations was over.[74]

In 1888, the recent deadly winter combined with a new, Democratic administration in Washington to make Wyoming stockmen uniquely interested in a state-run system to provide certainty in water rights. Now painfully aware that they had to raise hay for their cattle, they had already enmeshed themselves in machinations to ensure that, despite federal investigations and claim cancellations, they would own the land under their ranch headquarters and irrigable hayfields, while their livestock still grazed the public lands in summer. Hiring a bright young engineer to create a local system of secure water rights for their lands, protected by state law, seemed an obvious next step.[75]

Smaller ranchers and farmers meanwhile had embraced the idea of growing hay for winter feed, and they too saw the need to protect their access to water with legal water rights. The number of water rights people set out to claim in the decade of 1880–1889 was more than seven times larger than the claims made in the decade before.[76]

Given the prospect of real demand for water—and, therefore, real competition over it—the territorial legislature had begun to write more water law. In 1886, the legislature had set up a self-help system for allocating water and keeping the peace in dry times. People who claimed to own water were told to file their claim in the local court, which would sort out disputes (based largely on the size and date of the ditch built). Upon request of the irrigators, a water commissioner in each major drainage would divvy up water in dry times based on filings at the courthouse and any court rulings. In 1887, water users were required to install measuring devices, and the job of territorial engineer—Mead's future post—was created to measure streams and ditches, give technical advice to water commissioners, and (the part that appealed to him) recommend new water law. The moment of change to a different economy and society, when small ranch and farm operations with irrigated lands were increasing, was the moment Mead arrived.[77]

———

In many places in Wyoming, people were laboring to plan and dig a ditch and attempting to level-out rough land so ditch water could flood across it—all with only a mule and scraper for help. At the far-flung county seats, poring over the files of claims that were supposed to identify and protect those people's right to water, Mead was appalled and bemused by what he found. He compiled lists of the water claims, including some that were both enthusiastic and imaginative.

"The first thing that was manifest was that the virtue of self-denial had not been conspicuous on the part of claimants," Mead recalled as an old man. "If the amount of water claimed had existed, Wyoming would have been a lake."[78]

One man had claimed the right to take, from one river, sixty thousand cubic feet per second—more than all the water in all the rivers in Wyoming, as Mead pointed out. Meanwhile, "the ditch to divert this volume is just two feet wide and six inches deep."[79]

Conflicts in court based on such filings resulted in disastrous rulings. The district court in Cheyenne in the 1880s, for example, had arrived at a fantastic and potentially destructive result. Addressing the claims of little more than 10 percent of the water users on one creek, the court gave some irrigators twenty times as much water per acre of land as others got. In several instances, the court granted the water directly to individuals to use as they might like, rather than tying the water right to either ditch or land.

Mead was expected to enforce the decree after he arrived. When it came time to do so, Mead told the judge that the irrigators would lynch him if he did. The judge promised to jail Mead if he didn't. In the end, Mead managed to avoid both noose and jail—but only by investigating and making his own list of irrigators and water uses on the stream and getting all the irrigators to agree to it. Then, he invoked the aid of the territorial attorney general to sidestep the court decree.[80]

In the confusion besetting water rights, there was nothing to stop—and everything to invite—speculators in water claims who would exploit the people who actually wanted to put water to work. Mead argued to the Wyoming readers of his reports, and to the patrons who brought him to the territory, that his plans for active state ownership of water would avoid such exploitation and promote real development. Wyoming could become, Mead said, what it promised to be but was still short of in 1890: a place of stable irrigated agriculture, a state of towns with schools and churches, the home of communities.[81]

That promised to come with statehood, the goal Mead's patron Warren had been working toward in recent years. When the constitution was written for the new state, Mead wrote his water management ideas into the constitution with the help of Warren and his allies, including the leading law firm in Cheyenne.[82]

The constitution read then and reads now:

Water being essential to industrial prosperity, of limited amount and easy of diversion from its natural channels, its control must be in the State, which, in providing for its use, shall equally guard all the various interests involved.[83]

The water of all natural streams, springs, lakes or other collections of still water, within the boundaries of the state, are hereby declared to be the property of the state.[84]

Priority of appropriation for beneficial uses shall give the better right. No appropriation shall be denied except when such denial is demanded by the public interests.[85]

Right along with that, the state engineer and the Board of Control were enshrined in the constitution too.[86]

Everything Mead wanted followed from the constitutional provisions. People could get a right to use water, but it was a right limited by the constitutional premise that the real owner of the water was the people of Wyoming.[87]

Contemporaries offered the best summation of Mead's system. "Radical change . . . it reverses the present system," said Charles Burritt, describing the water plan to the constitutional convention. He saw that as a good thing. A lawyer with a blue-blood education, Burritt was the lead spokesman in the convention debates for the new constitutional language on water. Radical change, Burritt said, was exactly what was needed—the "present system" was a disaster.[88]

Burritt was also mayor of Buffalo (and one of the backers of the diversion from Clear Creek to French Creek). There were plenty of people in that part of the state who agreed with him that Mead's system was radical change. They did not think, however, that it was change for the better. There were many who were astounded to have the state suddenly claim the right to review and restrict how much water they could use, and when. They had their rights, attested to by their old water filings, and they expected to go file on more water—whenever they wanted it.

The opposition was vocal and active. The first time Mead denied a water permit, he remembered years later, "handbills were distributed in Cheyenne, which had the lurid heading, 'Do you want to live under a czar?'"[89]

Mead understood. "Those early irrigators had built their ditches and diverted water without having to ask the consent of anyone," he recalled much later:

They had taken and used streams just as they used the grass on the public range, and they fought control of the stream just as they fought all leasing laws for the governing of the range.

They looked on their water right as they did on a homestead filing, and they thought the claim which they had recorded [the water claims filed in county courts under the old territorial law] gave them a title to the amount of water stated in the claim.

The idea of absolute right to the water claimed went even further. They looked on the stream as they did on the air, as something to be enjoyed without any limitation from a public authority, and to be taken just as they shot game or caught fish.[90]

Mead traveled hundreds of miles around the state to explain his new system in person. Forty years later, farmers in Hyattville, a tiny settlement on the western slope of the Big Horns, some four hundred miles northwest of Cheyenne, wrote him a tribute memorializing his visit to preach the virtues of the state water system. "We [re]call those days when with team and wagon you visited our ranches and taught us the 'Mead doctrine' of water rights. Your contribution to our state is an obligation which we can never repay," the Big Horn Basin Pioneers Club wrote to him in 1930, inviting him to their picnic that summer. (Mead was then commissioner of the US Bureau of Reclamation and immersed in plans for what became Hoover Dam. But he went to Hyattville for the picnic.)[91] Remembering those wagon trip days, Mead said, "I became the voice of John crying in the wilderness for a more adequate public control."[92]

Mead's superintendents, dispersed among the four major drainage divisions of the state, were on the front lines. They regularly reported to him the resistance and resentment they encountered. Opponents recognized the new system for what it was: an assault on the concept of absolute private property rights. They said it violated the US Constitution's protection of private property.

By the mid-1890s, superintendents were yearning for a court ruling to settle the question, yes or no. The decisive court challenge came finally in Buffalo— which, in its way, was the natural, and perfect, setting.[93]

———

Buffalo, and Johnson County surrounding it, epitomized the volatility of Wyoming in the early 1890s. The open range stockmen were frustrated by their inability to manage the grazing lands for their own use, even as they tried to reorder their reduced operations in light of the grim losses that had bankrupted so many of their fellows after the bad winter just five years earlier. Around them, new small ranches, some of them run by former company cowboys, were steadily increasing. Conflict between the waning old and the waxing new worlds exploded in 1892, just south of Buffalo.[94]

It was less than two years since Wyoming had won statehood and celebrated the prospects for "joy, wealth, prosperity" and the coming of the science and art of civilization, as the poet of the day had put it. Leading Wyoming stockmen pulled together a trainload of gunmen from Texas plus some locals and sent them north. Switching to horses and heading to Johnson County, the group trapped and shot down two men who were described variously (depending on the source) as rustlers or as small settlers. As was revealed later, these two were only the first on a list of people in the area that the stockmen considered rustlers and had marked for death. Ultimately, however, the "invaders," as the stock-growers' force was popularly called, wound up besieged by enraged citizens from Buffalo. The gunmen, who had certainly never expected to find themselves surrounded by a citizens' militia, made it out alive only with the aid of the US Army sent by President Benjamin Harrison. Warren, with friends among those who had organized the assault on Johnson County, was fortunate to be out of the invader action, being in Washington as one of the new state's US senators. But he woke President Harrison in the middle of the night to order in the army, rescue the besieged gunmen, and quell what was called the Johnson County War.[95]

The blazing guns in Johnson County spotlighted a West-wide situation that a Public Lands Commission appointed by President Theodore Roosevelt a dozen years later analyzed this way:

> At present the vacant public lands are theoretically open commons, free to all citizens; but as a matter of fact a large proportion have been parceled out by more or less definite compacts or agreements among the various interests. These tacit agreements are continually being violated. Violence and homicide frequently follow, after which new adjustments are made and

matters quiet down for a time . . . but an agreement made to-day may be broken to-morrow by changing conditions of shifting interests.[96]

The commission also pointed out the results for the once-fabled land of free grass: "The general lack of control in the use of public grazing lands has resulted, naturally and inevitably, in overgrazing and the ruin of millions of acres of otherwise valuable grazing territory."[97] The commission recommended a remedy—federal management of grazing via a leasing system—that took thirty years to be enacted.[98]

Once the invaders had been arrested, Johnson County leapt into a murder prosecution against the Wyoming cattlemen and the Texas gunmen they'd brought with them. Warren's right-hand man in Cheyenne, Willis Van Devanter, a hard-nosed attorney, the Republican party chairman, and a Mead ally getting the water language into the constitution, went to work. He engaged in skillful maneuvering that, combined with the county's poor finances, ultimately led the Johnson County attorney to drop the prosecution in disappointment. Warren later managed to get Van Devanter appointed to the US Supreme Court.[99]

Wyoming voters, however, did not hesitate to convict. Voter anger at the failing big stockmen's view of economic justice was eloquently expressed in the statewide elections that followed that fall of 1892. Warren was up for reelection to the US Senate. The uproar of the votes against his Republican party resulted in a legislative deadlock (at the time, legislatures elected US senators). Warren was temporarily stripped of his Senate seat—the only interruption in what became a thirty-eight-year Senate career.[100]

Buffalo was, to say the least, not likely to be friendly to Republican initiatives from Cheyenne. Even before the invasion, in fact, Johnson County had barely approved the new constitution. Neighboring Sheridan County, to the north, had flat rejected it. Burritt, the lawyer mayor of Buffalo who had backed the water provision in the constitutional convention, acted as a confidential informer for Van Devanter and other lawyers for the invaders, betraying details of the county's strategy and its difficulties in prosecution of the faltering murder case. Though Burritt's letters were confidential, people in the county likely suspected his true allegiance. As Burritt wrote to Warren in understatement, in the aftermath of the invasion, stockmen supporters like himself had to keep quiet in Johnson County.[101]

Mead, his system, and the water language in the new constitution were undeniably associated with Republicans, stock-growers, and government from faraway Cheyenne, where for the most part the stockmen had held sway. Mead was like early progressives elsewhere in the United States: he had not hesitated to ally himself with those "big business" interests to accomplish what he wanted for planning and efficiency. In the 1890s, Mead joined Warren in first urging the cession of federal lands to the state. When that failed, he ultimately joined Warren in promoting the leasing of rangeland. He saw it as helping livestock ranches become stock farms, which he saw as Wyoming's future—ranches where privately held irrigated bottom lands would be associated with grazing leases on nearby federal ranges. That pattern, already in the making, was as friendly to the small settlers as it was to larger ranches. Grazing leases, however, were initially anathema to would-be small settlers, who saw the high plains as places to be farmed and believed that both federal land cession to the states and leasing proposals were plots by big stockmen to lock up the public land for their own use. In the early decades of the twentieth century, small settlers established "dry farming" homesteads on the grasslands, particularly in the eastern part of the state, after the best lands in the valleys had been settled. Some of those homesteads failed quickly, but many persisted into the 1930s.[102]

Mead had started to put his water system into action just a little over a year before the invaders went into Johnson County. The legislature passed enabling legislation in 1891. Mead appointed the water superintendents and set them to work on stream adjudications across the state to provide expert review of water claims and determine what water rights should be protected and listed. It sounded fine. But it was a Herculean task. Mead's reports of the 1890s bulge with the detail of stream-by-stream adjudications, along with data from the streams measured, and the permits issued, by his tiny staff. The vault at the engineer's office still has on file the stacks of postcards certifying that the superintendents had notified each water user of upcoming adjudication hearings. (Sometimes today, people who dream of challenging those original adjudications come in to check the postcards.)[103]

The adjudications were a significant means of preaching, stream by stream, the value of the state system in fostering orderly water use and water-using communities. Starting with the very first adjudication, Mead noted, irrigators resisted the idea of being restricted to only the amount of water they actually put to use. But then they got the chance to examine the engineers' official measurements of how much water was available in the stream, and they could look through the compilations of claims filed over the years in county courthouses. After that, they stopped resisting, realizing that if everyone's exorbitant claims for water rights were not cut down to actual use, there would be water enough for only the first few claims on the stream.[104]

Mead launched an adjudication in Buffalo as he did all over the state. Then he hit an awkward stumbling block: his first water superintendent joined the invaders in the Johnson County War. Right there was the evidence, for anyone who had doubted it, that Mead's system was part and parcel of the program of the invaders.[105]

After the invasion, Mead had to start over in Buffalo. When the superintendent had joined the invaders, he lost all the records of the adjudication hearings. The proofs of water use had to be retaken. To finish the job, Mead looked for new people. He ran through one short-timer and eventually, in 1895, named as water superintendent the popular young railroad engineer Edward Gillette. Gillette stayed on the job supervising water and promoting the state water system in northeast Wyoming—with remarkable success—for the next five years.[106] When Gillette became superintendent of Water Division II, it was just five years since he had first led a railroad survey crew into Wyoming from the Black Hills. In that short time, however, he had acquired some significant political connections that could aid Mead's cause.

The girl Gillette married in Sheridan was Hallie Coffeen, daughter of the most prominent man in town. Her father was, in fact, the man who was elected to Congress as a Democrat on an anti-stock-industry platform in the fall 1892 election. Henry Coffeen was a former college teacher from the Midwest who, once he had landed in Sheridan, had become a merchant and an early irrigator, a member of Wyoming's constitutional convention, and a supporter of the Knights of Labor. A major street in Sheridan is still named after him. As a statewide politician in Wyoming, he turned out to be short of judgment and staying power. By the time Mead appointed Coffeen's son-in-law Gillette to the water superintendent's job, Congressman Coffeen had lost his bid for a second term. One reason was that he blunderingly prompted the army to close the fort that had been Buffalo's bread and butter.[107]

Despite Coffeen's problematic congressional record, Gillette's job was probably easier in disaffected northeast Wyoming because he had a Democratic father-in-law. Certainly, that would distance him from the image of a water superintendent who joined the invaders. Gillette also had his own solid reputation as the man who brought the railroad to the Powder River basin and to Sheridan (to Buffalo's disappointment, the violence and turmoil of the invasion had led his railroad managers to bypass that town).

Gillette himself was known to be cool headed and competent. He became a trusted lieutenant for Mead. They agreed on what the community benefits of irrigation could be. Some twenty years later, after having done a stint as Wyoming's state treasurer and run a private engineering practice of his own, Gillette helped Mead again, with inspections and critiques of federal irrigation projects in Wyoming.[108]

Gillette took his seat as superintendent of Division II on Mead's Board of Control in time to finish, in early 1895, the adjudication proceedings in Buffalo. He proved himself an excellent superintendent for northeast Wyoming, not only in the details of the adjudication he finalized but in the way he implemented the results of that adjudication.

In Buffalo, relations between people using water—like those on French Creek and Clear Creek—had become strained in the years after the town was founded. In 1889, some had gone to court to sort out water rights. It had been a dry year in Johnson County, and the town of Buffalo particularly suffered as upstream diverters took advantage of their position on the stream to take what they needed. The court case started up but was interrupted by statehood and the new constitutional provisions. Given what the constitution of the new state said, the local judge ruled that the state water administrators should take over the case.

What with the new state law being written and then the water superintendent joining the invaders, it took several years before state law could be applied to determine water right priorities around Buffalo. Moreover, it was a complicated job. On French Creek, for example, by 1894, it was not easy to delineate the water sources supporting irrigation in the valley. Was a certain field, for instance, irrigated by water natural to French Creek or by water diverted from Clear Creek? A state official would have to trust irrigators as to the source and then measure the rights by examining the irrigated land.

In the end, Gillette finalized adjudications around Buffalo in fall 1894 and spring 1895, enforcing them in the 1895 irrigation season. But he made tough rulings. On Clear Creek, for instance, Gillette decided he could certify water rights to only about 10 percent of the volume of water that most people in the Buffalo area had claimed. Meanwhile, 1895 and 1896 were dry years.[109]

French Creek ranchers chafed at an adjudication awarding them rights to less water than they had claimed. But they found in the end that their survival was ensured by receiving rights to and protection for the amount of water they had been using. The adjudication confirmed the crucial priorities. It officially listed the French Creek ranches' rights to divert from Clear Creek with a priority date just ahead of the cattle company's big diversion. The actual amount of water they had used, to which French Creek ranchers now had a right officially affirmed by the state, turned out to be enough to support their valley through all the changes of the next century.[110]

Meanwhile, by imposing a limit on the upstream claims on Clear Creek, the adjudication also seems to have aided the water supply situation of the town of Buffalo. Overall, the adjudication had arrived at a workable accommodation in water rights, which allowed the town and its surrounding ranches to survive and grow. The ranches on French Creek enjoyed workable amounts of water over the next century, even in dry years, to operate satisfactorily.[111]

Water users up and down the stream, Gillette found, saw the value of the new adjudication. He had been surprised and much relieved to find that it was the irrigators along the streams who mutually enforced the new state adjudication, because they saw it as "the only just solution of the problem." The irrigators, not part-time water commissioners alone, kept water users within the limits of their appropriations and thereby made it possible for most appropriators to use enough water to sustain their operations.[112]

Old conflicts and attempts to speculate in water had been virtually eliminated, Gillette wrote. In the Buffalo area as elsewhere, more recent appropriators had typically suffered from the actions of some of the earlier water claimants who wasted water, extended it to new lands, or loaned or sold it to someone—all for their own profit and at others' expense under their claim to "prior appropriation rights." Now, Gillette said, the water rights list implemented under state law allowed water commissioners and vigilant neighbors to limit early appropriators to the water that had been used, on the land for which the water was originally appropriated.[113]

The ban on transfers of water rights to different lands "is simply the salvation of later rights, and checks in a measure the vast advantage which the earliest rights on a stream possess," Gillette wrote. Anyone who used water on lands not described for that water right wound up with only a new water right with a later, junior date—and did not dare to object. "In times when water was extremely scarce," Gillette reported, "there has been such an intense feeling among his neighbors against such misapplication of water . . . that no other determination would long be tolerated by his neighbors."[114]

No one in Buffalo, then, seriously challenged Gillette's adjudication. Yet one entity from elsewhere did: a mortgage company from Colorado. The company objected to Gillette's adjudication and went to court over it, launching a challenge to Mead's new water system that went directly to the Wyoming Supreme Court.

The mortgage company was called the Farm Investment Co., based in Fort Collins, where Mead had started his western career. When the financial panic of 1893 rocked the nation with bank and railroad failures, agricultural prices also hit bottom, and dispossessed people marched on Washington. Buffalo had seen hard times then too, and Farm Investment had picked up several properties there by foreclosure. In the subsequent Buffalo water rights adjudication, the company successfully proved water use for the water rights accompanying several foreclosed-on properties, and rights for the amount of water used on those lands were recognized in Gillette's adjudication.[115]

But the company had one property, with an 1879 water claim on French Creek, for which it had offered Gillette no proof of water use. Possibly, the company feared that scrutiny would show the water hadn't been used. Not surprisingly, Gillette did not list that claim in the final order of the Buffalo adjudication listing certified water rights in priority.[116]

After a few years, Farm Investment could see what the adjudication meant in action. Company officials didn't like what they saw. In 1899, Farm Investment sued to stop the use of water on French Creek in the manner set by Gillette's adjudication. The company, in fact, sued a long list of people on French Creek, including everyone from the operator of the first general store to the widow of the former fort commander. One of those defendants was John Fischer, a native Bavarian who had started Buffalo's first brewery with locally grown barley to serve thirsty soldiers from the fort. Fischer's land and water right served the only ranch still making a living through cattle on French Creek a hundred years later, in the 1990s.

Farm Investment's lawyers challenged water use by all the French Creek users via a wholesale attack on Wyoming's new water system. The company used all the arguments and epithets likely standard for years among opponents of Mead's new water law. Farm Investment, like modern "property rights" groups, charged the state government with violating the Fifth Amendment. Mead's system, Farm Investment claimed, violated the Fifth Amendment's prohibition on the taking of private property for public use without just compensation. So the Wyoming courts, as the twentieth century came on, had to answer, for their own time, the familiar question: When is regulation of private property for the public good a "taking" of property—a serious invasion of private rights that the public must pay for?

In 1899 Wyoming, Farm Investment thought the answer to that question was clear. The company portrayed itself as flabbergasted to see a potentially valuable 1879 water right effectively disappear. A preexisting property right could not be wiped out merely because its owner did not come forward to participate in some new state system for listing rights, the company argued. The Wyoming water system improperly sought to have engineers be "invested with . . . the absolute control over the most valuable property interest in the state," the company's lawyers argued.[117]

Mead's new system was a bold-faced effort "to take away private property without due process of law," a piece of obviously unconstitutional retroactive legislation, the company charged. The system's limits on water rights, tied to actual use and actual measurements, were an attempt—"under the assertion of state ownership of the waters within the state"—to provide for "almost denial of property rights in appropriations of water." The purpose of the engineers' plan was "not to regulate, but to destroy. It therefore cannot be sustained, so far as it operates to defeat vested rights," the company said.[118]

Mead's plan was "an ingenious combination of provisions supposed to be adapted for the advancement of an enlightened public policy" but fatally "intermixed with others in conflict with the fundamental law and constitutional principles," the lawyers wrote.[119] The new water laws marked a change from prior appropriation law in the rest of the West, "developed by the experience of half a century." It should and no doubt would, the company's lawyers said, suffer the fate of other sets of statutes that attempt to depart from popular custom; such efforts are "but the invention of the theorist . . . a proposal to try an experiment which is generally rejected on the trial."[120]

Farm Investment was confident in the dedication of nineteenth century US courts to the sanctity of property rights. The company had not reckoned, however, with the man leading the Wyoming Supreme Court, and his dedication to Mead's new concept for water.

Charles Potter, chief justice of the Wyoming Supreme Court, was no stranger to Mead's ideas. Only six years older than Mead but an 1870s transplant to Wyoming, Potter had been a constitutional convention delegate and a partner in Van Devanter's prominent Cheyenne firm. As such, he was privy to the legal thinking of one of Mead's prime allies in promoting the new water language for Wyoming. After adoption of the constitution, Potter had become first the attorney general and then a Supreme Court justice for the new state.

When the Farm Investment case came to him, Potter was in full swing in what would become a thirty-year career on the state's high court. A good portion of his career was, in turn, a thirty-year effort to interpret and implement Wyoming's unique water system. Potter had seen the territorial conditions that had prompted the effort to write new water law, and he knew Mead's philosophy and the goals of the new system. He embraced both. In opinions written in a long series of cases, he excelled at defending, explaining, and firmly establishing the water system. On only one occasion did he deviate notably from Mead's views. After his death, it became painfully clear that the Wyoming Supreme Court had lost its link to the philosophical foundations of the state's water law. Increasingly in water cases, starting in about the 1930s, the court drew on traditions of land law that Mead had sought to avoid.

In the 1890s, however, Potter set the course to support Mead's system. In 1896, in one of the first water cases to come before him, Potter had already laid the groundwork for defeating any challenge to the new structure. He made this initial case an occasion for broadly endorsing the idea that in Wyoming, water law could and must be dictated by the necessities of Wyoming society in peculiarly Wyoming terrain.[121]

In the face of evidence to the contrary, Potter in 1896 rather boldly announced that unlike other western states such as California, Wyoming had never countenanced the old British concept of water rights, the "riparian" view that rights to water in a river were to be shared among all the owners of land along the river. By that pronouncement, Potter denied a good deal of ambiguity in Wyoming's past and expelled it from the state's future. He wrote out of existence both the probable history told by early Wyoming ditches serving riparian lands, and 1875 territorial statutes that had appeared to embrace whatever ideas various Wyoming settlers might have had about water including, for some, the idea that they owned water flowing through their land.[122]

Potter had good reason to skirt the contradictions of the past. He, like Mead, saw Wyoming's disadvantage in low levels of population and development as an advantage when it came to adopting better water law. He saw the appeal of a blank slate. And like Mead, Potter was well aware of the litigation and sometimes violent conflict that the recognition of two competing water rights theories had encouraged in California and its neighboring states.

In his library (volumes of which now rest in the basement of the University of Wyoming College of Law), Potter kept a classic treatise on California water law that documented the costly struggles between riparian and prior appropriation water rights. In 1896, when that first Wyoming water case came to his court, the facts of the case were nearly identical to those of a California case, which Potter marked in the margins of the treatise. Potter seized the opportunity then to write a decision, on the same fact situation, countering the California court's decision. The California courts had given riparian rights consideration. Potter, by contrast, ruled riparian rights out of Wyoming.[123]

Potter found in 1896 that the "imperative and growing necessities" of irrigation and other water uses in an arid state like Wyoming had required that only prior appropriation be recognized. In language that would serve in later cases to justify the development of Mead's new system, he emphasized the need and ability of the people of Wyoming to craft law to suit their own place.[124]

California's experience also inspired Potter, as well as Mead, to make room for public concerns in Wyoming's new water-law system. J. N. Pomeroy, the distinguished law writer who authored the treatise Potter studied on California water rights, believed that the use of water is, in the dry states of the West, properly a matter of public concern, and water law must recognize that.

Pomeroy argued eloquently that prior appropriation, in the usual practice in the West, did a much poorer job than riparian water law in acknowledging that the proper use of water is and must remain an issue in which a fundamental public interest is involved. He explained that riparian law, built on centuries of experience with the effects of competing demands for water, had a tradition of considering the needs of other people and of the general countryside dependent on a flowing stream. That was the origin of riparian law's requirement that landowners make "reasonable" water use, Pomeroy wrote. Riparian law treated water as a resource in which everyone had a common interest. Prior appropriation law, by contrast, in its pure form had little concern about the impacts of water use on others, he said. In fact, Pomeroy noted, prior appropriation theory gained popularity in frontier societies—like the early West—partly because it unabashedly allowed a first user to dry up a stream if he could. This potential for the prior appropriation doctrine to encourage a rapacious exploitation of water, early noted by Pomeroy, was later made notorious by twentieth century critics of western US water law. It was one of the problems Mead had sought to correct by writing public ownership into Wyoming's constitution.[125]

When the Farm Investment Company challenged Wyoming's new water system in 1899, in Buffalo, Potter had the opportunity to back Mead's effort—to emphasize and to validate public ownership and public control of water. He seized his chance. His opinion in the case, *Farm Investment Co. v. Carpenter*, written in 1900, has stood for more than a century as the keystone supporting Wyoming's system against attacks on its constitutionality. His opinion has been routinely cited, in Wyoming and in other states, whenever a Wyoming-style water-law system came under fire.

In his decision, Potter firmly repulsed two lines of constitutional attack. First, he rejected the company's argument that Wyoming was improperly taking property rights without compensation by use of the constitution's pronouncement that water is public property and its subjection of private water filings to state-imposed limits. Next, he rejected a second major constitutional claim the company had put forward, that the legislature had usurped judicial powers by giving the state engineer and his superintendents the authority to adjudicate and determine water rights.

Overall, Potter ruled firmly for the state on the core issue in the case. He upheld the Buffalo adjudication completed under Gillette. He held that the Farm Investment Company's early-date water claim—so sacred in the eyes of its owner—could indeed be lost if the company refused to participate in the state adjudication system. Potter ruled that if the holder of a water claim dating from Wyoming's territorial days failed to come forward now, under the state system, and offer proofs of water use, the old claim would have no place on the official state lists of water rights listed in priority. The old filing would have little practical significance.

How?

First and foremost, Potter declared, water *is* a public resource—not simply because the constitution of Wyoming says so, but because of a long and steady tradition among the people and in the law. The new constitution merely restated that traditional view.

Second, Potter said, the state must have authority to regulate such a public resource. The state must be able to institute proceedings to determine the relative rights of private parties in this public resource—to avoid chaos on the streams. He cited the well-recognized "police power" that government must have to keep the peace. From that he argued that governments can require preexisting rights, as well as new ones, to submit to a reasonable system of listing rights and keeping order among them.

As a result, Potter ruled, a preexisting claim held by an individual or a company like Farm Investment, if that entity fails to participate in the state's reasonable adjudication procedure, will very properly be left off the new state list of water rights. And the state's water commissioner, when dividing up waters in times of scarcity, will very properly not allocate any water to that old water claim since it is not on the state list. So as a practical matter, the old right becomes meaningless.

Legally, Potter noted, Farm Investment still had its territorial claim. The legislature hadn't put into the law any penalty—like forfeiture of a claim—for failure to participate in the state process. Potter carefully declined to read such a penalty into the law. He pointed out that the company could still make a legal claim—that it could, for instance, try in court to sue neighbors for damages when they got water and the company didn't. Perhaps the company might get money damages. But it wouldn't get water.

So the company's legal claim survived, but it had limited practical value. And it did not survive for long, even in that condition. Potter suggested that the legislature might want to write in a penalty, such as forfeiture of a claim, for failure to participate in an adjudication. And Wyoming legislators did so, at their very next opportunity.[126]

Potter laid this all out with some eloquence, but he kept the tone mild rather than impassioned. The art of his opinion was that he managed to give a resounding endorsement to the revolutionary change Mead's system had brought to Wyoming water but make the system sound not revolutionary at all.

Of course, no one else had seen it that way. The one thing that both sides in the Buffalo water case agreed upon was that Mead's system was a dramatic departure from any water law in Wyoming's territorial experience. Not only the Farm Investment Company, on the offensive, but the French Creek ranchers, defending the new law, and even their mayor Burritt, back in the constitutional debates ten years before—they all thought it was radical change. They differed only on whether it was a good thing.

Potter performed quite a tour de force. The chief justice of the new state clearly saw his task as one of solving the problem of how to uphold the new system without putting the court in the position of endorsing drastic (and, as Farm Investment would have it, potentially unconstitutional) change in property rights matters. He accomplished that by portraying Mead's system as a natural outgrowth of both territorial law in Wyoming, and the theory of prior appropriation itself.

On how water was viewed in mid-1880s Wyoming, as the territory grew toward statehood with small settlers increasingly populating the landscape despite the big range cattlemen, Potter wrote:

> The cultivation and even the occupation of the lands within the territory had been attended with the expenditure of much capital and labor, and the very existence of the homes of a large class of citizens, as well as the productiveness of the soil, depended upon the security to be afforded the appropriations of water which had been made; and in view of the many rights already accrued, and the inception of new ones which would necessarily accompany the continued growth of the territory, the welfare of the entire people became deeply concerned in a wise, economical and orderly regulation of the use of the waters of the public streams.[127]

By contrast, one of Potter's peers in the 1889 constitutional convention—a member, like him, of the convention's water committee—had characterized territorial water legislation as a worthless hodgepodge:

> The legislatures of this territory have attempted to deal with this (water) question, and from time to time demonstrated their ignorance of the whole matter, legislating in one direction at one session, and undoing all their work the next.[128]

That was a somewhat common view of what the territorial legislatures had wrought. Nonetheless, Potter blandly asserted in 1900, in the *Farm Investment Co.* case, that territorial water law had envisioned an overall goal. That goal, he said, was to protect the economic interest built on water, "give stability to its values, assist in a desirable conservation of the waters, and avoid confusion and difficulty in their distribution."[129]

Since Wyoming territory began, Potter concluded, in water law, "the significant feature of the changes and additions from time to time has been the principle of centralized public control and regulation."[130] He moved right on to quote the water provisions of the constitution of 1889 and the subsequent statutes with their elaborate structure of centralized public control. He saw no need to mention

Mead and the labors of 1888–1890, which created the new water law. In sum, the Wyoming Constitution "would seem rather to declare and confirm a principle already existing, than to announce a new one," Potter declared.[131]

He found, remarkably, traditions of public control of water not only in Wyoming's territorial law but in prior appropriation law itself. That theory, he noted, developed on the public domain and from the needs of the arid country and therefore was based on public ownership of water. "Under the doctrine of prior appropriation, it would seem essential that the property in waters affected by that doctrine should reside in the public. Such waters are, we think, generally regarded as public in character."[132]

Moving on again, Potter found that it is appropriate for government to regulate a resource that is "public in character." Where water supplies are short, the state must exercise its power of public control, regulating who gets water first according to a comprehensive list of rights, in order to allow the proper functioning of individual rights to use water. And just as with deeds to land, where the state can require all transactions, old as well as new, to be recorded publicly in order to ensure valid land transfers, so state water regulators can require all water claims, old as well as new, to submit to the state process of proof and adjudication in an official water rights list. Potter continues:

> Where various rights are connected with the same stream or body of water, a
> subsequent claim cannot be successfully regulated without including, in the
> regulations, all rights. The water to which the use of each attaches is public, and
> the people as a whole are intensely interested in its economical, orderly, and
> inexpensive distribution. It is a matter of public concern that the various diver-
> sions shall occur with as little friction as possible, and that there shall be such
> a reasonable and just use and conservation of the waters as shall redound more
> greatly to the general welfare, and advance material wealth and prosperity.[133]

To be effective, the regulators must determine the relative priorities of all the relevant claims, old and new, on a stream. To do otherwise "must result in practical failure, in times when official intervention is most required. In fact, that had been demonstrated under our former system," Potter declared.[134]

Finally, Potter backed Mead on his choice of whom to put in charge of sorting out these water rights to create the state list. It was properly not the job of the courts but a task for technical experts—the state engineer and his staff:

In the development of the irrigation problem, under the rule of prior appropriation, perplexing questions are continually arising of a technical and practical character. As between an investigation in the courts, and by the board, it would seem that an administrative board, with experience and peculiar knowledge along this particular line can, in the first instance, solve the questions involved with due regard to private and public interests, conduct the requisite investigation, and make the ascertainment of individual rights with greater facility, at less expense to interested parties, and with a larger degree of satisfaction to all concerned. In the opinion of an able law writer upon this subject, the powers of the Board of Control in this respect constitute one of the most praiseworthy features of our legislation.[135]

Here Potter quoted a contemporary writer on water law:

He says, 'In the State of Wyoming, at least, there will no longer be the ludicrous spectacle of learned judges solemnly decreeing the right to from two to ten times the amount of water flowing in a stream, or in fact amounts so great that the channel of the stream could not possibly carry them, thus practically leaving the questions at stake as unsettled as before.'"[136]

———

Dedicated to "the principle of centralized public control and regulation"—with Potter's endorsement of that core concept in 1900—Mead's water system was solidly enshrined as a distinguishing feature of Wyoming government.

With the principle of active state ownership, administered by a cadre of technical experts, Wyoming's system could distribute rights to water in a way that both protected and limited those rights. It could, ideally, provide to a growing new society both security and the flexibility to accommodate change.

"The Wyoming system" was the latest thing for water people all around the West. The concept was promoted (by Mead and by others) as a model for other states. Several states did institute some form of the Wyoming system—but few were willing to follow Mead and Potter all the way.

Most states resisted the idea of administrative experts adjudicating water rights. Even when they saw the advantages of taming pure prior appropriation and therefore embraced the ideas of state ownership and state permits for water, most states balked at giving a board of experts the power given to Wyoming's board to determine and limit private rights. So those states that adopted part of Wyoming's model typically left out the administrative adjudication of rights. Often, under pressure from lawyers and judges who saw water rights as their business—a matter of property rights, much like rights in land, and therefore the domain of the courts—they declined to adopt that portion of Mead's ideas.[137]

Potter's ruling confirming Mead's system in its entirety, however, authorized Wyoming to pursue its experiment in water management. Key content of the water law, like the tying of water rights to the land, had yet to be tested in practice. The unique authority Mead's system gave to a board of practical experts to interpret and implement the idea of state ownership of water, however, proved itself a critical factor in the working of the Wyoming experiment.

With Potter's decision on the Buffalo case, the work of Mead and of his superintendents like Gillette was justified. Their efforts had paid off. Next came the problems of seeing the water system implemented day to day and seeing whether it could go on as it was intended. Dreams of development, and the practical frustrations of trying to make those dreams come true, brought about changes in the original system. Indeed, by the year of the court decree in 1900, those changes were already at work. In 1900, however, what seemed most important was that Potter had affirmed the Buffalo adjudication and thereby validated the idea of public ownership and control of water.

Fred Bond, a Mead lieutenant, called Potter's decision a "subject for especial congratulation." By 1900, Bond had become state engineer, since the fame of the new water system had taken Mead to a job in Washington, DC. Bond wrote in his report to the Wyoming people that the question of whether water is publicly or privately owned was now settled in Wyoming, in favor of the public, as was only right. Any further debate on the topic could be relegated to other states "still struggling to reach the high plane occupied by the Wyoming standard," Bond said.[138]

Gillette, meanwhile, had gotten the most satisfaction from the way Johnson County people had accepted the system long before the Wyoming Supreme Court upheld it.

"It has been somewhat of a revelation to me," Gillette reported to Wyoming readers back in 1896, "and strengthens me in the opinion that our water laws are exceedingly just and effective, and that more than ordinary care and foresight was used in their framing than our citizens ordinarily appreciate."[139]

2. PUBLIC ORDER vs. PRIVATE GAIN

> I was told later that two men were guarding us kids from
> the creek bank 200 yards away with Winchesters.
>
> —JESSE SLICHTER, RECALLING HIS FAMILY'S
> STRUGGLE FOR LAND AND WATER.[1]

Places: Box Elder Creek, central Wyoming; Crazy Woman Creek, northeast
Wyoming's Big Horn Mountains.

Time: 1890s.

Careyhurst and ditch, Box Elder Creek. J. E. Stimson Collection, Wyoming State Archives.

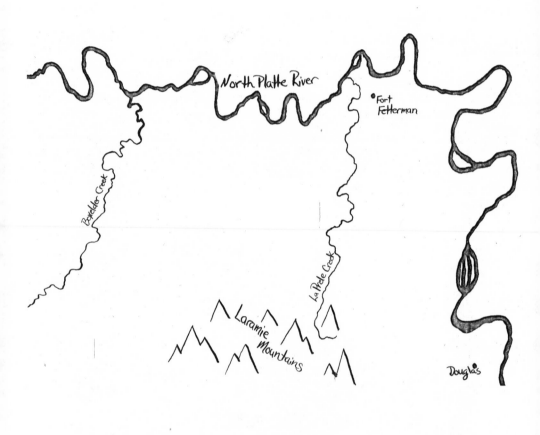

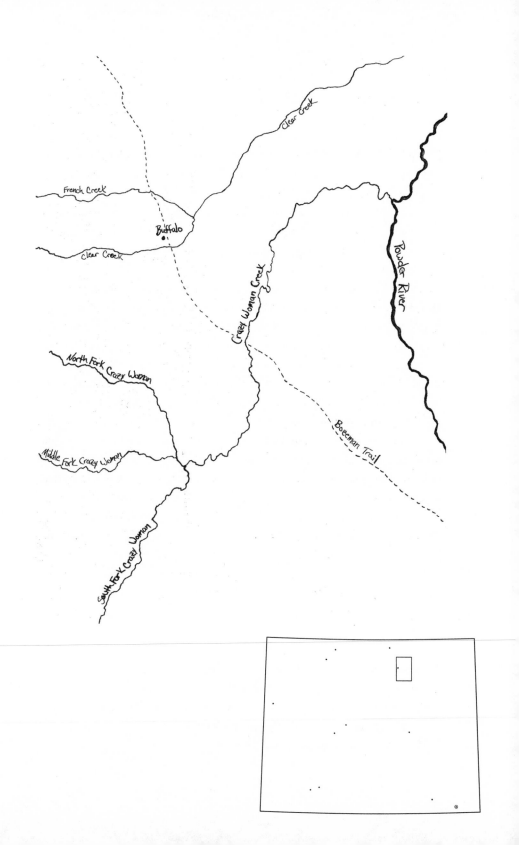

THE WORLD IN WHICH Mead sought to work shared, of course, very little of the logic with which he shaped his new water system. Wyoming in the late 1880s and early 1890s was an oddity in nineteenth-century America, an anachronistic "frontier" between the two developed coastlines symbolizing the Gilded Age. As such, it played host to two worlds. For many, it was a place to build a cabin or a dugout, to start a new life. For a few others—those with enough money to play—it had been a gaming table made of grass; it was not a place to live.

As Mead set up his water system, those two worlds were still in conflict, and the friction produced sparks that cast a lurid light, as in the Johnson County War. Mead's vision of a public interest governing the distribution of water made headway, but his water system inevitably bore the imprint of the times. Power contests that used water as a weapon could not be stopped by Mead's system and his few lieutenants working across a vast territory. Most people, as Mead well knew, believed that water like land could and should be the basis of private property, and therefore potentially of wealth. There was plenty of temptation and opportunity to make one's own rules.

Yet Mead's work had an impact. His system cut down to size the dreams and sometime pretensions of the 1870s and 1880s; it made it possible for a more stable society to take root. Gillette's adjudication in Buffalo demonstrated that. Mead's concept of an overarching public interest in water and its orderly use slowly shaped people's reality. The frontier nature of nineteenth-century Wyoming affected Mead's system, but his logic also changed the Wyoming that moved into the twentieth century. Each left its mark on the other.

Events in two locations demonstrate how that happened. The first was on Box Elder Creek, a small tributary of the North Platte River in central Wyoming.

———

The story of Box Elder Creek is dominated by one figure: Joseph Carey. He was originally an ally of Francis Warren and even did Warren's campaigning for him in late summer 1890 when Warren was ill during the race for governor of the new state. Mead was well acquainted with Carey.[2]

Carey and Warren became two of the "grand old men" of the state's political folklore. They were the vivid, long-remembered, nineteenth-century originals, emblematic of the ambitions for personal wealth from Wyoming's resources—water, grass, or (later) minerals. With their identity and their wealth rooted in the open-range stock business, they carried the state's nineteenth-century roots several decades into twentieth century politics. Warren, Mead's sponsor, was territorial governor, state governor, and US senator for nearly forty years. Carey was territorial delegate to Congress and author of statehood, briefly US senator, and a one-term governor, running against Warren's faction, just before World War I. Mead worked with them for most of his life, as he too went to work on a national level at what became the Bureau of Reclamation. Warren and Carey were not to be evaded; their goals and their way of operating were a major feature of the world in which Mead's system had to make its way.[3]

There is perhaps none better than "Judge" Carey, as he was known much of his life, to represent the attractions and contradictions of the Wyoming world where Mead worked. A fellow Republican leader (later a federal judge) said of Carey years later:

> Carey was perhaps the most astute and effective stump speaker Wyoming has ever produced. While his political statements were not strictly based on recorded facts, he was able to convince the voters in an unusual way of the logic of his arguments.
>
> I recognized Carey as a man of pre-eminent ability and frequently re-marked that any citizen of Wyoming might well feel proud of him when he appeared as he frequently did among the Governors and Statesmen of the Nation. He was a big man in big things but he was possessed of a peculiarly vindictive disposition and temperament oft-times akin to a school-girl. He took to heart any opposition to himself and punished those who had op-posed him in an exceedingly icy manner.[4]

Carey appeared to have viewed politics and political office as matters of personal sway. By contrast, Warren took a pragmatic approach, focusing on party unity and a fine-tuned patronage machine, to make federal money an important driver of Wyoming's economy. Warren had by far the longer political career.[5]

Joseph Maull Carey came to Wyoming as a bright young man from a well-established East Coast family, looking to make his fortune. He had sat out the Civil War and came to the new territory of Wyoming as its first US attorney in

1867 when he was twenty-two, just out of law school. The job was his reward for working back home on Republican campaigns. Four years later, he was on the territorial Supreme Court, where he got his lifelong title as Judge Carey. But in 1876, just before veteran Civil War officers launched the last assaults on native tribes in Wyoming, Carey lost his court post in a factional party fight, and he turned his attention to cattle. He had fifteen thousand head brought to the North Platte River the next spring after the great grasslands north of the Platte became open for cattle herds to enter.[6]

Further west, on the Wind River, the Shoshone and Arapahoe were also turning to cattle. Until the mid-1870s, the Shoshone had prospered by hunting and by selling hides. The 1868 treaty creating the Wind River Reservation intended to cast the Shoshone in the role of farmers. They received initial government support in learning to farm, with some success. Their real livelihood, however, was based on hunting, until buffalo and other game became scarce. As the game herds disappeared, as their people became more impoverished, and as whites made homes and mines on the reservation, cattle herds emerged as their persistent goal. In 1872, the Shoshone sold some reservation land south of the river to the government in exchange for cattle, first running them effectively in communal herds, but then forced by the reservation agent to parcel the herd into individual ownerships. The process enabled nearby whites to buy out the herds of individual Shoshone cattle owners for cash. Congress, opposed to "perpetuation of distinctive national characteristics in our midst" and sure of the benefits of small individual farms, increasingly pushed for the native people to develop private farms on reservation land while "surplus" land on the reservation would go up for sale to new settlers. The Arapaho were forced to live on the reservation starting in 1878. Proposals and negotiations for cutting down the reservation continued for years as the world of settlement and commerce grew more active at the reservation's borders.[7]

Carey, of course, was at ease in that world. Back on the Platte River, where Carey planned a cattle kingdom, he made himself a place near the river crossing at the army's Fort Caspar. He established it personally, taking a hand in building the first one-room log ranch house himself. He also set up operations on a place about thirty miles east, near the next army outpost, Fort Fetterman. The spot he chose, where the creek called Box Elder met the river, had meadows that had been harvested for native hay in a few previous years by men who made a living as freighters and supplied beef and hay to the army and its horse herds. Carey headquartered there, south of the river, eventually naming his place Careyhurst. In 1877, he turned his cattle out to the north—swimming the herd across the sometimes treacherous river to reach the good grazing.[8]

Others also brought in cattle, and the new industry grew quickly, attracting investors told to expect 150 percent profit in five years from the "free grass." Markets for beef were growing in a hungry postwar United States, and on its frontier were free pastures. That brought Wyoming its brief decade of fame, from 1876 to 1886. On the edges of the grasslands, a few mansions sprouted, where guests enjoyed multiday hunts; champagne was served at the cattlemen's club in Cheyenne.[9]

Carey spent much of that decade organizing the stockmen to pursue their interests. It was a heady challenge for a lawyer, guiding a new industry working in new terrain. Helping to recast a fledgling Wyoming Stock Growers Association (WSGA) in 1874, Carey became its president at the industry's peak in 1884–1885, just before the crash resulting from the disastrous 1886–1887 winter.[10] He became a Republican national committeeman and in 1885 was sent to Congress as the sole representative of Wyoming Territory.[11] Though the grazing lands were public domain subject to laws encouraging settlement and land ownership, the cattle companies' "home ranges" extended far from headquarters without any ownership. Boundaries existed, but they ran unmarked, known only to the people in the business.[12] Accompanying the joint roundups and cattle brandings, the companies and their association focused energy on blocking competition from small ranchers who were sometimes former company cowboys. The WSGA represented a majority of cattle on the range, but not the majority of cattle owners. They wanted to keep non-member operations in check and discourage the emergence of new ones, and they needed to draft strict rules to do so—association rules or, when they could swing it, official legislation of the territory. Carey and close associates did that work.[13]

First, the WSGA adopted an internal blacklist rule that forbade members to hire a cowboy who owned his own brand or a herd. Then, in 1884, as Carey became president of the association and then the territory's delegate to Congress, the organization successfully shunted a "maverick law" through the Wyoming territorial legislature. The law gave the WSGA sole control in disposing of unbranded calves on the range that were found without their branded mothers—the "mavericks." The WSGA then made sure that most of those calves went to its own members, making it all but impossible for smaller ranchers to buy the mavericks and build themselves a herd. The transparent goal was to ensure that cowboys remained wage earners, not property owners (and that anyone who had his own small herd might therefore be conveniently labeled a thief). In 1886, association members joined in a wage cut, resulting in an unprecedented cowboy strike. Not all WSGA ranches resisted striker demands to restore wages to the pre-cutback levels, but Carey and his "CY" ranch near Fort Caspar did. The WSGA's blacklist and its maverick law—"class legislation," in the view of the Buffalo correspondent of one Wyoming newspaper—are what led to a slow-burning discontent and ultimately a level of violence in Wyoming, demonstrated in the 1892 invasion in Johnson County, that did not mark the cattle industry in nearby Montana or Colorado.[14]

After the bust of 1886–1887, the surviving cattlemen, trying to become better businessmen, were haunted by the memory of carcasses piled under melting snow. Some investors discovered that their herds had been magnificent only on paper, and the surviving operations began to think seriously of feeding hay to real cattle, in winter. Native hay had been harvested for years in creek-bottom lands like Carey's on Box Elder, but the new idea was to go further, to water whatever could be watered to raise hay and feed the herd through the winter. The companies that had pulled through the bust began to think more seriously about irrigation, on land they could own. The federal Desert Land Act, enacted a decade earlier, had made construction of irrigation ditches one way to obtain land. The law was often used by Wyoming settlers to obtain land they irrigated for farms, with a few cows and a garden. Cattle companies, too, had sometimes put it to use—for real, or with fraudulent "dummy entry-men" to claim lands laced with ditches that stayed remarkably dry, there for nothing more than to claim title to the land. Fences were sometimes strung across the public range illegally. When young Elwood Mead came to Wyoming, he reported seeing networks of dry ditches. The outsized claims to water he recorded were also sometimes the work of stockmen trying to lock water away from small settlers. The federal government's General Land Office gathered evidence suggesting that some Wyoming stockmen had made fraudulent land claims. Carey was no exception. In 1885, he expanded his holdings on Box Elder Creek by buying out three men who had made Desert Land claims two years before; he had a contractor dig the ditches for them, paid them $48,000 for their claims, and paid the federal government its $640 fee for the 640 acres. A General Land Office investigator called foul and canceled the claims and with them Carey's title to that land.[15]

———

Carey saw a role for small ranches and farms in Wyoming—in certain places. In 1883, well before the bad year of 1886–1887, he had joined other WSGA members in starting the prototype water development company that aimed to profit by selling irrigated land to colonists. Apparently not designed as a water monopoly of the kind Mead deplored, the company sold both water rights and land to settlers. After the crash, Carey managed to hold on to his ranches—and also, in 1888, worked with an oncoming railroad to plat some of his CY ranch into city lots for sale to make some money on the creation of the new city of Casper. As one of the authors of the WSGA's toughest legislation, however, Carey was not friendly to settlers infringing on his world. In 1886, three would-be settlers eyeing Box Elder Creek had filed the complaint against Carey that led the General Land Office to investigate and cancel the Desert Land Claims he had bought that covered key bottom lands on the creek. The settlers moved right in on those lands the next year. Carey wouldn't stand for it.[16]

One of those settlers was John Slichter, from Iowa—"very tired of the Iowa mud," his oldest son said. Slichter came to Wyoming in 1883 with his wife and three children and settled on a small place on the North Platte River. He dreamed of breeding and selling horses and pursued that with modest means. Slichter also hauled freight, to and from the new railroad line stitching west to the new town of Casper. But Slichter's original place on the North Platte was not ideal. The river was an unreliable neighbor, not easy to tap for irrigation water. Fed by mountain snowmelt, the Platte would surge into violent flood stage in May and occasionally cut itself a new channel—and then dwindle to a listless near-nothing in fall and winter. Box Elder Creek, flowing into the Platte a few miles upriver, had good bottom land and timber and was a far more manageable water source than the river. Slichter saw the advantages and led two other men in filing the complaint against Carey over the Desert Land Act claims Carey had bought up the year before. Slichter had the land surveyed, and he testified to federal land officers in Cheyenne, seeking to show that creek bottom land was never "desert" and then to file on it himself under the Homestead Act. Carey was still Wyoming's delegate to Congress, but although it took a while, his title was eventually canceled. After a year in a nearby little coal town (a century later, the site of a major power plant), Slichter moved his family onto 160 acres on Box Elder. He built a small frame house there and a ditch, and he even started delivering to nearby ranchers coal dug along the creek. He grew a crop of oats. He figured he could water some one hundred head of stock and horses on the creek and irrigate to grow hay for feed. In October 1889, he began to dig a ditch. But, as he complained to Mead's superintendent nearly a decade later, soon, "a part of my land that it was constructed for was taken from me."[17]

Slichter had, of course, run afoul of Carey, who wasn't about to lose any Careyhurst property without a fight. Carey, in turn, unwittingly ran into a witness who left a rare record of what the Judge would do to someone who got in his way.

Slichter's oldest son, Jesse, was about ten or eleven years old at the time. The struggle between his father and Carey, in about the year 1889, was seared into his memory. Many decades later, when asked to write about family history, Jesse Slichter wrote about Carey.[18]

As could have been expected, the story he told is one-sided. But it is suggestive of what was said of Carey by his smaller neighbors. Carey, wrote Jesse Slichter, pushed out settlers who had come in on the canceled Desert Land entries near Careyhurst on Box Elder. The method was straightforward and nasty: Carey had his men go with wagons to the homestead cabins and dump everything people had into the creek. If possible, they did the job when the families were away.

In the case of the Slichters, Jesse wrote:

> I well remember the day our time came. Two wagons and four men and the foreman, Edward David, drove into the yard. [He] ordered the wagons backed to the door but the men refused on finding a woman and children. Lots of angry words were exchanged before they left. A short time later . . . Father and Mother were arrested and ordered to appear in Glenrock, 12 miles away. Father's lawyer advised him to leave us three kids at home alone. The same wagons appeared, but when they saw us playing in the yard, they got scared away. The judge dismissed the case. I was told later that two men were guarding us kids from the creek bank 200 yards away with Winchesters.[19]

Jesse Slichter took care all those years later to name Carey's foreman because the foreman, like Carey, was not just anyone. Edward David, grandson of a wealthy family in upstate New York, had come west about six years earlier following an uncle who had been the first surveyor general of Wyoming Territory, in 1868. The surveyor general's daughter had married another early territorial officer—Carey. Edward David was therefore a young cousin of Carey's wife. David ran a ranch in northwest Wyoming for Carey until the 1886–1887 disaster and then came to the Platte to superintend all of Carey's ranch interests in central Wyoming, including on Box Elder and in Casper. He soon helped organize a new county around Douglas, was elected to the territorial legislature, and became a trustee of the new University of Wyoming. To Jesse, however, he was the man who came to push the Slichters out of their new house.[20]

Carey won in the end, for the most part. He eventually got the key Box Elder properties back, titled to him. He was in Congress, and a Republican— and the General Land Office investigations were started by Democrats who had taken power only briefly in Washington and lost it in 1889, the year Carey got the 640 acres on Box Elder back. Carey summarized it all quickly, nine years later: "Slichter squatted on the land while it was in contest by the U.S. and afterwards had to leave it when contest was decided in our favor." By the time he said that, Carey had built Edward David and his family a big stone house on the Careyhurst ranch, a house that was still standing nearly one hundred years later.[21]

John Slichter, meanwhile, didn't give up. He had his mind set on getting a better water source than the unruly Platte. He still had his ranch on the river. He tried lengthening the ditch he'd started on Box Elder Creek so the ditch could take the creek water back to his river ranch. It took seven years to get the whole ditch dug (he had all kinds of trouble on the surveying work and nobody but young Jesse to help run the scraper for the ditch), but he did it. That was when Slichter and Carey came into conflict again. This time it was over water rights, and this time Carey did not get all his own way.[22]

This new confrontation started in 1898, the year Mead's new state-supervised stream-wide adjudication process reached Box Elder Creek. James A. Johnston, the superintendent of Division I, gathered evidence on oath in April as to how much water people took and where they used it. He took most evidence in Glenrock, west of Box Elder; but he took Carey's evidence in Cheyenne, for the Judge's convenience. Wyoming was a small society, and Carey was a big man (and by then in a standoff with Warren as part of a bitter Republican Party feud).[23]

Johnston knew Carey well; he had recently been president of the irrigation colony company that Carey and the other cattlemen ran.[24] Johnston was also a friend and colleague of Mead's, having helped get Mead to Wyoming. A Democratic territorial legislator in 1888, and then a delegate to the state constitutional convention, he helped write Mead's water language into the Wyoming Constitution. In the 1890s, Mead made him superintendent of the water division of the North Platte basin, which included the capital, Cheyenne.[25]

Johnston took testimony from Slichter in the 1898 dispute with Carey. Slichter declared that the ditch he had built to reach his North Platte River ranch was part of the original ditch he built from the creek to water the place on Box Elder, where Carey's foreman came to push him out. Slichter started digging the ditch on October 14, 1889. That date would put his water rights on Box Elder ahead of some other ditches, including some of Carey's ditches on Careyhurst. Without the 1889 date, Slichter said, he couldn't even get enough water to keep a garden going on his lands by the river, much less to grow hay for his stock.[26]

Carey, for his part, stated simply that Slichter had "squatted" on Careyhurst land back in 1889 and had simply left. Carey claimed that he himself had used the ditch Slichter once built, ever since the land ownership was settled in Carey's favor and Slichter was back on the river.[27] Johnston ruled for Slichter, though he cut back the water volume in both men's water rights. His ruling meant Slichter got reliable water from Box Elder Creek to use on his place by the North Platte. Carey didn't lose much in terms of water for Careyhurst—he had plenty of other water rights running through a welter of ditches.[28]

Between the big cattleman and the small-time settler, it was not the federal land office but Mead's water-law system—and its elaborate process of testimony and onsite inspection—that ultimately had some impact on what happened there along the Platte. Carey got rights to land and to water, but not to the exclusion of the new people coming in.

————

The Slichters, it turned out, were among the people who had come to stay and to build a town serving small farms and ranches, people whose families would stick around longer than the Careys. John served on the first Wyoming jury after statehood. Jesse and his brother Charley became county officers around the time of World War I and after.[29]

The town these people built, Douglas, took over as the local center from the remnant of Fort Fetterman, whose buildings for a time had served as the store, saloon, and hospital for the surrounding region. In those days, the area had a deputy sheriff, named Malcolm Campbell, just about Carey's age. A Canadian farm boy, Campbell had made it to Wyoming after the Civil War and wound up a freighter, driving bull teams of supplies for the Wyoming forts while the troops were active there. In the 1870s, he too had lived on Box Elder Creek, in a little adobe house, and harvested hay there to feed the army horses at Fort Fetterman. As the wars ended and the cattle business came on, he sold out. Campbell's place ultimately became part of Careyhurst.[30]

Campbell signed up as deputy near Fort Fetterman on the Platte under the county sheriff who was one hundred miles away, so he was pretty much on his own. His job was to keep some kind of order in the sudden little settlements on the edge of the great new cattle ranges. Hundreds of cowboys were soon making big wages on the range, and they would come to "town" at the old fort to spend their money. They were "young and carefree farmer boys up off the farms of the eastern states," Campbell said. A "happy go lucky lot . . . few among them were deadly or unfriendly"—until they started drinking, Campbell recalled. Then the gunfights started. Jesse Slichter said his mother and aunt were terrified whenever John Slichter went to old Fort Fetterman for supplies. The drunken brawls could entangle bystanders.[31]

It sounds like the setting for an old-fashioned Western, and it was. The straggling town at old Fort Fetterman was the model, for Owen Wister, of the town where he set *The Virginian*. In the summer of 1885, Wister spent two months as a guest on an area ranch owned by "Major" Frank Wolcott, a Kentuckian who had fought on the Union side in the Civil War, come to Wyoming to work for the General Land Office, and become a cattleman later investigated by the General Land Office for fraud. On his visit, Wister noted that Wolcott was battling squatters nearby—trying direct intimidation and blackballing them with the Stock Growers Association so they couldn't hire out part-time with cattle companies to bring in money while proving up on their homesteads.[32]

In 1885, unknown to Wister, Wolcott was beginning to take out heavy loans that, after the 1886–1887 drought and hard winter, he wouldn't be able to pay back. Foreclosed upon in 1892, Wolcott became bitter and desperate. It was he who led the Invasion of Johnson County. A cattleman colleague described him as "a fire-eater, honest, clean, a rabid Republican with a complete absence of tact, very well educated and when you knew him a most delightful companion. Most people hated him, many feared him, a few loved him." A neighbor said, "I never knew a man more universally detested." Deputy Sheriff Malcolm Campbell described him as honest, "a polished gentleman," and also a "bantam rooster," and summed him up: "Wolcott was a man of very positive convictions, and his experiences with small ranchmen around his VR Ranch had often been of a violent nature." Wolcott joined Edward David as one of the first commissioners of the new Converse County in 1888. He had also been, for four years, a justice of the peace. Campbell was elected deputy and was in office when Edward David took Carey's men to push the Slichters out of their cabin.[33]

Years later, Campbell set down his record of the world Wister romanticized. He described "really vicious" criminals he had to deal with, but added, "there were many thousands of other cowboys who were loyally serving their employers, doing their daily duties willingly and well, and helping with the example of their fine, strong personalities, to build up the self-respect of the country."[34]

Campbell felt, however, that the thousands of cattle on the range "offered a great inducement to cattle thieves, as the cattle were free and roamed for hundreds of miles, uncounted and untended. At the termination of most of the drives, the cowboys were paid off and left to shift for themselves. Many lost their money in the first gambling joints, and found themselves footloose in a land teeming with cattle which could with very little risk be driven away and sold." And, Campbell said, after the cattle bust following the winter of 1886–1887, cattle stealing became a major problem, and he was frustrated by what he saw as the impossibility of getting local juries to convict for cattle theft.[35]

At the peak of the boom of the early 1880s, meanwhile, the big cattlemen— particularly the British—were, according to Campbell, "often arrogant, unfitted for the west and its demands, and careless of the country and its people. The cowboys in their employ knew them only with amusement, and there was built up then from the first a class contempt which was to reap its whirlwind in blood ten years later" in the Invasion of Johnson County.[36]

Campbell dictated all these memories at age ninety to Robert David, the son of Carey's ranch superintendent Edward David. Robert, adopted from Chicago in 1897, lived in Douglas as a teenager. His father built a house in town where, in the World War I era, the family dressed in white tie and tails for dinner and were served by a Japanese cook. Small ranchers, like those Edward David had ousted from Box Elder, resettled and raised their families nearby. Robert married, clandestinely, "beneath him" into such a family, who ran sheep. He found that he could breathe easy in Douglas only after his father retired and moved several hundred miles away to Denver.[37]

———————

Some 120 miles northwest of Box Elder Creek was Crazy Woman Creek, in the foothills of the Big Horn Mountains, about thirty miles south of Buffalo. The story of water rights there illustrates the impact of frontier Wyoming on Mead's system, rather than the other way around.

Malcolm Campbell, down in Douglas, knew a wide stretch of the range. He counted "only two ranchmen of limited means on all the land between the Platte and the Missouri" in Montana—an area of several hundred square miles. Both those small ranchers were on Crazy Woman Creek. One was named John R. Smith. Smith did not like being pushed around on either land or water rights.[38]

Like Campbell, Smith was a former army freighter. Entering the Civil War at seventeen, he was a color-bearer for an Indiana regiment with Grant at the siege of Vicksburg. Immediately after the war, he came to Wyoming, went on a hunt with a portion of the Cheyenne tribe, got work as a freighter and a scout for the army, found a wife, and started a family of his own. He sent the family back home when the fights with the Sioux got hot in 1875–1876, and then he worked as a scout and freighter for the army. After that, he settled in. Using the Desert Land Act, he got title to land close to where Crazy Woman Creek intersected the Bozeman Trail on its route north from Fort Fetterman to Fort McKinney and the future town of Buffalo.

Crazy Woman is not a particularly generous stream. In its neighborhood, in fact, it's the byword for dry. "If it's dry anywhere, it's dry on Crazy Woman," say state water commissioners on the east of the Big Horns. Smith's spot on the main stem of the creek, below the confluence of three tributaries, was probably the best place for water on Crazy Woman. His choice echoed the classic pattern of first settlers on Wyoming creeks: the earliest water rights, with the highest priority, are typically on the lowest, best spots, while later settlers make places for themselves upstream.[39]

Smith began digging irrigation ditches in 1878 to meet the requirements for getting title to the land under the Desert Land Act. He brought his wife and children to the ranch in 1879, cut native hay, and grew what vegetables he could (cabbage did best) for sale at the nearby trail crossing station and the fort. The territorial governor appointed him to help organize Johnson County in 1881. In 1888, he helped petition the legislature about the need for schools. The Smiths stuck it out and did all right—the family kept the ranch until the 1950s.[40]

As Campbell noted, however, Smith's place was small, atypical in that portion of the Powder River Basin. Big cattle companies predominated. In May 1883, five years after Smith started his ranch, Crazy Woman Creek saw hundreds of cowboys and two miles of wagons in the big roundup of May 1883. Soon after, the forks of the creek in the hills up above the roundup were the setting of considerable water maneuvers—resulting in a tangle of water claims that long resisted straightening out under Mead's system.[41]

In 1884, a big British-owned cattle company ordered its foreman to start focusing on "doing all we can to take up water" on the North Fork of Crazy Woman Creek—one of the small mountain basins above Smith. The company manager, Moreton Frewen, was a Brit, later known to his peers as "Mortal Ruin," who had built what locals called a "castle" on the Powder River. Frewen said he was harried by new settlers "claim-jumping" on what he had regarded as his company's range. He ordered his foreman to take up claims on whatever land and water he could.[42]

The ranch foreman was named Fred Hesse. He came from a middle-class British family, had once studied law, and wound up working for Frewen's Powder River Cattle Company in Wyoming and running the big roundups on Crazy Woman in the early 1880s. Around 1885, Hesse made land and water claims, as Frewen ordered. One of his best friends did much the same, making water claims on the Little North Fork of Crazy Woman to form an irrigation company planning to sell water. Hesse's friend was called Frank Canton. Born Josiah Horner, Canton changed his name after being convicted of bank robbery in Texas. Landing eventually in Wyoming, Canton fooled the people of Buffalo into electing him sheriff and the WSGA into naming him a stock detective in 1886. Hesse and Canton became infamous in northeast Wyoming a few years later as the local agents who helped fire up and guide the stockmen in the Invasion of Johnson County in 1892.[43]

In 1885, when Hesse and Canton filed on water, the Wyoming ranges were still in a "wet cycle" that made the land look lush. But the next year, 1886, brought the drought summer that preceded the famous bad winter. No one had been there long enough to have seen a drought. Getting by on Crazy Woman was tough. Smith wasn't getting water, and he didn't believe the problem was simply drought. He'd kept an eye on all the activity up above him: "They were all using it above me, and they were all taking it out," he said, even though he had gotten there first.[44]

Coincidently, 1886 was also the first year of the territorial water law—probably intended to help livestock companies shore up their water claims—which called for people to file any water claim, new or old, in the county clerk's office rather just in a notice posted beside the ditch. In Buffalo during that hot dry summer, all kinds of filings came in, usually testifying to a ditch-digging effort that had begun a year or two before. People seemed to have remarkable memories of the exact day on which they'd started to dig a couple of years before. The exact date would, after all, set the priority of their right. Their signed statements of claim were accompanied with reports on the dimension of their ditches. Smith, as well as Hesse and Canton and plenty of others, all filed papers declaring what day they had first dug their ditches along Crazy Woman.[45]

Smith didn't get much more water in his ditches the next year, or the next, and by 1888, he was mad enough to go to court. Into Buffalo court files came Smith's testimony and the records of all the considerable effort that had been put into water—or at least into water claims—by Hesse and Canton and others above Smith on Crazy Woman since the mid-1880s.

As it happened, the year 1888, when Smith went to court, was also Elwood Mead's first summer as territorial engineer. Part of the law creating his job, after the disaster year of drought and blizzards, allowed people to ask the territorial engineer to measure their ditches. Smith, preparing his court case, had Mead survey the capacity of his ditches. The court didn't require evidence on anything else—like how much land was actually watered by those ditches. Rather, as the law provided, the district court judge in Buffalo in 1889 set

priorities according to the date and size of ditches and awarded the water
rights by taking into account not only what people were actually irrigating
but also what, in the official legal language, they "proposed" to irrigate.
Proposals? Everyone on the creek had proposals. Lots of them. And proposals
alone were the basis of the judge's decree on Crazy Woman. That decree
illustrated why Mead soon replaced district courts with a board of practical
experts who could assess actual water use.[46]

Smith came out just fine, though, in the decree from the court in Buffalo. He
got what he wanted—a big water right dated 1879, the best big water right on
Crazy Woman Creek. Based on the size of his ditches, he got an award of water
for 1,200 acres, the amount of land he "proposed" to irrigate. His claims weren't
even the most ambitious. Up and down the creek, people got awards of water
rights aimed at irrigating amounts of land ranging from modest (150 acres) to
preposterous (21,000 acres). The amounts of water that the judge awarded per
acre, meanwhile, varied wildly, so the total of water rights had no logical
connection to the acreage involved or to what the stream could provide. Canton
and Hesse got rights awarded too, big ones, upstream. But their rights were
established as junior to Smith's. Smith's was potentially the controlling water
right on the creek; it gave his ranch one-fifth of the creek's usual spring flows and
all the water in the creek once the high spring flows went down.[47]

Smith had beaten Canton and Hesse in water. In return, Hesse soon took
pains to label Smith as a rustler, thus ensuring that cattle Smith had shipped off
to market would be seized as stolen goods. The label didn't seem to stick well,
however, on a man who everyone knew was a first settler, entrusted early on
with public business in the county, and owner of a longstanding place everyone
passed on the old trail heading to Buffalo.[48]

The Invasion of Johnson County started soon after, and soon the invaders
were holed up, besieged, on a friendly ranch on the North Fork of Crazy
Woman Creek. Characters from all around the range had taken part in the
invasion, one way or another. Edward David, Carey's foreman down in Douglas,
cut telegraph lines so Buffalo wouldn't know the invaders were coming; his
neighbor Frank Wolcott had helped conceive the scheme and joined in leading
it; acting Wyoming governor Amos Barber, who had been a much-trusted
doctor at old Fort Fetterman, told militias around the state not to answer local
sheriffs' calls for help.[49]

Smith, on Crazy Woman, did what he could to hinder the invaders and bring the posse from Buffalo to arrest them. He also helped capture the invaders' supply wagons, which, along with dynamite and many rounds of ammunition, contained the list of seventy men whom the invaders had targeted to be killed in Johnson County.[50]

Just a couple of months later, the air had cleared somewhat. Smith and others, once smeared by Hesse and Canton as rustlers, gathered and branded their cattle peacefully alongside the outfits of ranchers that had joined the invasion. But tensions and resentments lived on, and so did the 1889 court decree on Crazy Woman water rights. The decree was long impervious to Mead's system, because the careful stream-wide adjudications by Mead's board of practical experts could not, Wyoming courts said, touch water rights in a stream already covered by a decree made by a court, however nonsensical the outcome. Fortunately, there weren't very many such decrees. And for a long time, people on Crazy Woman worked out a "good neighbor" agreement to ignore their decree, at least in terms of the volumes of water awarded. It seems that John R. Smith's ranch, for instance, never did use all the water he had "proposed," and so the other ranches got summer water.[51]

But in the 1970s, with the coming of the national energy crisis, water began to look valuable for power plants to burn Wyoming coal, and people on Crazy Woman started worrying that new owners of the old Smith ranch might try to sell the water right, which still looked so big on paper in the court decree, to an energy company.[52] Their alarm made it clear that the water right decreed to Smith had not seen full use for years. People upstream were relying on using water that the ranch had never called for. They went to court.

Mead's system finally came into play. The Board of Control—the practical experts—found a way to put its stamp on the old Crazy Woman court decree by "interpreting" it. That meant putting more realistic numbers on some of the rights prioritized in the decree, including officially allocating less water to the old Smith ranch. The board managed to banish the specter that had scared the neighbors— the possibility of the whole old Smith claim suddenly springing to life in new hands and completely disrupting everyone else's water use on the creek.[53]

Still, water on Crazy Woman Creek continues sometimes to function in a world of its own. The board had to leave untouched the water rights held by people who hadn't joined the complaint against the old Smith ranch rights. The best way to run the creek, therefore, has been by agreement rather than a legal list. Ideally, everyone would agree to take water under the per-acre standard that would have been in place with a stream-wide state adjudication. Calling in the state water commissioner would mean regulation of only some people under the old decree, with its odd, non-standard amounts. No one wanted that. So people on the creek were conditioned never to call on the state to resolve conflicts but to just rely on "self-help." In summer 2002, self-help meant that one old man on Crazy Woman was collared by an upstream neighbor and intimidated out of using his water. Rather than complain, the old man put his place up for sale. The state water division superintendent knew what had happened and saw it as a characteristic of life on Crazy Woman. He didn't interfere. He followed a long-standing practice of the water board Mead had established: unless a water user called for state help, it was not the place of the state to intervene.[54]

In fact, the self-help rule about water has prevailed all over Wyoming since before statehood and long after. It is the practice everywhere, not only the practice on the few creeks covered by old court decrees. Statewide, people can run water management on their creeks as they like, if no one complains. There is a sense that people should take care of their own water rights. Irrigators sometimes work out a way to rotate water use among themselves, with or without local state engineer's office approval, particularly when water is short. Sometimes there's a consensus agreement among neighbors, and sometimes there's not. If things don't go right, people have to muster the courage to speak up.

Before statehood, self-help was an official rule. Territorial law of 1886 said that officers in water administration should not intervene on a creek unless two owners on that creek asked them, in writing, to do so. That law stayed in place for a few years under Mead's new system. But even after it went off the books, the principle was well established, reinforced by the vast spaces of Wyoming and the small size of state water staff. Modern Wyoming water commissioners typically still don't step in to divide up water on a creek according to priority unless someone complains about not getting water. The current law says any *one* water user can ask a commissioner to step in, with a request in writing. On Crazy Woman in the 1880s, John R. Smith had certainly not been afraid to complain. But that isn't always the case.[55]

———

The 1889 Crazy Woman decree captures, as if in a photograph, the world of the 1880s—its bravado, its pretensions, and its jostling for power among the likes of Fred Hesse and Frank Canton versus John R. Smith, somewhat like Joseph Carey and Edward David versus John Slichter on Box Elder.

Mead walked into that world with a dual role to play. The stockmen who brought him in, just after the disaster of 1886–1887, wanted both security and flexibility. They wanted to protect what they'd scrambled for, like claims on Crazy Woman. They also wanted an avenue to something new, since the drought and winter they'd just been through suggested that the open range business of the past was not the model for the future. The constitution and laws Mead wrote meanwhile were true to his beliefs, requiring him to "equally guard all the various interests involved." That included the smaller farmers and ranchers, who wanted to find a foothold in Wyoming and help make the future.[56]

Mead offered a way to secure water claims, but only with scrutiny of water use. With that, he managed to provide both security and the flexibility to accommodate new people and new growth. His system could bestow a new imprimatur of legitimacy on existing water claims—but only after forcing those claims through a test. Old claims to water could make it into the new calf-bound, laboriously maintained state record books, inscribed as rights that the state would protect against all comers. But they would not necessarily be enshrined there in the size and shape of their progenitor's fondest dreams. Mead believed that proof of actual use, of an amount of water actually necessary for the crop, was the narrow gateway through which the ambitions and dreams of a Carey or a Slichter, and ultimately of a Smith, should be forced to pass before they could get the protection that the new state system offered. That was the way to make things different from what they were under the old order as represented by the Crazy Woman decree.

Amid the ruins of an industry where paper cattle and dummy entry-men had been common currency and no one cared to ensure that grass or water would be there to serve a common future, Mead's great feat was to give the state government the power to insist on evidence of genuine activity, produced by investment of cash, sweat, or both. On those terms, the state system would offer its protection to all, laying a foundation, in water at least, on which a new economy could perhaps be built.

The net result was that the small men were no longer necessarily dwarfed by the claims of the cattle companies. In the years following 1886–1887, despite the infamous drought and bad winter, more and more people came to try settling in Wyoming. They ignited the stockmen's anger and fear that erupted in the Invasion of Johnson County. The pages of the histories compiled years later, the histories of Box Elder or Crazy Woman creeks, echo with the voices of the people who came, and kept coming, to take out a homestead claim all the way into the 1920s. Some cowboys or foremen from the 1880s made good with their own ranches; some emigrant families like the Slichters managed to keep going on small stock-farms; some settled their families on ranches that are still working today. In Laramie, a Norwegian sailor turned railroad worker found a Norwegian seamstress to marry and managed to buy a small ranch. Stories like theirs multiplied. Of course, people going for grandeur were still part of the picture. Carey worked to add to the Box Elder ranch. His son Robert listed the ranch, still called Careyhurst, as his residence when he became a US senator in the 1930s. John Kendrick, who had come to Wyoming in the 1880s as a cowboy, accumulated cattle and land near Sheridan (some of it by buying up the land allotment rights of Civil War veterans who never saw Wyoming; Carey did the same). Kendrick became a popular governor and then US senator. In 1915, when as governor he sat on the state land board, he added to his holdings nearly ten thousand acres of former state land controlling access to creeks near Sheridan.[57]

For all these people, Mead created a set of pragmatic rules for who could get what water in a dry, tough land, rules on which people could begin to build their lives. In the Wyoming that entered the twentieth century, some of the frontier persisted and left its traces on Mead's water system—like the role that self-help retains on the creeks and how that shapes what water people really get to use. But a key Mead principle—the public supervision for overall public benefit, insisting on actual use of water—plus the structure of dedicated practical experts he put in place as supervisors, stayed in place and put their mark on Wyoming.

3. MAKING THEIR OWN WAY

Allow me to explain right here that the whole country is not farming land,
but only along the margin of the streams.

—JOHN GORDON, DECEMBER 1881.[1]

Places: Little Horse Creek, southeast Wyoming plains; Wind River, west-
central Wyoming valley; Greybull and Shoshone Rivers, northwest
Wyoming badlands.

Time: 1900–1920.

Excavation of Coolidge Canal, Wind River Indian Reservation. Courtesy of Edith
Adams and WRIR Photographs, Randy Shaw Collection, American Heritage
Center, University of Wyoming.

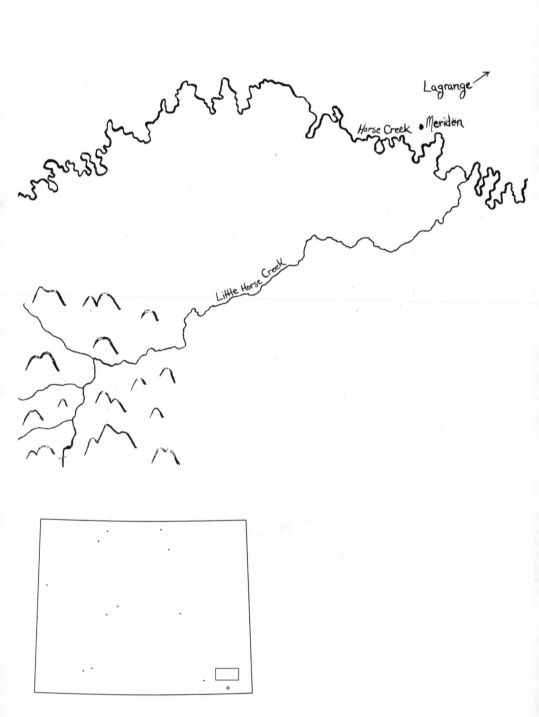

Lagrange

Horse Creek • Meriden

Little Horse Creek

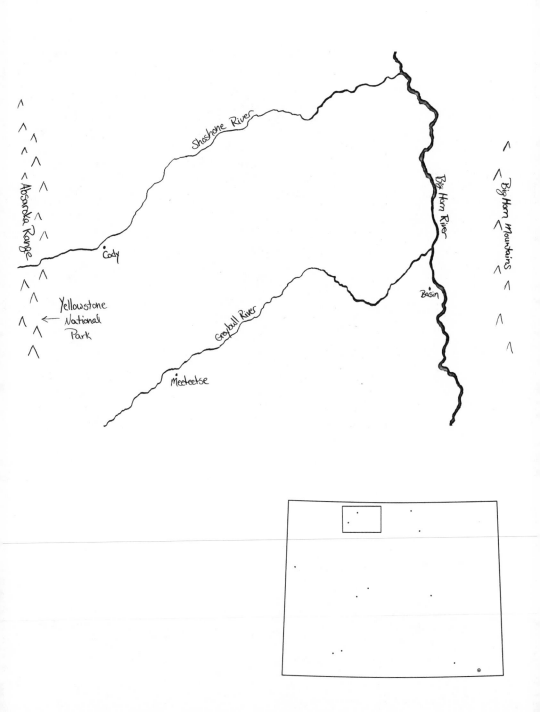

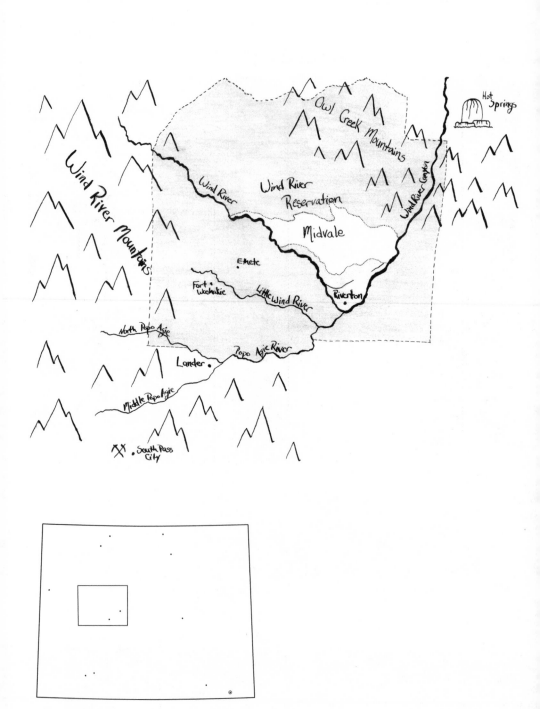

JOHN GORDON WAS IRISH, a fiddler who made his own violins. He came to the United States soon after the Civil War and was a carpenter in a Connecticut factory for several years. Then, with a Scottish friend he had met in the factory who was an avid reader of Horace Greeley's newspaper and its call to "go West, young man," he decided to try his luck in the West. He moved first to Greeley's irrigation colony in Colorado and then in 1878, with his wife and three small children, moved to a cabin in southeast Wyoming. Three years after the family arrived in Wyoming, in December 1881, Gordon wrote to a cousin in Ireland who was thinking of following him. Gordon figured that man needed to understand how different Wyoming was from Ireland. He sent his warning: "The whole country is not farming land."[2]

Elwood Mead, as state engineer, regularly told readers of his annual reports that with irrigation, Wyoming could become celebrated for agriculture. He soon agreed with Gordon that Wyoming would not boast farm products; rather, Wyoming would be known for raising prime stock fed on native pastures and hay or grain grown on irrigated bottomlands. Fifteen years after Gordon wrote to his cousin, Mead assured his readers:

> A union of farming and stock raising is the form of agriculture best suited to our conditions. If . . . the arid uplands [were] utilized to supplement the irrigated valleys, one to furnish summer pasturage and the other the winter's food supply, it would make Wyoming one of the most attractive and prosperous stock raising districts on this continent.[3]

But it turned out to be very hard even to transform strips of bottom land along the creeks into green hayfields; dry benches up above the big rivers might also someday be irrigated, but they were beyond the reach of individual efforts. Mead had told readers in his first annual report as state engineer in 1892 that major investment of capital would be needed for Wyoming to realize its potential through irrigation. Much of the North Platte river basin, where John Gordon and John Slichter lived, did not attract the number of settlers the way midwestern states did, Mead wrote. Bench land near old Fort Fetterman, for instance, "is not occupied, because as a rule, the pioneer is a poor man; the diversion of a great river is an undertaking he cannot compass. Until this is done the land has no value except for a grave, and is not attractive for that."[4]

Men like Gordon and Slichter labored to make irrigated fields, in tough terrain, all over Wyoming. Their struggles, and the welcome Mead and Wyoming gave to federal investment in building water projects on the big rivers, had a good deal to do with what became of Mead's water system.

In the first ten years of Mead's system—from 1890 to the turn of the new century—Mead and his small staff worked on establishing the state government's role as the gatekeeper for water rights, as demonstrated on Box Elder Creek and in Buffalo. They succeeded in convincing people of the principle that it was the state, not individuals, which would decide whether and how much of a water claim would persist with legal protection. Except in the case of water claims covered by troublesome territorial court decrees like the one on Crazy Woman Creek, water superintendents reviewed old claims for water rights from before statehood or new proposals made after 1890. They and the rest of Mead's board would decide, after careful examination, whether an old claim could be recognized with state protection as a water right and how much water that right covered. In the case of a new plan to use water, it was solely the state engineer's job to decide, after scrutinizing an application for its practicality, whether someone with a plan for water would get a permit as the first step to getting a water right.

But as the new century came on, new questions arose. Some of those were implicit in the water tangle that Gillette had cleared up with his adjudication in Buffalo. They became continually more pressing. Just what did you get when you were granted a Wyoming state "water right"? Was a water right something you could sell to someone else, somewhere else, to use the water as they liked? Was a water right something you could slowly make use of over time, maybe holding off for years before putting to use all the water involved? Did it depend on what kind of project you had, or what your neighbors thought?

Arriving at the answers to those questions involved considerable experimentation and debate. It took about twenty-five years, from 1900 to 1925, to work it all out, and the result was a considerable change from Mead's original plan. While Mead had envisioned straightforward state ownership and control of water, set by experts working for the good of the people, by 1925, Wyoming's water management system had moved in a different direction. Actions of the state water agency and the water users resulted in users joining the state agency as key decision makers. The agency and the users worked in symbiosis, acting as a community. The state remained the gatekeeper and recordkeeper but, as the history of Crazy Woman Creek suggests, the users—just as much as the agency—could set the rules for water use and help determine what enforcement occurs on an individual creek.

One reason that happened was simply the difficulty of enforcing Mead's idea of state ownership and control over the vast and varied topography that is Wyoming. Mead had a small staff and often had to dig into his own pocket to pay for their travel. In 1898, he left for Washington to work on national irrigation policy. His successors, imbued with his ideas, continued to deal with small staff and huge distances. The result was that people on Wyoming creeks got to experiment. Wyoming terrain and climate, for that matter, required experiment; it was not obvious how to make irrigation work. Facing all kinds of obstacles, people gradually adjusted and changed the water rights system.[5]

———

John Gordon tried ranching in a couple of places in southeast Wyoming. His first spot, in the 1870s, was on the Laramie River, and there, as the cattle business grew, he began to keep his house as an inn on the road north. Judge Carey often stayed with him, and Gordon liked to say that he first pointed out to Carey the potential for watering the flat bench lands above the Laramie River. These lands became the irrigation colony that Carey and other cattlemen launched (joined by Gordon's Scottish friend). They optimistically called the place Wheatland. In a peak year, before the disaster drought-summer and blizzard-winter, Gordon sold his own place in the area to some other big investors—who ended up among the Invaders of Johnson County eight years later.[6]

With money in his pocket, Gordon took his family on a pleasure trip back to Ireland and then brought them back to Wyoming to find another place to ranch. "We had decided we had enough roughing it on the frontier, and so looked at an old settled section on Little Horse Creek. I purchased three small settlers, and made an extensive ranch, built fine house and barn, refenced, reditched [*sic*] all the land, got all in fine shape to handle small bunch of fine cattle." He called the place Springvale.[7]

In 1891, southeast Wyoming—known as part of Water Division I to the new State Engineer's Office—was the most heavily populated area in the state. The superintendent of Division I was J. A. Johnston, the man who later judged the claims of Carey and Slichter on Box Elder. Johnston did his very first adjudication of water claims on Little Horse Creek, where Gordon had settled. There, a territorial court had started the work of sorting out water rights, but Johnston finished it. Little Horse Creek was indeed the "old settled section," as Gordon had put it. It was near the capital, and water from the creek had been at work irrigating fields for some fifteen years. Reviewing the evidence Johnston gathered, the state engineer and all four division superintendents, meeting as the Board of Control, issued an adjudication in 1892 delineating the water rights (and commenting on the difficulty of the task). The adjudication confirmed that Gordon, based on the places he had bought from earlier settlers, held the earliest and best rights on the creek for his Springvale Ditch Company.[8]

Neighbors immediately challenged the board order. There were a lot of claims on Little Horse Creek—sixty-seven, in fact. Hoping to overturn the board's decision, the holders of those claims went to court. As the case dragged on, Gordon eventually found himself in financial trouble—"I finally got enlarging myself too much," he said—and sold his place. First, however, to bring in some cash and to appease the major opposition, he made a deal to sell a "half-interest" in one of his key water rights to a bigger neighboring ranch on the creek.[9]

Gordon sold his half-interest in a water right to an operation called Little Horse Creek Irrigating Company. It was headed by George Baxter, who invested in cattle after coming to Wyoming with the US Army as a young West Point graduate. Later, Baxter was appointed territorial governor (to F. E. Warren's disgust), was ousted after forty-five days by President Grover Cleveland for illegal fencing but was elected to the Wyoming constitutional convention (where he argued for women's voting rights). He competed unsuccessfully with Warren for governorship of the new state. Baxter had joined the challenge to the Little Horse adjudication just as he was heavily involved in the planning of the Invasion of Johnson County, and the subsequent successful legal defense of the Invaders.[10]

The deal Gordon soon cut with Baxter, however, said that Baxter's Little Horse Irrigating could use Springvale's water right, with its early priority, every other week—when Springvale was not using it. Baxter was satisfied, and with that water purchase he dropped out of the court fight over the Little Horse Creek adjudication.[11]

But there was a place on Little Horse Creek between Gordon's lands and Baxter's. It was owned by the James R. Johnstons (no documented relation to Division I superintendent J. A. Johnston), a solid and prosperous family with Little Horse Creek ranches dating from 1883. The family founders were two brothers from Ohio who had gone to California in the 1849 gold rush and were smart enough not to look for gold but to ride the gold boom as merchants and lumber dealers. After switching to stock raising, the Johnston brothers eventually came back to Wyoming, nearly forty years after they had passed through on the Overland Trail. Settling on Little Horse Creek, they had extensive cattle, horse, and sheep herds. Characterizing themselves as people of "pluck, perseverance, and integrity," they were disturbed by the Gordon-Baxter deal. Not only did they have a ranch between Gordon and Baxter, but some of their water rights were between Gordon's and Baxter's in priority. Those water rights would get less water because of the Gordon-Baxter water sales contract. The Johnstons went to court to fight the water sale.[12]

In settling Little Horse Creek, people had inevitably developed a pattern of water use. Gordon's water rights, attached to the lands he bought from earlier settlers, amounted to 10 cfs; he had been diverting that much for irrigation every other week in the summer. The plan under the Gordon-Baxter deal was that Gordon would keep using his 10 cfs as always, but Baxter's company would use 10 cfs in the off-weeks, that is, every second week all summer.

Before the deal, when Gordon didn't use his water in the off-weeks, more water had been available for the in-between junior water right holder—the Johnstons. But with the Gordon-Baxter water sale, that water would no longer be available to them.[13]

On paper, Gordon had simply sold off half of his water right. But in effect, the Gordon-Baxter sale doubled the amount of water used under Gordon's high-priority right. Gordon's neighbors, the Johnstons, were outraged, and the State Engineer's Office was horrified. It was a classic example of what Mead had sought to prevent. The water commissioner at work on Little Horse Creek refused to honor the contract between Gordon and Baxter. The Johnstons meanwhile asked the courts—still reviewing the board's adjudication of the creek—to defeat the water sale contract. As the Johnstons and the engineers saw it, Gordon had sold off something he had never used and therefore never had. The Board of Control refused to recognize the sale.[14]

The Laramie County District Court overruled that board decision. The court held the sale valid and described a right to use water as a property right that could be sold like any other property. The Johnstons appealed to the Wyoming Supreme Court. While the high court decision was pending, Mead, no longer in Wyoming, was disturbed to see the district court ignore the tie of water to land that he believed was fundamental to the water law he had written. He wrote bitterly of the implications of the district court ruling:

> It is not believed . . . that (the district court opinion) will be sustained by the supreme court. If it is, water rights acquired during the Territorial period will become personal property. The water of the public streams will become a form of merchandise, and limitations to beneficial use a mere legal fiction. If water is to be so bartered and sold, then the public should not give streams away, but should auction them off to the highest bidder.[15]

The Wyoming Supreme Court ruled in 1904 to uphold the Gordon-Baxter water sale. Chief Justice Charles Potter wrote the court's opinion. He was the man who had worked with Mead on the water language in the constitution and just four years earlier had so deftly and stalwartly upheld the overall water management system in the Buffalo case. When it came to the question of selling off a water right, however, Potter parted ways with Mead. He thought it was a matter of course that water rights could be sold to a new location.[16]

Potter's decision landed like a bombshell on the State Engineer's Office. Superintendents of the water basins protested in public reports. Mead wrote ever more bitterly, saying that allowing a water sale like the Gordon-Baxter deal meant the court was dragging Wyoming down to the level of other western states, with the laws he believed "belong to the lower Silurian period."[17]

Potter saw the situation as a relatively simple one. He interpreted Gordon's 10 cfs water right as a right to 10 cfs of continuous use all summer—just the way it looked on paper. The irrigators, the State Engineer's Office, and Mead had pointed to the pattern of actual use of that 10 cfs of water—only every other week—and urged that the pattern of actual use determined the water right and should be protected. Potter did not grasp that, though. Taking the water right at face value as it stood on paper, the chief justice said that half such a right could be sold off the land and used elsewhere.[18] Potter's judgment made reality of Mead's nightmare—water being "bartered and sold," like any other kind of property. The portion of a water right that Gordon had sold could be used elsewhere. The Wyoming legislature deferentially followed the court ruling, enacting a statute allowing water transfers away from the original land.[19]

But the high court's decision and that new statute did not last long. Water users and the State Engineer's Office soon managed to overturn it. In 1905, the state engineer was Clarence Johnston, son of J. A. Johnston, who had been Division I superintendent. Mead had brought young Johnston on staff in the 1890s and taken him to Washington for a few years. Back in Wyoming as state engineer, Clarence Johnston convinced the legislature to appoint a special commission on water law, with himself as a member.[20]

Johnston was sometimes regarded by later members of the State Engineer's Office as someone who could overstate facts with a flourish—"Oh, Clarence!" members of the Board of Control would say, as late as the year 2000, when they found errors in water permits he had signed. But he was dedicated to upholding Mead's water law. The new water law commission, with Johnston on board, took the unusual move of sending out written surveys to water users statewide. The survey revealed user views on a variety of topics, including support for the common practice of rotating water use on a creek in times of shortage, allowed generally by territorial law, to be officially recognized with rules set by state engineer staff. Most important for the Little Horse Creek case, a strong majority of users responding to the survey opposed water sales. The commission convinced the legislature to enact new laws in 1909, adopting the users' views on issues like rotation and, most significantly, overturning the Little Horse Creek court decision. The key new law explicitly forbade the sale of water rights to use on other lands, on pain of loss of priority—and losing priority, of course, meant in effect losing the water right. A companion statute set an exception, so water for towns or other "preferred uses" could be changed from their original uses, paid for, and moved to the new use, without loss of the priority date.[21]

The new law meant that the options of someone who had a Wyoming water right had limits that were very clearly stated. Water rights generally could not be sold away from their setting; irrigation rights, for instance, could be sold only along with the land, except for the special case of a "preferred use" buyer, like a town. And what a water right consisted of was limited. The new law specifically provided that a water right held by anyone in Wyoming was defined by the actual characteristics of water use—how much water was necessary and used for what purpose, on what lands. "Beneficial use shall be the basis, the measure and the limit of the right to use water at all times," the statute read, and is still the law today.[22]

State engineer Johnston saw this statute he had helped draft as a fundamental step in clarifying the limits of a Wyoming water right—the limits that he, his predecessors, and the Board of Control had consistently applied. He laid out all the details of what it meant in a triumphant essay titled "What is a Water Right?," published in his 1910 public report after the struggle was over. The key principle, he wrote, was that the needs of the public and the community on a stream must prevail over the desires of an individual. The pattern of water use on which a community had been built and daily relied on should not be disrupted; even a town, a "preferred use," could obtain only from an irrigator, for instance, a right to the volume and timing of water that the irrigator had employed. The proper idea, Johnston wrote, was to "afford protection to every claimant yet . . . no man is given a weapon whereby he may destroy the prosperity of his neighbors. All that need be borne in mind is that the right to use water should be limited in accordance with the beneficial use made and the right should belong to that use rather than to the user."[23]

Baxter, the former governor who had bought water from Gordon, got to keep the right to that water as a result of Potter's ruling. Still, he must have been chagrined to see the general concept of water sales defeated for the future by the voices of other users. He, and likely his peers, wanted a straightforward rule allowing water transfers so someone with a water right could sell water to a new location—to anyone else who could find a use for it, wherever that use might be. Yet the idea didn't stick in Wyoming that individuals could own water as private property that could be moved, bought, and sold. The reverse happened. Wyoming wound up with a statute firmly stating the principle Mead had intended to be part of the original water law. Baxter and the high court had run into not only the strong beliefs about water in the State Engineer's Office, but also the on-the-ground experience of Wyoming water users. Experience told users that water was something different than land.

Mead's successors in the engineer's office were men he had trained and so imbued with his views that they can fairly be called his disciples. They believed what he believed: that allowing an "absolute property right in water" would encourage speculation that would destroy orderly water use and people's efforts to build settled communities.[24]

Little Horse Creek, a small stream, had fostered just that kind of community. The people living along it came to know their creek, their soils, their climate, and each other well. They found out how much each one's water use was tied to what others did along the creek. Water users around the state also heard of the deal Baxter had cut with Gordon, as the case went through the courts and the state engineers' reports highlighted it in outrage; people could talk it all out with each other and with the water superintendents, who felt passionately about it. Enough time passed for water users and superintendents to unite in a new understanding: that people working with a limited water resource were interdependent in a way peculiar to people living in arid regions. Water, in a place like Wyoming, could not be thought about in the same way as land.[25]

When Edward Gillette was superintendent in northeast Wyoming back in 1896, he wrote about what that notion meant on the ground. A dry year, he said, meant "intense feeling, theory and pet laws for the government of water are cast aside. The condition is an angry farmer, the half-matured crops on his land, which gave promise of an abundant harvest, are rapidly burning up." An irrigator had to be stopped from simply moving water wherever he wanted, Gillette wrote. "No other determination would long be tolerated by his neighbors."[26]

Justice Potter wrote in 1904 that to disallow the Gordon-Baxter sale would be "to deny the element of property in the water right itself." That he refused to do. But the water users and the engineer's office, followed by the legislature, went right ahead to do just that. They denied "the element of property." They saw water as so different from land that they had to try something new and make new rules.[27]

As Mead told national readers in 1902, when the Baxter-Gordon contract was heading to the Wyoming Supreme Court, people who worked with water often saw things differently from lawyers and courts. The legal system was all too likely to apply to water the rules about ownership of private property in land:

> The speculative value of the personal ownership of running water is so great that every argument which the ingenuity and intellect of the best legal talent of the West can produce has been presented to the courts in its favor. Organized selfishness is more potent than unorganized consideration for the public interests.[28]

The two figures in the saga of "organized selfishness" that had played out in Wyoming met different fates. George Baxter retreated to his native Tennessee, where his wife had come from a wealthy family. He had had a moment of fame in the society pages "from coast to coast" in 1900, when his daughter left her Denver fiancée waiting in church and ran off to marry a much older "wealthy socialite" in San Francisco. After John Gordon sold Springvale, he moved to Cheyenne and ran an experiment farm for the state's Department of Agriculture. As an old man, with a penchant for reciting Robert Burns, he moved to California to live with his daughter's family, and there at last he wrote, "I have a garden which I love to cultivate and make plants of all kinds grow to perfection."[29]

———

Most water users in Wyoming were more like Gordon and his Johnston neighbors than Baxter. They were small ranchers and farmers, and they had a personal relationship with their superintendents that was significant in shaping policy. The superintendent in Division III, northwest Wyoming (the basin of the Shoshone and Wind-Big Horn Rivers) wrote about that in 1910:

> The people are coming every day for advice and information that they can get in no other place. In this respect, it is one of the most important positions in the State. It deals more directly with the people, knows their needs and conditions better than any other place, and is of the greatest help to them, all of which they fully appreciate. In connection herewith I want to express my appreciation of the help and good will extended to

me by the people. They have given me every encouragement in the performance of my duties and without their hearty co-operation the administration of the work of this office would be extremely difficult. They have
always been consulted before any great change has been made and their
advice has always been good, founded as it is, upon actual experience and
observation.[30]

There was more water policy ahead to be made by the state engineer, the
superintendents, and the water users. By 1910, some questions had been settled:
the state water office, representing the public, would determine who got what
water right, whether by investigating old claims or issuing permits for new ones;
and once a person had a water right, it was defined by its use, and in general, he
could not sell it away from the place or purpose for which it was used. Just a few
years later, the high court affirmed the State Engineer's power to limit, in a new
permit, the size of a proposed hydropower dam to accommodate the future
value to the public of the only practical railroad route to connect northwest
Wyoming's Big Horn basin to the rest of the state.[31]

But it was not yet obvious how soon a person with a state water permit once
in hand must put the water to use. And that was a question with major
implications. Could you hold off, wait some years to put the water to use, but
still end up with a water right for the volume of water you first envisioned, dated
when you received the permit—with that valuable priority date, so that you
could always get that water before someone who got their permit later?

The State Engineer's Office believed the answer to that question was clear:
absolutely not. To acquire a water right, there were many obligations to meet.
Obtaining a permit was merely the first step toward securing a water right. To
get a valid right to use water, the next step was to put to actual use, and in a
timely manner, the water for whose use the state had given merely an initial
permit. Under the water laws of 1890 that Mead wrote for Wyoming, state water
permits had requirements written into them, clear deadlines for the start and
finish of water facility construction, and for getting water into actual use (with
irrigation, that meant onto the land). After water had reached farm fields and
proof of the water use matching the permit was presented to the local
superintendent, he was to inspect. If he could verify that the water was truly in
use, the Board of Control would issue a "certificate of appropriation." That

certificate was what conferred a valid water right, and it would be a right for only the amount of water the inspection showed was being used.[32]

Successive state engineers all shared the same view: a water permit once issued should be regarded only as a contract with the state. That contract merely gave a would-be user his opportunity to put water to use—under certain conditions, including firm deadlines. Only compliance with all the contract conditions, including deadlines, would result in someone's obtaining an actual right to water. The penalty for failing to comply on time was severe: the permit and the opportunity for a water right with priority position would be canceled. State Engineer Johnston, trained by Mead, put it this way in 1904: "If the applicant does not sufficiently appreciate the value of the water right sought to comply with terms of the permit, it should be canceled in order that others who have shown proper diligence may be protected."[33]

It was a matter of basic policy. The concern was fundamentally the same as in the water sales debate on Little Horse Creek. To allow applicants to proclaim themselves "users" by obtaining a state water permit but doing nothing more was to allow speculators to obtain Wyoming water rights and potentially hamper the growth of self-supporting settlements. Further, the growth of prosperous communities demanded that unsuccessful ideas make way for new, more promising proposals. That again meant that no one should be able to obtain a state permit, fail to use the water, but keep the permit and a claim to the water. If someone could do that, he could create for himself the ability to bide his time and keep the permit priority date, preempting others who could put the water to use right away.

Mead, early in implementing the state water system, had after all pointed out that sometimes he would issue a new permit, with a new date, to irrigate lands already covered by an earlier permit. He would do that if it appeared that the holders of the initial permit simply weren't progressing with their project. Mead informed one laggard permit holder that "the fact that you have had a year of unrestricted opportunity with no visible result, as I am informed" meant it was more important to the state of Wyoming to issue a new permit to someone with new plans than "that development should be retarded in order to protect your prospective profits." Mead similarly warned Judge Carey's Wheatland project that it would lose water right priority to later users, in adjudication, if project lands didn't use the water claimed in 1883.[34]

Mead had a similar goal when his state water-law code adopted the rule in Wyoming Territory—the standard rule in prior appropriation law—that even a water right with a certificate of appropriation could be lost if it were not used. A water right could be challenged and lost, considered "abandoned," if it could be proved that the water involved had not been used for a few years. In Mead's mind, water rights had to be placed in the hands of people who proved successful, in order to put Wyoming water to use quickly, keep it in use, and build communities in arid country.[35]

But, put Wyoming water to use quickly? Putting Wyoming water to use *if possible* turned out to be the more practical goal. In the end, in fact, that turned out to be so hard to accomplish that a Wyoming water permit came to mean something quite different from what Mead and the early engineers who followed him had imagined. A series of events in the first twenty years of the twentieth century led to that outcome. These events unfolded in a place that reeked of impossibility—a very different part of the state from Little Horse Creek. That place was the Big Horn basin of northwestern Wyoming, adjacent to Yellowstone National Park.

Yellowstone's wild landscape had been consecrated in the 1870s as the first national park, and it had attracted well-heeled eastern visitors and the construction of luxury hotels. Nonetheless, the Big Horn basin, east of the park's boundaries, remained frontier country in 1902. Isolated by physical barriers from more settled parts of the state, the region had the harshest terrain, sparsest population (less than five thousand, or about one person for every three square miles), least economic activity, and least social stability in Wyoming.[36]

The Big Horn basin also had in spades what any newcomer who tried irrigation found almost everywhere in Wyoming. Whether armed with cash or with just a mule and a scraper, everyone—including the state engineer and his staff—met surprises trying to put their ideas about water into practice. They were all profoundly ignorant of the land and climate in which they were working. The people before them, who had moved through there for thousands of years, had not seen it as a place to grow crops. The new inhabitants, coming from farming country, joined Mead in envisioning profitable agriculture—some even imagined row crops, in addition to the hay that eventually dominated most irrigated fields in Wyoming. The new people had little idea what would be required in finances and engineering to irrigate the terrain or to grow crops at elevation. They faced a steep and painful learning curve. Many private irrigation ventures died.[37]

The Big Horn basin was the home of the superintendent who wrote in 1910 about the importance of responding to the "actual experience and observation" of water users.[38] Mead described the basin in 1898, just before he left the state, as:

> An immense bowl entirely surrounded by lofty mountains. It has not, however, the appearance of a valley as that term is ordinarily used. The greater part of the surface is hilly and broken. Some of the bad-land ridges rise almost to the dignity of mountains and present a picture of aridity and desolation which disappoints and discourages many of those who visit this section for the first time. None of the broken country can be reclaimed. The limits of irrigation are restricted to the bottoms and table-lands which border the water courses. Outside of this the country is neither adapted to the construction of ditches nor fit for cultivation even if water could be carried to it.[39]

Nonetheless, there seemed no end to big ideas for testing the "limits of irrigation" in the Big Horn basin. Some dreamers brought money, some brought know-how, and some brought neither. One with money had the fine name of Solon Wiley, with a career behind him as a public-utility engineer in Omaha. Starting in 1895, he formed a company and obtained Wyoming water permits for a plan to build an irrigation colony on flat bench land above the Gray Bull (now the Greybull) River, a sizable stream that heads up in high mountains and had for some time watered ranch lands in pleasant foothill valleys before it crossed the more desolate country that Wiley hoped to make bloom. Wiley's permits included both a new permit, dated 1896, and some old permits he bought, dated 1893, which had been issued for a project on the bench land envisioned by Andrew Gilchrist, the Scot with whom Gordon had emigrated West. Gilchrist had become a major figure in Wyoming, acquiring significant ranch lands and a major share in the Stock Growers' National Bank. He died, however, before he could do anything with his idea for a Gray Bull project.[40]

Wiley also got federal public lands set aside for his project, under a new law to help water projects, a law Judge Carey, in his last year as US senator, authored in consultation with Mead. The Carey Act, enacted by Congress in 1894, was intended to boost the chances of success for private irrigation developments. The idea was to let state governments set up contracts with private developers so the states would supervise major irrigation projects to which federal lands would be allocated. Those "Carey Act lands" would be "segregated" from the general public lands for project use only—they couldn't be homesteaded or otherwise claimed by others while the state-approved project was getting underway.[41]

Wiley took up the Carey Act challenge, came to the state, accepted state supervision, and got federal lands designated for his irrigation project. He was not an absentee entrepreneur but a very hands-on one. He personally supervised construction of his canal, and his wife cooked for the laboring crews. But the planned network of ditches was ambitious, and progress was very slow. It looked like Wiley might not meet the construction deadline in his state water permits—a deadline of the last day of 1902.[42]

As his canals began to reach some of his lands, however, in 1901–1902, Wiley wanted water for the few farmers he had convinced to settle there already. A major drought had struck in 1902, one of the worst many Wyoming settlers had ever seen. Wiley got into an argument with downstream neighbors using the Farmer's Canal, a group of experienced Mormon irrigators who had figured out how to farm formerly desolate lands and had a state certificate for their water right. Their priority date was 1894. Wiley, who had gotten a state water permit for his venture only in 1896, nonetheless claimed an 1893 date for priority to take water from the river, based on the 1893 permits he had bought up. Those permits had contemplated watering the lands now beginning to be reached by Wiley's canals. So now Wiley believed the date of those permits had become valuable.[43]

Back in 1895–1896, when Wiley started his project, Mead was state engineer in Wyoming. Mead had issued a new permit for Wiley's more ambitious project to water the lands covered by the old 1893 permits, under which no land had yet been watered. Consistent with his policy of encouraging the new, Mead had issued a new permit in 1896 to cover substantially the same lands as the 1893 permits. The new permit had the 1902 construction deadline. As he recalled later, Mead also made it clear to Wiley that his was a new project and could get no earlier priority date than his filing of 1896; the old permits Wiley had acquired would not give him an earlier priority.[44]

Now, in 1901 and 1902, as Wiley argued with the Farmer's Canal and claimed

1893 priority for getting water, Fred Bond, Mead's successor as state engineer, had to decide what to do. The problem was Wiley's proposal to take water in a priority earlier than the Farmer's Canal. Bond appreciated Wiley's construction difficulties, which meant the looming 1902 deadline set by Wiley's 1896 permit was not likely to be met. He also had some sympathy for the eager farmers at Wiley's colony, who were living in sod huts, waiting for water in drought, hoping to raise a crop and maybe build a cabin. The engineer was totally unwilling, however, to give Wiley's project any priority over the Farmer's Canal, where the farmers were equally or more deserving. They had worked hard and successfully watered the land and "proved up on" their water right, getting it officially adjudicated.[45]

The state engineer asked for advice—twice—from the state attorney general. He did not, however, get the answer he wanted. The attorney general insisted that Wiley should indeed get water with the 1893 priority position he sought, ahead of the Mormon colony. The attorney general also went further and said an unmet construction deadline on Wiley's project would be no problem and should not affect his permit or his priority right to water, whenever he did get everything built and settled.[46]

This attorney general was an interesting character, quite an obstacle for state engineer Bond to encounter. Named Josiah Van Orsdel, he had come to Wyoming from Pennsylvania in about 1892 and was a lawyer for another Big Horn basin promoter. He knew Mead, because that promoter had had Mead survey his irrigation project plans. Van Orsdel eventually became head of the Wyoming Republican Party, succeeding to the position earlier held by Willis Van Devanter, who had been Warren's right-hand man. Then Van Orsdel was appointed as the state's fourth attorney general, a post he filled for seven years, longer than anyone before him. In 1905, he became briefly a Wyoming Supreme Court justice, and then, like Van Devanter, went on to legal work in Washington, first as an assistant US attorney general and then as a judge on the federal court of appeals there.[47]

The first problem State Engineer Bond presented to Van Orsdel was whether water should be delivered to land still being developed under a water permit in a river already in heavy use. When other users wanted the water, had their rights adjudicated, but had a later priority date than the permit claimed for the partially developed project, did that project get some water first? Van Orsdel responded, quite naturally, that someone with a water permit who was working to put water to work under it had to be able to get water under the permit priority date, ahead of others' later rights, in order to have a chance of meeting deadlines for getting the water put to use. In any given irrigation season, he said, it was up to the local water commissioner to decide how much land was truly ready for irrigation under the permit and so how much water should be delivered under the permit.[48]

Van Orsdel accompanied that statement with a rather expansive new view of water rights in Wyoming, declaring that state water permits were much more than Mead or his successor believed. State water permits should, he said, be regarded as documents that conveyed a germ of a property right (an "inchoate" right, he called it). The permit holder had obligations: to nurture the right by putting the water to use. The State Engineer's Office, however, also had obligations: to protect that right at the earliest stages, which could require delivering water before construction was complete. Van Orsdel noted that questions about delivering water to land still being developed, to the detriment of later-date water holders, wouldn't come up very often in the case of small plots of land that individuals could develop in a timely fashion. But it would and had come up, he said, in the situation of a big project like Wiley's, where getting settlers and water on the land would be a lengthy process.[49]

The problem was that the Farmer's Canal adjudicated rights were *not* later but earlier than Wiley's 1896 permit from Mead. Farmer's Canal had priority over that permit. Wiley's demand for water had a major hitch: the only permits he had of an earlier priority date were the 1893 permits that had never been used. In 1902, Bond asked again, with more urgency, for legal advice on Gray Bull River water, highlighting that the basis of Wiley's claim was those 1893 permits so long unused. Van Orsdel responded by expanding on what an "inchoate water right" meant. The "inchoate" rights received under a Wyoming water permit couldn't be canceled by the State Engineer's Office—only the Board of Control could do that, if the water user sought to show that water was used, in the adjudication of his water right. The germ of a property right granted in a Wyoming water permit gave a permit holder not only the right to use water but also the right to change plans for reaching the same lands with water, he said. If someone like Wiley, with a new plan, bought up older permits, he was merely changing old plans and could claim the old permits' priority date. The 1896 permit Mead gave Wiley had been for the same amount of water and land as under the old permits. It should be viewed as just an extension of time, to the end of 1902, for the water to be put to use under the 1893 permits and their priority date.[50]

Van Orsdel went further: water projects like Wiley's were a special case for Wyoming's water office. Under the Carey Act, developers such as Wiley got federal lands set aside for irrigation projects that the state approved and provided with water permits. The state had to certify that enough water was available for the project. State law implementing the Carey Act had declared that water certified as sufficient for the project would go to the lands set aside for the project. And a big project could take years to develop. Given the probably long time until project completion, it was only logical that "sufficient water must be reserved at all times to irrigate the lands segregated." No matter how long it took to build the project's irrigation canals, the state couldn't turn around and give priority to subsequent water rights, just because the water reserved for the Carey Act project had not yet been used. Wiley need not worry about the December 31, 1902, construction deadline; the construction deadline in the state water permit simply could have no force for a Carey Act project.[51]

Van Orsdel's conclusions caused an uproar—along the Gray Bull River, in the State Engineer's Office in the capital, and beyond. Van Orsdel had said that Wiley's project should get water ahead of the Mormon farmers, no matter how long it took Wiley to get the project going. People were outraged. Van Orsdel wrote to Mead about it. Mead was now heading up the Division of Irrigation Investigations for the federal Department of Agriculture in Washington, DC. From Washington, Mead told Van Orsdel he recalled the sequence of permits involved in the Wiley project and the Farmer's Canal very well. He declared flatly that Van Orsdel was wrong. Van Orsdel's pronouncements, Mead wrote, "would cause unending confusion, would be unjust to other appropriators on the stream," and, overall, violated the Wyoming constitutional language on water that Mead had spearheaded. Mead similarly dismissed Van Orsdel's view that the Wyoming statute accepting the Carey Act negated state water permit deadlines. Mead had been one of the people drafting that statute. Van Orsdel responded doggedly that he knew that his view was contrary to that of everyone involved in Wyoming's acceptance of the Carey Act, but he believed his view would prevail in court. And as for his interpretation of the "inchoate" rights conveyed under Wyoming water permits, Van Orsdel concluded, "I am aware that I have departed from some of the principles upon which we were always agreed, in my opinion, but I still believe that my opinion to a large extent, will be upheld."[52]

Van Orsdel's prediction was both right and wrong. Initially, the State Engineer's Office had little choice but to implement the attorney general's opinion and give Wiley water. The Farmer's Canal immediately filed suit, represented by another Cheyenne lawyer who had been both a Wyoming Supreme Court justice and US attorney for Wyoming, a prominent Democrat challenging the arguments of prominent Republican Van Orsdel. The state district court ruled for Farmer's Canal, and against Wiley—overruling Van Orsdel's opinion. The district judge said Wiley could get some of the water he needed—but only *after* the Farmer's Canal got all its water. The court ruling stuck. Both projects on the Gray Bull were successfully irrigated under that order of priority for decades to come (and they eventually worked together to get reservoirs built on the river to improve water deliveries to both). The State Engineer's Office, meanwhile, gave settlers under Wiley's project extension after extension, into the 1950s, to get the water onto the ground.[53]

All this was, however, only an initial wrangle over what was involved, what had to be done, to acquire a protected priority right to water in Wyoming. There was more to come—and Van Orsdel's thinking in the Wiley case managed, through various twists, to have an impact where it mattered: in future policy, nationally and in Wyoming.

———

The turn of the century was a time of considerable turmoil and significant action in national policy affecting the West. The pressure of a growing population in the industrialized East merged with western aspirations for more land settlement and agricultural production and together produced new national policy. In 1902, just two weeks before Van Orsdel sent his advice on the Wiley case to the state engineer, Congress had passed the National Reclamation Act. It made its way through Congress with considerable fanfare and controversy, with Wyoming's delegation—first Senator Warren, then Congressman Frank Mondell—much in the fray. The new law provided for federal construction and maintenance of major dams and irrigation works across the West. Mead, with Warren's ear, had pushed for a more state-centered program, but that idea lost out. Mondell, whom Mead privately considered ignorant about irrigation—"opinionated and somewhat bumptious"—pushed the successful language for expansive federal involvement in irrigation that President Theodore Roosevelt wanted. Promotion of the final Reclamation Act played on the widespread recognition that private ventures had proven incapable of taking on the big projects, and the belief that not only the West but the nation would benefit greatly from irrigation to create small western family farms, producing solid citizens along with fruit, vegetables, and grains. Agriculture was seen in the nineteenth century as the universal first step in development of "civilized" societies, and the ideal of the small family farm, however inappropriate to the topography and climate of the West, drove policy, investment, and boosterism well into the twentieth century. The passage of the Reclamation Act was watched closely in Wyoming. The years ahead showed what faith in farms might mean for people on the ground there.[54]

The turn of the twentieth century also produced a congressional consensus for an increasingly hard-headed approach to a fundamental social issue in the West—the relations between the native people and the whites who now dominated the region. Initially, national policy attempted to keep the two very different groups apart. Surviving members of native peoples in the East had been forced West in the eighteenth and nineteenth centuries; mid-nineteenth-century treaties with native peoples in the West had typically, as on Wind River, reserved large tracts of land to tribes while "opening" to settlement certain lands that the tribes formally agreed to relinquish. But as white settlers kept pushing into the tribes' lands—seeking land or gold, massacring buffalo, and bringing on the army—the tribes became, in the national mind, a defeated former enemy to be handled by special policy. As historian Frederick Hoxie describes it, in a close study of federal Indian policy, 1880s policy became "assimilation." On land previously held by the federal government for the tribes in communal ownership, native people should become private landowners and family farmers. Their children would be educated to a white Christian model (hence the infamous shipment of children to a harsh life, and sometimes death, at boarding schools far from home). In a generation, they would be "civilized" according to the standards of their conquerors, perhaps becoming professionals as well as farmers, mingling with the rest of the population in the pursuit of happiness extolled by Jefferson as the right of all men.[55]

The idea that agriculture was an appropriate occupation for native peoples had been a feature of many mid-century treaties like the one in 1868 creating the Wind River Reservation for the Eastern Shoshone. Farming became the official national goal for native people in the 1880s, to be accomplished by allotting to individuals, once they were deemed "competent," small plots of private property in the original reservations. It was a policy originally intended to be benign, to bring on "civilization" on the white model. No magical transformation of native people into farmers occurred, however. By the turn of the century, the assimilation goals of Congress and the Department of Interior supervising reservations hardened, sanctioned by the US Supreme Court, into what another historian describes as a "coercive and vicious" policy: native people were to be put onto their own small farms, ready or not; the "surplus" land in those nineteenth century reservations was to be taken and opened to new non-native settlers; and native families, now considered capable only of rudimentary education and manual labor, could just learn to survive in the school of hard knocks.

Wyoming congressman Mondell embodied that policy in the late 1890s. He had arrived in Wyoming in the late 1880s at the age of twenty-seven, helping to open coal mining in the northeast of the state in the wake of Gillette's rail survey. After being elected to Congress in 1898, he promoted manual training for native children, to teach them "how to perform the work they would be called upon to do, particularly in housekeeping and agriculture, after they left school."[56]

Historian and Wind River area rancher Molly Hoopengarner and geographer Paul Wilson have traced in detail what that policy did in agriculture and water use on the Wind River. Federal negotiations with the Shoshone and Arapaho on the Wind River Reservation began soon after Congress passed the 1887 Allotment Act aimed at putting native families onto surveyed lots and selling "surplus" lands to new settlers. The tribes' people were starving as the term of the annuities promised by the 1868 treaties was running out. After years of talks, the tribes sold, for cash and cattle, ten square miles of a hot springs area in the north of the reservation in 1897. Mondell made that happen, and the same year, he pushed for new surveys "with a view of preparing for the cession of a portion of the Wind River Reservation later," as he put it. He saw the Wind River as a "fair and fertile region," and he wanted a large area north of the Wind River opened to settlers. He also noted that the Department of Interior was slowly making allotments there for native families, "evidently with the view of justifying a grandiose Indian irrigation plan." He and Warren accordingly got the allotment process speeded up—and restricted mostly to the south side of the river—while pushing for more land cession from north of Wind River despite opposition in both Interior and Congress.[57]

By 1904, though the reservation's agricultural economy had begun to improve, with people hand-digging their irrigation ditches, the Shoshone and Arapaho leaders had seen enough bouts of impoverishment and starvation that they agreed to a final cutback of the reservation. The land cession covered a very large area north and east of the Wind River. Altogether, the reservation was cut to about one-fourth its original size of 1868, thus accomplishing the goal of the first Wyoming territorial legislature. The transaction was supposed to include assistance for the two tribes including cattle and an irrigation system. The final 1904 deal, however, was not itself a land sale. Congress and the federal negotiators managed to ensure that the United States would be not the buyer but the broker, simply offering the land for sale to white settlers. Mondell promoted inflated estimates of what such sales would bring in. Those estimates went far toward assuring the tribal leaders that the revenue would provide what they needed in the way of livestock, schools, supportive payments, and a full-fledged irrigation system that could raise hay for cattle. Ultimately, however, the estimates were proved wrong. White settlers took up less than a quarter of the ceded lands—the best quarter. What was left remained unsold yet out of Shoshone and Arapaho control for about the next thirty-five years. The newcomer whites paid for the land they took up, but those limited proceeds couldn't pay the real cost of the projected benefits for the reservation. In turn, those costs—particularly for the irrigation system—turned out to be, as usual for most irrigation projects in that era, woefully underestimated.[58]

The US government began work right away on the irrigation system for the lands remaining in the reservation but never completed it. That is true of every other irrigation system ever built by the US government for native people nationwide. Congress's spending on reservation irrigation systems built under the Indian Service was dwarfed by what it spent on projects built on non-reservation lands by the new Reclamation agency launched in 1902. At Wind River, the expected cost of the entire irrigation system on the reservation system was eaten up by construction of just one major ditch, and that was all that revenue from the land cession sales could cover. Construction continued steadily, with water being brought to more and more acreage—twenty thousand acres by 1920. But after the first new ditch had been completed with cession sales money in 1907, irrigation became a new cost of farming. Native people on newly allotted "family farms" on the reservation found that they had to pay out cash just to try to make the land produce. The federal government charged water fees to cover construction, operation, and maintenance of the new irrigation system (the government refrained from charging interest on those costs). The new system meanwhile made some of the old hand-dug, free ditches inoperable. For people who had been told that they would get an irrigation system in return for ceding reservation lands, people who were just beginning to save seed and money for the next crop season and who had been pulled away from their fields to help build the new system, those terms were discouraging or impossible. Shoshone and Arapaho use of irrigated lands on the Wind River Reservation dropped steadily, while some white settlers took over allotments on the reservation by sale or lease and began to dominate the irrigation done with the new reservation system. The land base in use by native people, already drastically shrunk, became increasingly fragmented.[59]

———

The white settlers had a lot of frustrations driving them to help create that fragmentation. Lands held under allotment by Shoshone or Arapaho individuals were attractive for lease or purchase because they could be served by that slowly growing reservation irrigation system. The land that white settlers could buy on the ceded portion needed irrigation (though that was not much highlighted in the sales promotions), but a functioning irrigation system was even slower to arrive there.

Since 1898, Wyoming's secretary of state and governor had been seeking cession of those lands with an eye to settlement. They and State Engineer Johnston had been fully aware of the need for irrigation to support any agriculture on the ceded lands. At the time Johnston opposed speculation in the Little Horse

Creek case, he had thrown himself and his rhetorical flourishes into pushing for a huge irrigation project to cover over 330,000 acres on the ceded portion of the reservation. He described the project as one of several that would deliver Wyoming's dream future, as he assured Wyoming readers. The projects, he promised, meant that "Wyoming is to witness a development along agricultural lines within the next few years which will place her among the first western states in the variety and volume of crop production." Secretary of State Fenimore Chatterton, no rhetorical slouch himself, joined Johnston in pitching the glowing prospects for such a project and its future community of prosperous small farmers (presenting the usual overestimated profits and underestimated costs).[60]

Brandishing a State Engineer's Office preliminary survey, Johnston and Chatterton managed to attract investors, a group led by a successful Chicago businessman who had grown up on a Nebraska farm and kept a hand in farm and irrigation ventures. Chatterton (while still in office) became the investor group's lawyer, and in August 1906, Johnston issued to Chatterton, as the company's lawyer, a very ambitious permit for use of Wind River water on the ceded portion of former reservation land.[61]

Johnston also undertook to vet and regulate the company's construction work and its contracts with settlers, though the state had none of the official regulatory role it could have had under the Carey Act. The Interior Department, brokering the ceded lands, refused at the outset to make them a Carey Act project. Nonetheless, Johnston persisted in attempting to manage the development. To his great irritation, the Chicago businessman hired his own engineers to survey the lands and subsequently refused to build the big project, concluding that it would never be an economic success because of porous soils which would require expensive concrete lining on canals. As an experiment, the Chicagoan did build a small pilot canal system of about fifteen miles near the new town of Riverton that had sprung up in response to the prospect of land deals, irrigation on the ceded lands, and the glowing picture state officials had painted. The pilot ditches carried water, but the Chicago investor concluded that the settlers themselves were a key problem, because they "could not have made a success of any farming operations. Most of them were speculators who had come in when the reservation was first opened to settlers in the hope of some easy money. Real hard work in the fields was the last thing they were looking for."[62]

The pilot system eventually watered only a little over ten thousand acres—about 3 percent of Johnston's proposed grand total. The whole business was much in the press, with one newspaper accusing Johnston and the former secretary of state of betraying their trust as public officials by working with and for the Chicago investor group. They had aided a "colossal conspiracy" to create a monopoly on water rights via the grand 1906 permit, the paper charged. Johnston canceled the 1906 permit in 1910, then left the state in a huff. The next state engineer reinstated the permit with its valuable priority date intact and put it in trust in the hands of the state's top officials in hopes of finding new investors.[63]

Finally, two groups of landowners organized on their own, one taking over the pilot ditch that had been completed, and the other getting Wyoming congressman Mondell's help to take over and extend another ditch to water about twelve thousand acres. That ditch had been started by a Shoshone family on their land on the north side of the river before the 1904 land cession. State officials gave both groups portions of the big water permit from 1906, with arrangements for first dibs on water under its prized priority date. People who had waited ten years or more finally got water. Meanwhile, the rest of the big, grand project languished; it didn't look profitable to any investor. The federal Reclamation office finally took it over in 1919. Wyoming congressman Mondell had meanwhile grown adept at tapping annual federal "Indian appropriations" bills for money to pay for irrigation infrastructure for white settlers on the ceded portion of the Wind River Reservation. Construction got underway with those funds. The project included, as had been originally envisioned, conversion of Bull Lake on the reservation into a reservoir feeding the Wind River for the settlers' project. There was also a dam across the Wind River to raise water levels enough to divert it into a long canal running ten miles to reach project irrigated lands for settlers.[64]

All those ins and outs translated into mounting frustration for the new people who had come to the new town of Riverton following the government offer of land. Some, therefore, looked for places to make farms on lands in what remained of the reservation, where they could see the Indian irrigation system under construction. It was legal to lease or buy those reservation lands that had been parceled out as allotments, and so these newcomers did that. [65]

———

Individual native people accordingly came under pressure to lease or sell their allotments. And then the issue of water rights for those allotments became an issue that intensified the pressure to sell.

As the opening of the ceded lands to settlers was pending in 1905, the reservation superintendent had filed permit applications for state water rights for the lands on the reservation that were being allotted to individual members of the Eastern Shoshone and Northern Arapaho tribes. The next year, the 1906 state water permit for the ambitious state-endorsed project on the Wind River began its long life. Under Wyoming water law, in order to keep their 1905 state priority rights alive ahead of that permit and its potentially massive water withdrawals, the native people had to meet permit deadlines to start using and keep using the water.

In Montana, much the same situation had developed. Native people on the Fort Belknap Reservation saw settlers up and down river starting to irrigate with state permits, before their own irrigation was fully underway. The federal agent in charge there, however, succeeded in getting a federal court to rule in 1905 that federal law ensured that the native people at Fort Belknap had the best water right on the river and could put that water to use at their own pace, independent of state law. In 1908 the US Supreme Court upheld the decision in the case, called *Winters v. United States*. The top court said that the tribe at Fort Belknap had rights to water, reserved to them with a priority dating from their treaty with the United States, for present and future needs, with no set amount or deadlines to be imposed by state law.[66]

As a historian of the *Winters* case has pointed out, the idea of such an "inchoate, unquantified, flexible reservation of water," endorsed by the US Supreme Court, was attractive not just to native people but to some outside the reservation. Some white settlers in Montana welcomed the decision, believing it would attract federal investment for major water development in their area that would serve them as well as reservation lands. Certainly, the concept of an inchoate right, in these arid lands where projects could take a long time to reach full size, had been useful to Van Orsdel in supporting investment in Wiley's Carey Act lands in Wyoming.[67]

Van Orsdel, remarkably, had a role in that Supreme Court ruling in the *Winters* case. In 1907, he was in Washington as an assistant attorney general and landed on the team arguing the case for Indian water rights in the *Winters* case before the US Supreme Court. As part of its decision upholding the Indians' rights, the court agreed with an argument Van Orsdel made: that the United States could reserve land for its purposes, including Indian reservations, and water for that land would be reserved as well, independent of state water law. It is hard not to hear an echo of his argument a few years before in Wyoming, where he said state water law did not apply to land segregated and reserved by the federal government for a special purpose—in that case, irrigation under the federal Carey Act. [68]

Once the *Winters* decision was issued in 1908, the US Office of Indian Affairs and the US Department of Justice put it to work on the Wind River Indian Reservation. On a tributary of Owl Creek, near the northern border of the reservation, a Shoshone family named Duncan had previously been determined "competent" to take their own land, and so they had an allotment. In 1911 and 1912, the Wyoming water commissioner closed the Duncans' headgates to ensure water would go down to whites who had been settled outside the reservation, near the mouth of the creek, since the 1880s. In response, the superintendent of the reservation told Washington that there needed to be a "test case" on reservation Indian water rights in the Wind River. To the "considerable annoyance" of the Wyoming water superintendent for the Wind and Big Horn basins, the US attorney filed a complaint in federal court in 1911 and again in 1912 against the water commissioner. The federal lawyer ultimately argued that based on the *Winters* case, Wind River Reservation lands and allotments on them, like the Duncans' land, had water rights dating from the treaty of 1868—the best priority in the Wind River basin.[69]

The goal of federal officials on the reservation, following congressional policy, was, however, still to push native people on the Wind River Reservation onto small private farms. So in 1912, even as the case from Owl Creek got underway, the reservation superintendent and a special inspector from Washington met with both Eastern Shoshone and Northern Arapaho members in council. The officials wanted tribal members to sell "extra," unused allotted lands to white settlers. They said flatly that the state owned the water, and everyone had to follow state water law. The water right could be lost if the water was not used and the state permit deadlines, extended once, were now 1916. To get the water on the land by then, the people would need horses and plows—and, the officials said, they could raise the money for that by selling off land and reducing their family's allotments to a workable forty acres. The reservation officials did not mention that at that moment government lawyers were challenging the power of state law over Wind River water for the reservation. They emphasized only state law deadlines.[70]

The meeting went on two days. Tribal members were skeptical that they would get much for their land and doubted they could move water onto remaining land in just four years. They noted that they had been pulled off their fields in just the past few years to build the new federal irrigation system to serve the allotted lands—including lands they were now urged to sell.[71] The Arapaho wanted to go to Washington to confront the secretary of the interior about it, though others discouraged them.[72] Big Plume, a member of the Arapaho council who had been on a delegation to Washington a few years earlier, had this to say (as translated by agency staff transcribing the meeting):

> About eight years ago there was a government officer and he was setting just like you are now and he told us how we were to live. Two head men of the government here raised their hands to God saying they had done everything straight and that there would be no cheating of one another. That government man when we ceded that north portion of the reservation he told us himself that we should get one and a half million dollars for the other side or the ceded portion of the reservation. And we remember what he said about these ditches at that time. There was to be a certain amount of money set aside for these ditches and he told us you work on these yourselves and get the money back and nobody can say anything about these ditches, that is what he told us. These promises that were made to us were or have never been fulfilled as yet.[73]

Another tribal member commented:

> And about these ditches, I cannot quite understand it, it goes hard on us all. This water I supposed was free to everybody because you all know that we could not live without water. These four years I cannot quite understand and it makes me feel bad to think that if we don't get down to doing something this water will be taken away from us and our lands. When we ceded that other side we supposed we were going to get something out of

it and why should they compel us to pay for these ditches. After they have taken the land away from us now they want to come and take our water away from us and after while we will have nothing left that we can call our own.[74]

The reservation officials responded that the secretary of the interior recommended that the native people sell more land simply because he was trying to do all he could to save their water in the face of state laws. "The water," said the inspector from Washington, "appeared to be free as long as there was no demand for it. It is the demand by the white people who have made these laws that has brought these conditions about."[75]

After this meeting, allotment sales were "successful," the officials reported to Washington. Though they also told Washington that they were anxiously awaiting the result of the Owl Creek case, they appear never to have told tribal members about that until four years after that council meeting, when the federal court in Wyoming ruled on the case in June 1916. The ruling by Wyoming's first federal judge, himself a former member of the state's constitutional convention, upheld the Wind River Reservation's 1868 treaty-based right to water based on the *Winters* case. A new superintendent immediately told agency staff to explain the decision to the Shoshone and Arapaho, to "assure them that their right to this water must not be interfered with" by state water officials. He told staff to watch state engineer staff to ensure that they followed the court order. By the mid-1920s, another case came before the next federal judge for Wyoming, Blake Kennedy (the longtime leader in the state Republican Party who drew a sharp picture of Carey in his memoirs). Kennedy too upheld 1868 treaty rights for the reservation, based on *Winters*. Citing *Winters*, Kennedy wrote, "The Government has reserved whatever rights may be necessary for the beneficial use of the Government in carrying out its previous treaty rights," on the Wind River Reservation, and those necessarily included the rights to water for irrigation. These federal court decisions went unpublished, however, and can only be traced in government archives.[76]

Wyoming's Congressman Mondell meanwhile railed against treaty-based rights for any tribe's present and future needs. After the 1908 *Winters* decision, he described as "monstrous" the legal theory that he characterized as saying that in an arid country, there could be any power to "stay development until the crack of doom because there is somebody too indolent or too indifferent to develop." And in all the decades to come, congressional action essentially echoed Mondell's antipathy to treaty-based rights that could enable tribes to develop water over time. Congressional spending on reservation irrigation projects, which could have been a major aid in putting native water rights into use, steadily dwindled nationwide.[77]

At Wind River, the result was probably much as Mondell would have hoped. Over the next half-century, white settlers bought out forty-two thousand acres of land originally allotted to Shoshone or Arapaho people and leased thousands of acres more, all of which tended to be land reached by irrigation. The white settlers ended up dominating use of the new reservation irrigation system. Native people tended to dominate lands on private ditches on the reservation, however, because those ditches were built independent of the new irrigation system and were therefore not high cost. Shoshone and Arapaho people also had better access to grazing lands on the reservation kept for shared use (though managed by the federal government). Cattle herds supplied by the cession terms did well until a federal official supervising the reservation again forced communal herds into ownership by individuals, who sold them off.[78]

Haltingly, however, the idea Van Orsdel had formulated, of reserving water from the ordinary operation of state water law for certain places and people and their future development, continued to make its way forward. Even on the Wind River Indian Reservation, it eventually did so in the late twentieth century, based on the *Winters* case—to the chagrin of those who thought like Mondell.

––––––

Elsewhere in Wyoming, the US government, with a host of eager settlers behind it, soon pushed against the strictures of state water law that had been dear to Mead. In doing so, federal officials made full use of the views Van Orsdel had expressed.

Not on the Wind River but on the Shoshone River in the Big Horn basin, north of the Gray Bull, the new US Reclamation Service of 1902 decided to build one of its very first projects. To do so, the federal agency acquired an unused permit from 1899. The 1899 permit was part of a venture of Buffalo Bill

Cody, who had made his career in the 1870s working as an army scout in Wyoming in summer and performing skits on the East Coast in winter about heroic army scouts. In the 1890s, Buffalo Bill, already famous for the world-traveling *Wild West Show* he would lead for decades, had a partner for development ventures: George Beck, who had a ranch near Sheridan and was the son of a US senator from Kentucky. Beck, a Democrat and perennial political hopeful for statewide office, had opposed the stockmen's invasion of Johnson County, in 1892, and warned fellow cattlemen not to join it. A couple of years later, he saw the Big Horn basin from atop the Big Horn Mountains, imagined it as akin to the Promised Land, and talked Buffalo Bill into joining him in irrigation ventures there. [79]

It was Beck and Buffalo Bill who hired Van Orsdel as lawyer and Mead as engineer to work for them as a team, several times in the 1890s, to lay out irrigation plans. Mead was not immune to the intoxication of potential development. In some ways, he was the biggest dreamer of all. Contrary to the grim realism of his description of the entire Big Horn basin for Wyoming readers in 1898, in 1896 he had written a flowery piece for a national irrigation booster magazine on behalf of Buffalo Bill's first canal project on the Shoshone River:

> Ever since the advent of the first emigrant this tract of land has been a source of longing to the homeseeker. As the possibilities of this region became better understood its attractions have increased until it has become generally known and regarded as the most extensive and desirable body of irrigable land in the state. [80]

Unfortunately, Mead was no better than any contemporary at foreseeing the obstacles to irrigation projects. For even the relatively modest first project that Buffalo Bill proposed for the hopeful "homeseekers" in his namesake town of "Cody," Mead's survey and projections for Shoshone canal projects proved woefully inadequate. [81]

Buffalo Bill also had a second, more ambitious plan: diverting part of the big Shoshone River to irrigate bench land above it. Despite their combined Kentucky charm and *Wild West Show* star power, Beck and Buffalo Bill never could pull funding together for that one. They did, though, secure a state water permit for it, signed by Mead in 1899. [82]

Though Buffalo Bill and Beck didn't build their dream project, its 1899 permit was tantalizing. Perhaps water for that imagined project could take priority over water for the various irrigators who had moved onto the Shoshone River, with smaller-scale efforts, soon after 1899. Under the Reclamation Act, the federal government had to have state water permits for its irrigation projects. The new federal Reclamation Service acquired Buffalo Bill's 1899 permit as a cornerstone for the service's own grand ideas for the Shoshone. It also applied for and got additional permits from the Wyoming State Engineer's Office for water storage and other aspects of the project federal engineers were designing, which was even more ambitious than Beck and Buffalo Bill had dared propose.[83]

The Reclamation Service met with its own difficulties, however. The federal plan was to build a dam to make a big reservoir and then put in a series of diversions downstream to take river flow and reservoir-storage water up out of the canyon to irrigate the dry flat benches above. The Shoshone is a serious river, fed by the snows and also by the sulfurous geyser pools of Yellowstone. Originally known as the Stinking Water, it had undergone a cosmetic name change, with Mead's assistance. Neither damming its steep canyon nor diverting it and sending its water to farm fields was easy. By 1910, the government's Shoshone project, using both direct flow and storage permits, was far from complete. By 1915, the State Engineer's Office felt compelled to say that the 1899 permit for diverting direct river flow of the Shoshone was void. The federal government had failed to meet that permit's construction deadlines.[84]

The Reclamation Service protested. Federal lawyers exhumed the Wyoming attorney general's opinion of 1902 on the Wiley Project, whose permit, Van Orsdel had said, would keep a priority unaffected by missed construction deadlines. The Reclamation Service demanded the same treatment. Van Orsdel's old opinion thereby took on a life of its own. Never mind the district court decision that had effectively overruled it on the Gray Bull River, not far from the Shoshone.[85]

The Wyoming attorney general of 1915 presented Van Orsdel's 1902 opinion to the new state engineer, James True. True came from an early pioneer family in Denver that had become a major player in civic life there. His brother, a renowned mural painter, would soon create the murals that adorned Wyoming's state capitol building, and a nephew, H. A. "Dave" True, later played a major role in the development of the oil and gas industry in Wyoming.[86]

James True reluctantly allowed the federal project on the Shoshone River to keep its 1899 priority date to divert unstored river flow. One consideration that might have weighed heavily with him was that by 1915, there were already plenty of people trying to live on the federal irrigation project lands along the Shoshone, people with high hopes of making farms and building new lives. They were the people who needed the water that the 1899 permit could provide.[87]

There was Charley Robinson, one of the first to arrive, in 1907. He was an Iowa farm boy whose Irish parents had met on an immigrant ship, married when they landed in New York, and drove an ox team out West at the end of the Civil War. Charley heard about the federal irrigation project on the Shoshone and brought his wife Maggie ("a shy Colleen") and three children—the oldest was six—to a cabin on the barren flats by the river. Maggie found the place lonesome, but seeing the train go by every day, a link to the outside world, was "the thing that helped me from getting too discouraged," she said years later. The first time Charley tried irrigating, he used so much water it ran down the road for five miles. But he got better at it. The Robinsons settled in, and Charley became a buddy of Buffalo Bill's. By 1913, when the oldest child was twelve, they managed a family holiday. Charley took them all on a "covered wagon safari" into the wooded beauty of Yellowstone, some eighty miles away, where they feasted on fresh-caught fish. [88]

There were William and Margaret Sedwick, who came in 1908 with two grown sons. They were from Iowa too, and they were tired of being flooded out each year at their farm along the Des Moines River. They came with $400 in cash and one hundred slips of elm trees to plant. The trees all died the next winter. But "believe me, pioneering did not seem so bad to me, everybody knew everybody else, and were all so neighborly," said their son Bill, years later. Both he and his brother Henry married girls from other homesteads. Except for a brief stint in 1918, when Henry, just after his wedding, found himself guarding a railroad in Siberia in World War I, the Sedwicks stayed on the Shoshone Project and farmed there all their lives.[89]

The Sedwicks and the Robinsons and others like them had moved onto a solid block of fifteen thousand acres that the federal government had opened to homesteading in 1907, to be served by water from the new federal project, with its big dam still under construction. This meant that the people who came there, no matter what their background, very soon had a lot in common. They all shared the same experience—surviving the winter in a rough cabin, eating jackrabbits, breaking up the dry sagebrush-covered ground, trying to figure out how to irrigate and what crops to grow, how and where to sell them, and how to get their children some schooling. They built shacks at first, and then homes, on the farm lots, and they opened stores and alfalfa mills in the town. A hardware store, two blacksmiths, a drug store, a barbershop, a bakery, and a confectionery were all there by 1910.[90] They shared hopes, disappointments, miseries, and successes. The neat houses they set down in their new town on the orderly streets drawn by Reclamation engineers today still reflect the bonds formed between them. The town was named Powell by the engineers, after John Wesley Powell, though he had argued unsuccessfully thirty years earlier for a different kind of development of the arid West. The settlers wrested the identity of the town away from the government as quickly as possible, when the town lots were put up for sale. All prospective businessmen bought the cheap lots on a narrow street and made that street the center of town—instead of the broad avenue, with bigger, more expensive lots, that the federal engineers had laid out to be the main street.[91]

The farms outside of town were on homesteads, but the families didn't own them free and clear after five years, as ordinary homesteaders did. These homesteads couldn't be farmed without water—from the big, expensive, federal dam and canals. Like the people on the Wind River Reservation, the settlers found that they had to pay "water fees" designed to pay off those construction works. In their case, until everything was paid off, the federal government had a lien on their homesteads. That, in turn, meant it was very hard for the homesteaders to come up with cash to buy livestock, machinery, or seed. With their farms burdened with a federal lien, the only loans they could find were typically set at high interest rates to compensate for the poor collateral.

That's what Elwood Mead found. He came back to Wyoming in 1915. He was consulting for the Reclamation Service, assessing the condition of farmers on federal irrigation projects. It was the very year the Reclamation Service and the state engineer were wrangling over whether to extend the 1899 permit on the Shoshone. Farmers on the federal irrigation project on the Shoshone and elsewhere in the United States had just succeeded in getting the Reclamation Service to extend time for payment of their water fees—the farmers had organized delegations to lobby for extensions in Washington. But in Mead's eyes, that time extension meant that the farmers would just suffer longer, with the federal lien stuck on their farms longer. He proposed that instead the federal government give irrigators agricultural training and, most importantly, a loan program with interest rates they could afford. He interviewed the Shoshone Project farmers. Even those who had arrived with cash in hand were now in debt at 10 or 12 percent interest, and Mead believed that the extension on water fee payments was likely to attract people with even fewer resources who would get into worse debt. He took seriously what one man deep in debt on the Shoshone Project had told him: "Simply, when you come here you put your foot in a trap."[92]

But there were those who wanted to stay on and who believed in the promise of a federal water project. "We like the country here and the people, Albert Shoemaker told Mead. "There is a good class of people here, industrious." Shoemaker had come from Nebraska in 1909 with his two grown sons and their wives. He wished that his relatives back in Iowa, who never could afford more than renting a farm, would come to the Shoshone so they could someday own a place. "Of course, we will never have millionaires but maybe we will have some people in good common circumstances," Shoemaker said. "I think it is a good thing, making homes for lots of good young men like mine."[93]

No wonder it was almost impossible for a Wyoming state engineer in 1915 to cancel a water permit for a project that hosted people struggling to make homes. In fact, no Wyoming state engineer ever canceled the old Buffalo Bill permit, though for another twenty years and more, state engineers kept raising the issue of permit expiration—and then giving up on it. It took a long, long time for the project to use all its proposed water, getting the water onto all its proposed acres. The people on the farms were able to "pay off" the costs of the project only because Congress forgave the interest owed nationwide on Reclamation project construction costs and also allocated some of the Shoshone costs to non-irrigation benefits. It would be some ninety years later, early in the twenty-first century, before the Shoshone project finally proved up on the old Buffalo Bill permit and received a state certificate for an adjudicated right to divert the flow of the Shoshone River.[94]

But it all worked out so no one along the river complained too often or suffered too much. Since the Shoshone River carries snowmelt from the heights of Yellowstone and surrounding mountains, the big dam the federal government built (eventually named for Buffalo Bill) wound up supplying so much water, so steadily, that everyone on the river benefited. Smaller-scale irrigators who settled a little later downstream initially protested the extension of Buffalo Bill's old permit to serve the big federal project. But ultimately, all that water drowned out those protests. People got what they needed.[95]

————

That was the Shoshone's story. Exceptions to the imperatives of Wyoming water law—specifically, to the deadlines and demands embodied by state water permits—continued. On the Wind River, State Engineer True applied the precedent from the Shoshone. When the federal government took over the ambitious project on the ceded portion of the reservation in 1919, he extended the life of the 1906 permit, still covering water for 330,000 acres, though it had been used for only 20,000-plus acres since it was issued. Farmers under the new project organized as the "Midvale" irrigation district. As they struggled with soils and climate, the land took decades to develop, and the project never reached the size once envisioned. Nonetheless, just as on the Shoshone, state engineers extended the big 1906 permit for decades, allowing water use on the project to expand over time as more land was put under irrigation.[96]

And other entities got their own special rules. Wyoming's growing towns, like the capital, Cheyenne, believed that they should have rights to more water than they needed just then so they could plan and provide for a population expected to grow someday. The state's Supreme Court agreed in 1913.[97]

In ordinary irrigation use statewide, however, the State Engineer's Office struggled to maintain the parameters of state water law. The office did manage to establish that the amount of water covered by an adjudicated right, or a permit, was only the *maximum* amount of water that could be taken from a stream. The maximum might not always be allowed if it would be wasted—if, for instance, a ditch could not carry it. The amount of water that could be diverted under a water right was the amount a user could prove was being beneficially used.[98]

The engineer's office continued, however, to be alarmed by water users missing deadlines on their water permits. Some water users saw no need to get more than just a permit, believing that one step was all they needed to be able to use water. The water superintendents remained chronically short of time and money to travel, inspect, and adjudicate all the new water uses made by permittees.[99]

Inevitably, some users naturally believed that their permits alone had given them a "water right of perpetual value," that they could rely on for the future, without undergoing adjudication. State Engineer Clarence Johnston had early on warned that users might see permits this way.[100] Users who took water based only on a permit that was never adjudicated could believe they had a water right, with the permit application date as their priority date. That was a twist on the rule of "relation back." The engineer's office and water users in Wyoming had long understood that when first steps toward obtaining water were followed by diligent work to get construction underway and put water to use, the priority date of the water right eventually adjudicated would "relate back" to the date of those first steps. The first-step date in Wyoming was the date a permit application was filed. *So if I have only a permit, but I did all the work and I use the water, I have a water right with the priority date of my permit application.* That was what people believed. They believed that even if they had never actually received the blessing of an inspection, adjudication, and official water right certificate from the Board of Control.[101]

Water users had a familiar model to draw on—though it was from land law, not water law. Homesteading under federal land laws was still very much alive in Wyoming, peaking in the 1920s. Under the homestead laws, a good faith effort to meet the terms of a government permit would secure the title to land. So it would make sense to settlers that a good faith effort should also be good enough to entitle them to keep using water for the long term in the pattern they had managed to create in a tough setting. That could mean that a state water permit would not only establish a priority date, but it would convey the seed of a private property right. Van Orsdel's view could prevail over Mead's original idea of an initial permit as a short-term contract. [102]

The State Engineer's Office fought this development. In 1916, State Engineer James True (fresh from the vexation of the federal Shoshone project permit) demanded a strict new statement of the original rule minimizing the significance of permits. The legislature promptly did his bidding. The new law said water permit deadlines were to be met, on pain of forfeiture of the permit and its priority position. The new law said water users must seek appropriation certificates. Water users had to submit proof of water use for superintendent examination within two years of getting water onto land (or, for permits that had already been pending for years, the deadline was to be 1919).[103]

It was an uphill battle. The state engineer's files from the time are full of handwritten letters explaining why permit deadlines couldn't be met. In cases of small water users, the state engineers sometimes declared water rights expired for failure to meet deadlines, despite good excuses. "I was sick all summer and was not able to work. I wish you would give me a little more time to finish this ditch," one would-be small irrigator pleaded with the state engineer in 1918. True replied firmly: No, the permit was expired, gone and off the books.[104]

But what appeared to be a victory for demanding adherence to the rules quickly dissolved into another defeat. In 1920, the next state engineer declared that the statutory requirements had to be relaxed. Many water users had not complied, and superintendents simply couldn't review all the water claims that should have been adjudicated two years after irrigation. The speed of development in other states also made for reluctance to cancel would-be early-date, high-priority Wyoming water rights, particularly as interstate litigation and interstate water agreements ensued in coming years. Further, the legal burden was on the state engineer's office to warn users of impending forfeiture if they failed to submit proof of water use for adjudications. Warnings often failed to go out, and with no warning notice, permits could not be canceled. Permits on the state engineer's books might even be technically stamped "expired" but would not actually be canceled and the water rights forfeited. This practice went on for decades.[105]

With limited inspection or intervention from state water officials, water users took over some of the supervision and enforcement role that the State Engineer's Office had once expected to perform. Users could determine for themselves what water use would occur, and the evidence is that they did so. Neighborhood monitoring of water use appears likely to have become the norm by the 1920s, judging by the evidence presented in the lawsuits of later decades. The engineers were still the gatekeepers—they determined who got a state water permit. Many users did meet deadlines and go through the inspection and adjudication process. But the local economy and the resulting value of water, high or low, could make a difference in whether water was used each year and what the neighborhood would tolerate. What happened with water also varied depending on the personalities and interpersonal relations on a creek—on whether, for instance, the local norm was accommodation or intimidation. No doubt then as now, successful intimidation could overcome a senior water right, as happened on Crazy Woman Creek.[106]

Ordinary water users in the first decades of the twentieth century learned from painful experience just how hard it is to make irrigated farms out of Wyoming prairies and dry canyon-side benches—even harder than it was on the creek bottomlands taken by earlier settlers. The superintendents from the state water agency, with their offices out among the people, witnessed that struggle. Together they transformed Mead's system. The federal government, as in the Shoshone Project case, was a powerful lobbyist. But even more important were local hopes, epitomized by the deference given to federal water projects— the hopes of development and home building in a nearly impossible place, of which the Big Horn basin was only one daunting portion.[107]

Dreams postponed, or never realized, weighed on Mead's system and reshaped it. Would-be water users statewide, no matter their power or their finances, all identified with the frustrations of trying to get ditches built and water onto land as quickly as Mead had imagined it should happen. As the years went by, the state legislature invested in trying to attract people and capital. Water users and engineer's office watched the population and economies grow in other states, downstream along the major rivers whose headwaters were in Wyoming's mountains. They feared what those states might demand, in priority claims to water, while Wyoming continued to develop so slowly. Opportunity, and then security, for Wyoming users became a priority.[108]

In the 1920s, State Engineer Frank Emerson acted on that concern when he once again extended the permit for the federal project on the Shoshone River. Emerson was one of several engineers whom Johnston had sent to help federal staff survey the Shoshone project, seventeen years earlier. When he extended the Shoshone permit, Emerson took pains to outline his reasoning, balancing the importance of protecting small appropriators who had met their permit deadlines against the need to foster water development in Wyoming given the faster growth in downstream states.[109]

Emerson was also Wyoming's negotiator for its first water agreement with other states, the Colorado River Compact of 1922. There he also sought to ensure Wyoming's right to reserve water to itself for development in an unknown, possibly distant future. Negotiations for the compact, among all seven states sharing the Colorado River, featured California and Arizona pushing for a dam to serve the flood control and irrigation needs of their growing economies. The upstream states, including Wyoming, wanted to ensure that a big project for those states would not preempt future water use for their more slowly developing states. On the Green and Little Snake Rivers, Wyoming's tributaries to the Colorado, Emerson attempted to ensure through the compact that Wyoming would have the right to put to use a portion of those waters, someday when it could. Other upstream states secured the same prospect for their populations. Wyoming voters rewarded Emerson by making him governor four years later. The creation of an interstate compact on the Colorado River, with the states getting congressional approval of a mutual agreement on how to share water for present and future among themselves (instead of letting courts divvy up water), encouraged later river compacts.[110]

––––––––

Over these years, engineer's office staff and water users became a community to themselves, sometimes feeling almost besieged. The superintendents and water commissioners were local people; often they, too, were water users. They and the other water users supported and worked with each other to fashion water management in a tough physical setting, where the best that could be done sometimes fell short of the best that was once envisioned. People wanted the ability to develop in their own way, to suit their own situation in their own time. They organized themselves, often into ditch companies, canal companies, or larger irrigation districts—organizations that, depending on their size, could hire their own ditch riders to monitor water delivery to organization members. By 1920, row crops like sugar beets and hay were being grown. Beet processing factories were built, and beets remained economically important to the Big Horn basin and the North Platte River basin for decades to come. Rather than the cornucopia of crops once dreamed of, however, most agricultural production in Wyoming was in hay, valuable for ranch consumption and for sale, just as it is now. Local efforts, rather than big federal projects, were responsible for irrigating most of the land. Local people, working with engineer's office superintendents and commissioners when they needed to, often determined how much water was really used, and in what patterns.[111]

Water law and management became the work of the community of water users and engineer's office staff. The needs of the people on the ground came to be reflected in the law, and they made the law different from the system Mead envisioned. Water users needed to be able to rely on getting the water they had won by the sweat and pain of digging their ditches. They had staked all on the promise of a place that could be changed to become the kind of place they wanted. They needed to believe that they could deliver on that promise, for their children if not for themselves. State staff saw that, and they had not the heart to defeat them. Neither did they have the money nor the people to enforce deadlines and inspect all irrigated land.

As a result, the thinking that came from land law ultimately made its way into thinking about water. On Little Horse Creek, Wyoming water users had very clearly distinguished water from land. They pushed for banning most water transfers because of the interdependence among neighbors that water uniquely creates. But when it came to what a state water permit granted, water users tended to adopt a land law approach, so that a permit alone could become in effect a water right. Along with that, water users (and eventually the local courts) began to see a person's right to water as more and more like private ownership of land: a right they and their children could rely on. Public ownership of water and the concept of serving the public welfare was still represented on the ground by the engineer's staff who were part of the local work on water management. Meanwhile, water users, however, found that they did not share Mead's antipathy to granting "private perpetual rights" to the public waters.

In 1902, when Mead had only recently left Wyoming for the national stage, he wrote of the western irrigation laws he had championed:

> A climate so different from that of the East as to profoundly modify the structure of plants and the colors and habits of animals required a corresponding modification of laws and institutions to bring human settlement into harmony with its environment.[112]

People and place in Wyoming, in their turn, changed the laws and institutions Mead created. Mead had admired other countries that allocated water for the greatest public benefit. Under western US conditions, his way to provide public benefit was by requiring water permits, enforcing permit deadlines, and making way for new water projects where old ones had failed. Others, though, saw another way to provide public benefit—by reserving water, perhaps perpetually, for uses still in infancy. So in early twentieth century Wyoming, water supplies were reserved into the future for cities, for big projects that required federal funding, or for the future development that might be chosen by the new settlers or by people in the native tribes. The reservation of water in the Colorado River for the upper and lower basins of the river through the Colorado River Compact cleared the way politically for the big Hoover Dam that created Lake Mead. That reservoir was named for Mead, who supervised the dam's construction as the head of the federal Bureau of Reclamation in the 1920s.[113]

Mead still had his doubts about the public benefit of the big federal irrigation projects as they operated in those years—failing to water much land, with costs unrecovered and fees too high for many settlers. In 1929, he sent a young researcher back to Wyoming to investigate living conditions on the newest farms under the slowly expanding federal irrigation projects on the Shoshone. He had already tried to get Congress to condition new Reclamation projects upon the creation of social and economic aid programs to settlers. He failed in that. Congress seemed unable to resist creating new irrigation projects, with fanfare and pork-barrel political rewards, regardless of the fate of the people who settled the projects.[114]

Mead worried about those people. To a friend back in Wyoming, he explained why he wanted an economic investigation:

> The idea underlying this [investigation] of ours is that while the people on our Government projects are doing as well as people elsewhere, we are not satisfied that they are doing as well as they could if the whole of reclamation were more carefully thought out and provided for . . .
>
> [It] costs a great deal of money to change a piece of wild land into a farm, and unless the land is occupied by settlers who are good farmers, who either have capital of their own or can be provided with capital to properly equip their farms, the enterprise will be an economic failure and the settlers will encounter so many obstacles that they will become embittered.
>
> We ought to know how that can be averted, and if it can't be averted,

then we should quit building more works. We should not go on as we have been in the past, filling a project with settlers and then having to answer the charge of having deceived and betrayed them.[115]

Mead's investigator soon reported her findings on the new Willwood Division of the Shoshone Project in Wyoming, near Powell:

Too much cannot be said or written of the poorly built structures that serve as homes for the Willwood settlers [in a new division on the federal Shoshone Project near Powell] . . . Single room, flat-roofed, tarpaper shacks. It takes courage to rear families in such surroundings where even the floor, for example, is so badly constructed that rattlesnakes, seeking the warmth of the stove, crawl through. In far too many instances . . . the burden of the hardships during the period of development falls upon the women. Although pioneer agricultural methods and equipment are now out of the question, the most primitive of pioneer conditions and methods are still to be found within the walls of the homes.[116]

Others saw those shacks differently. One of the women on the Willwood district was Anna Christofferson, who moved there just about when Mead's young assistant visited. Christofferson had good memories of life on the irrigation project—regardless of its burden on women. She and her family, despite all the hardship, saw only project benefits.

Christofferson remembered, for instance, joining other women to lobby for a mail route:

We went to every homestead on the Willwood, even the bachelors' places, and hung out dish towels and diapers on the lines and fences so when the man from Washington came, he saw evidence of lots of people. We got our mail route. My fondest memory is how people loved people. We were all in the thing together. We were all hungry together. We were all happy together. Anything that hurt any of us hurt all of us.[117]

Her daughter Betty was born on Willwood in 1930. When she was sixty-six, Betty said:

I've heard that some of these people that are so strict on ecology say they shouldn't have opened up this land, they should have left it how it was—not change the world. I think of all the people that started lives here. They sent kids out to be doctors, lawyers, nurses, and everything else. Why is that wrong? To me it is right, and it is a much prettier place. Don't change the world? Where would I be?[118]

4. LIVING THROUGH LEAN TIMES

The pioneer pattern of agriculture is impressed upon the Wyoming land.
—FRANK TRELEASE, WYOMING LAW PROFESSOR, 1966[1]

Places: Horse Creek, southeast Wyoming; Laramie River, southeast
Wyoming.

Time: 1915–1975.

Ersanch family, 1930s Hawk Springs, harvesting sugar beets.
Courtesy of Wyoming State Archives.

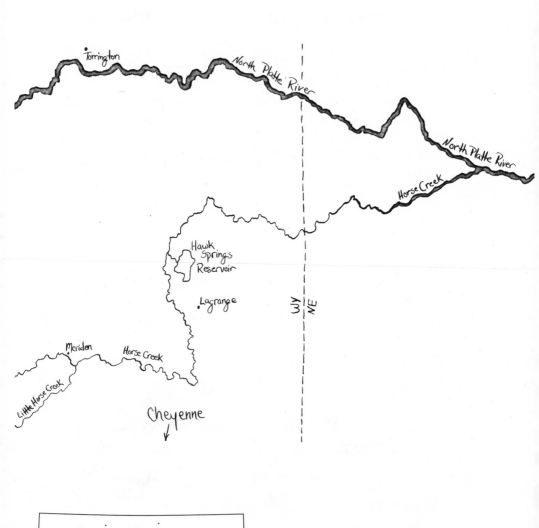

Torrington

North Platte River

North Platte River

Horse Creek

Hawk
Springs
Reservoir

Lagrange

WY | NE

Meriden

Horse Creek

Little Horse Creek

Cheyenne

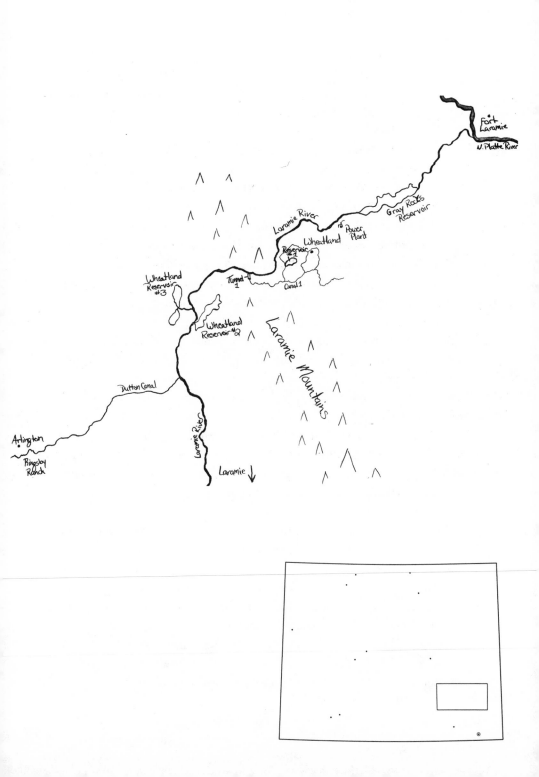

LITTLE HORSE CREEK, WHERE Johnnie Gordon had his Springvale Ranch, soon joins larger Horse Creek running northeast another twenty miles or so, to where the landscape of rolling plains begins to host some isolated big bluffs with striking rock formations.

In 1884, the Sherard family came out from Kansas and Missouri in two covered wagons and stopped near those bluffs. Nelson Sherard, "Nels," age fourteen, was driving one of the wagons. The ranches on Little Horse owned by Gordon and the Johnston family were two of only four settlers' ranches the Sherards had passed as they drove their wagons some sixty miles up from Cheyenne. More settlement was yet to come on Little Horse.[2]

But along Horse Creek itself, in sight of the big bluffs, there were already a few more settlers. A Nebraska-based townsite company even had people at work staking land and water right claims—anticipating profit if a rail line came through someday. Horse Creek was fed not by snowpack on high mountains but by local storms—it could run all year, off and on. Big cattle herds from Texas were driven north through the valley a couple of years after the Sherards arrived. A store became the core of a little settlers' town named LaGrange.[3]

At age fifteen, Nels started riding for cattle roundups and big ranches. He witnessed the drought and bad winter of 1886–1887 on the open range and eventually tried cowboying hundreds of miles away, near the head of the Gray Bull River in the Big Horn basin. He was a crack roper, lassoing even a gray wolf. He came back to marry a LaGrange girl; the Johnstons on Little Horse hosted the wedding. By 1902, Nels had moved his young family to a place near LaGrange, where he ranched for the next forty years.[4]

In 1902, the communities on the creeks north of Cheyenne were on edge over a murder the previous summer. The young son of a sheep rancher forty miles west up the creeks by LaGrange had been shot from ambush and killed. Tom Horn, known to be a hired killer for big cattle ranchers, was arrested and tried for the boy's murder. LaGrange-area men were on the jury. Nels Sherard went all the way to Cheyenne to hear every bit of the trial. "He was a nester," Sherard's son Don said of Nels, years later. "He was a small rancher. He wasn't in the class of the big boys that had the big ranches. He wanted to hear it all. That was a big deal in those days." Horn was convicted and hung.[5]

The next year, Nels Sherard took out a ditch to irrigate 190 acres on his new place, and Johnnie Gordon, now removed to Cheyenne, came and surveyed it for him. Nels and his wife put together a ranching and dairy operation. It was a satisfying life, but not an easy one. They had four boys but lost four others in infancy; their only daughter died at age eight after an appendicitis operation performed, too late, on the kitchen table. [6]

In 1916, when he was forty-six, Sherard himself sparked a "big deal" that caused a ruckus in Cheyenne. He complained that the townsite investment company, which owned a ranch and a big ditch dug in the 1880s that hadn't carried water for years, had lately started taking water through that ditch for its ranch. As Sherard understood the Wyoming water system, the old water right had been abandoned and should be gone. As a boy, just after reaching Horse Creek in 1884, he and his brother and father had helped dig the big ditch in question, he said. But by the early 1900s, it was "growed up with grass." There had been breaks in the ditch early on, making it hard to use, and it hadn't been running water for years.[7]

Others in the neighborhood said the same. There was a man who as a boy rode horseback to town for school (then held in summer and fall). There was the longtime rural mail carrier. There was a man who drove freight teams back and forth to Nebraska. They all said they had no trouble crossing the small swale that once had been a ditch. It wasn't a ditch anymore. It was full of dirt and grass, never water. That was, they said, until about the summer of 1909. The freighter, who grew up in the east and had never seen an irrigation ditch, remembered asking people in 1909, "It didn't rain, how did that water get there?" His wagon and team got stuck in mud and water where for years past there had been dry ground.[8]

These Horse Creek neighbors joining Sherard to describe what they'd seen were a solid bunch. One who was born in Prussia, taken on board ship with his parents at age five to make their way to Ohio in the late 1850s, had gone West in the 1870s. He worked as a freighter and a cowboy and finally made his own ranch near Horse Creek, where his son eventually also ran the ranch and went on the school board and the boards of various statewide associations. Another man had come from Nebraska in 1903 to manage his brother-in-law's ranch but then homesteaded his own place, helped build the church and the cemetery, and became a county commissioner. Another who came in 1901 to work for his uncles on a ranch and then got his own place was later deputy sheriff and a state legislator. Several of these men had joined Sherard in helping build a big new hall for the LaGrange store after it burned down.[9]

Horse Creek by 1916 was a busy place, with a lot of farmers and ranchers wanting water, so the state water office had a water commissioner working there, named Clint Donahue. When water was in demand and users would affirmatively "call" for getting water under their water rights, water commissioners would go by the official list of water rights and close headgates in priority order as necessary to get water to early right holders. Like other part-time water commissioners, Donahue was a farmer and rancher. He had come to the area in 1904, got a place on Little Horse Creek, and added the job of water commissioner to his work in 1916. He held that post for forty-five years. Donahue later testified that water started running through that old ditch in 1909 only because in fall 1908, the townsite company had acquired a new water right for its old ranch. That water right allowed the company to use the old ditch to transport water from Horse Creek to a new "reservoir" on the ranch—a small bowl of a pasture bounded by a little ridge plus some dirt berms. That meant, as Donahue put it, that the water suddenly flowing through the old ditch was "1908 water." The original line of the ditch was dug in the 1880s, but after years of non-use, it was now just a conduit for water with a 1908 priority to get to the new reservoir. Sherard and his neighbors agreed with Donahue. Sherard's own most recent water rights were dated 1904. Donahue's opinion meant Sherard should receive water before the ranch reservoir did.[10]

The townsite company's view was different. The company said that the water now running through in that big old ditch was "1884 water," because the ditch was built in 1884, for a water right dated 1884. The difference between 1884 and 1908 was of course significant. The 1884 right was big enough to take much of the water in the creek, water that otherwise would go to smaller farmers like Sherard that had settled and gotten their water rights after 1884.[11]

Sherard and one of his neighbors went straight to the state engineer to complain that the 1884 right had been long out of use and therefore should be considered abandoned and forfeited under Wyoming water law. The state engineer took the question seriously and launched a major investigation of water rights and usage on Horse Creek. The resulting report showed how little use there had been, over the past twenty-five years, of many big old water claims on Horse Creek and neighboring streams.[12]

Failure to use water could bring severe consequences—in Wyoming and across the West. Mead's laws had adopted the custom developed under prior appropriation that fit in with his view of water. If a water right is defined by water use, then users should lose their right if they don't use the water for years at a time. A water right could be declared "abandoned" because of non-use, and the water right itself, with its valuable priority date, would disappear. That makes room for someone else who came along later to get a new right to water and use it. "Use it or lose it" remains many people's shorthand version of the prior appropriation system, along with "first in time, first in right."[13] In Wyoming as elsewhere, whether and how a water right should be declared abandoned has, not surprisingly, been a question much debated, contested, and documented in state records. Nelson Sherard's complaint to the state engineer about Horse Creek started a chain of contests that significantly affected what happened to unused water rights in Wyoming.[14]

For Mead, abandonment was part of the mix of certainty and flexibility, a tool for allocating water for ultimate public benefit that he sought for the new system. People who failed in their attempts to put water to use should get out of the way and leave that water to be used by those who came after with perhaps a better idea. Just as no person should be able to transfer a priority water right to another place, so no one should be able to keep a right that is not used. The early priority date should vanish with the failed use, so that new priorities could replace it.

Mead's ideal of quick turnover after failure, however, suffered the same fate in abandonment policy as it did in permit deadline enforcement, as seen on the Shoshone. The ideal disappeared when theory was overcome by practice.

Wyoming's abandonment rule under Mead was strict, repeating what the territorial laws said: water users who failed to use water for only two years running could lose their right. Then in 1905, just after Nelson Sherard took out new water rights for his small ranch, some users got the legislature to change the law in Wyoming so that only water rights left unused for five years or more could be declared abandoned. That rule was more typical in the rest of the West. Meanwhile, Van Orsdel, the Wyoming attorney general, who tended to view water rights as property rights like land, said in 1904 that only the courts could declare a water right abandoned. It took until 1913 for the State Engineer's Office to get the legislature to modify the statutes so the Board of Control could declare abandonment after investigating formal complaints about unused rights.[15]

Nelson Sherard soon prompted the Board of Control to consider the possibility of widespread abandonment of water rights north of Cheyenne. Sherard complained that the townsite investment company ranch had not used its 1884 water right for many more than five years. That old right, like many water rights claimed in the 1880s, had been ambitious. It anticipated irrigating 4,500 acres of ranch lands from the 1884 ditch. Records in the county courthouse show that the townsite company, through its subsidiary, acquired the ranch lands in the 1880s through the typical pattern for major companies of the day, hiring dummy entry-men to take advantage of federal land laws. One part of the ranch to be served by the ditch, for instance, was acquired via a "desert land" patent issued to a New Yorker who had to pay a small amount and assert that he had brought water to otherwise "desert" land. In 1886, he handed over the property to an agent of the railroad's irrigation company who officially deeded it to the company the next year. In 1889, the water right for the ditch had been officially confirmed by a district court judge settling a priorities dispute on Horse Creek, just before statehood. That was one of the notorious territorial decrees Mead found so troublesome because it awarded wildly varying volumes of water with little attention to actual ditch size or acres irrigated.[16]

The state engineer who received Sherard's complaint in 1916 was James True, the man who just the year before had reluctantly extended the permit for the Shoshone project. As he considered Sherard's grievance, True began to wonder how much acreage along Horse Creek and nearby streams really was being irrigated by the water rights that ranches had acquired decades earlier, including rights confirmed by old court decrees. He got major funding from the legislature for a wider investigation on Horse Creek and Crow Creek (Crow Creek is a stream much nearer the capital that also had water rights set by territorial decree). True put a surveyor to work, and the results of that detailed field survey showed that on many ranches, as the owners had begun to use the water, the actual acreage they found it feasible to irrigate was much smaller than the acreage they had claimed or been awarded.[17]

True therefore launched a wholesale abandonment action against portions of nearly fifty water rights on Horse Creek and Crow Creek. He proposed that portions of most water rights be declared abandoned, to the point that most rights would shrink by well over 50 percent. The legal water right would then reflect the actual use of the water. A look at the topography on Horse Creek even today makes it clear that what True proposed made sense—the reduced water rights he outlined covered essentially all the land practical to irrigate in the area.[18]

True's eminently sensible idea led to turmoil. His massive abandonment action was an assault on overstated water rights held by wealthy individuals and companies that had considerable influence in state and even national politics. Owners of the land on Horse Creek and especially on Crow Creek included leading politicians. Warren, in his prime in the US Senate, was one landholder whose water rights on Horse Creek were affected. Former territorial governor George Baxter, prominent in the water sales contract on Little Horse Creek, had rights on Crow Creek that would be cut. When the Board of Control held a hearing on True's abandonment proposal for Horse and Crow Creeks in 1917, crowds of landowners, lawyers, and onlookers turned out. Their numbers were so great that the Board of Control had to relocate the hearing to the House chambers in the state capitol building.[19]

The Board of Control proceeded, heard the evidence and agreed with True. They declared as officially abandoned major parts of the big old claims, including the railroad land company's 1884 right that Sherard had targeted.[20]

That decision did not last long. The major landowners in the case hired the most influential lawyers in the state—including the firm founded by Warren's former lawyer Willis Van Devanter (by that time a sitting US Supreme Court justice). The lawyers did not argue the facts. They cited no details of water use or non-use. Instead, they simply challenged the procedure. They said the abandonment declaration was void because the state engineer and the other administrators could not launch a wholesale abandonment procedure. Each single abandonment charge, they said, had to be brought by an individual water right holder. They won with that argument; the district court ruled in favor of the landowners. At True's request, the attorney general appealed to the Wyoming Supreme Court. The appeal was withdrawn, however, with no explanation on record. Judge Carey's son Robert had been elected governor. He managed extensive family ranches, including lands on Horse Creek. None of their water rights had appeared to be unused, so they were not affected by True's abandonment action. But perhaps Carey sympathized with those influential neighbors who did not want to see their water rights abandoned. The entire Horse and Crow Creek abandonment effort died.[21]

Abandonment was, in a water supervisor's eye, a natural cleanup tool—a way to put failed, sleeping, or simply excessive water use claims out of the Wyoming water rights system and make room for the work of potentially more successful, active users with feasible plans. Horse and Crow Creeks, however, had proved to be hazardous cleanup targets. The buzzing of the crowds in the state House chambers echoed in the district court decision the next year. The owners of the big water claims on those creeks wanted to hold on to them, considering them a property right like a right to land. A water right gave land its value, and they wanted to hold on to rights covering as much water as possible. Their lawyers, arguing for a process limiting the scope of the abandonment searchlight, essentially argued that water rights, like land titles, deserved protection as property.

Users had begun to think water rights should become private property, but courts had the power to harden that growing idea into law. When a water dispute went to court, it was an opportunity for courts to import concepts problematic for water management. That was what had happened initially with the fight over water sales on Little Horse Creek.

"The law abhors a forfeiture" is a standard tenet of the Anglo-American common law of property in land. Starting with the 1918 district court ruling blocking the cleanup of Horse and Crow Creeks, that concept from land law pushed its way steadily into judicial decisions in Wyoming on water. As a result, "the law abhors abandonment" could almost be said of Wyoming water law today, even though abandonment remains on the statutes.[22]

Yet if "the law abhors a forfeiture," Wyoming water supervisors do not. Water users may not either, depending on how they sit. The decades after 1918 were hard. People seeking a self-sufficient way of life kept coming and homesteading, creating more farms and small ranches, mostly irrigated and a few dry-farmed. Yet a combination of mechanization and market prices that dropped quickly from highs reached in World War I meant a struggle to survive, particularly for irrigators who had been tempted into single-crop agriculture dependent on bank loans and market sales. Bouts of drought in the 1920s followed by severe drought in the 1930s made survival still harder. Through those years, water users would sometimes charge neighbors with abandonment, bringing individual abandonment cases (just as the lawyers in Cheyenne had conceded they could). The Board of Control, in turn, sometimes declared abandonment and sometimes did not. It was not that either the users or the water supervisors were avid for abandonment—it was simply a tool that had to be available. In the Horse Creek case, it could have protected the small farmers like Sherard against the opportunism of the railroad company. But whenever courts were invoked in the years ahead, they steadily restricted the conditions under which a water right could be forfeited for failure to use the water.[23]

———

In the end, the 1918 district court in Cheyenne helped to allow the big 1884 right on Horse Creek—impractical at the time, and under-used—to stay on the books for over half a century more, available to its owners to try to use as convenient. That created considerable tension and conflict in the neighborhood.

In the Great Depression Horse Creek questions came back, this time in the form of a single abandonment claim brought specifically against the townsite investment company's ranch. This time, an irrigators' association filed the claim, to protect the water supply for a reservoir they had built. The association had an interesting history. Their reservoir was built by a group of local ranchers on Horse Creek and nearby Bear Creek led by Frank Yoder. They had seen a chance to develop some upland prairie to the north into irrigated fields they could promote for new settlers.

Frank Yoder and his brother Jess, like Nels Sherard, had come to Horse Creek as teenagers with their families in the 1880s. The two started various ventures in the area, including the store in LaGrange and their own ranches. Twenty years later, in the early 1900s, water storage was an important topic of discussion everywhere for Wyoming—finding a way to beat the uneven pulses of nature that sent water by in torrents one month and left creeks and rivers nearly dry in the next. Catch the water in winter and spring and release a nice steady stream of it when and if you needed it. That was the idea. The Reclamation Service was building big reservoirs—on the Shoshone, far to the northwest, and on the North Platte. There they broke ground in 1904 for the Pathfinder Reservoir, which would provide water to a lot of land in Nebraska and some in Wyoming—lands a little to the north of Horse Creek.[24]

Groups of irrigators launched plans for smaller reservoirs. In spring 1908, Frank Yoder signed the application for water rights for a reservoir to be built on top of Hawk Springs, storing water from Horse Creek, to irrigate prairie to the north. His group well knew that their fledgling irrigation project could never get water from Horse Creek in summer, given all the active water rights in the area with early, higher priority dates. But Horse Creek often flowed in the winter, fed by occasional storms. The reservoir was intended to take advantage of that—to take water in the non-growing season, in the winter, and store all the water the new farmers would need in summer to irrigate the prairie lands north.[25]

Yoder applied for rights for the reservoir in May 1908 (with the help of Johnnie Gordon, acting as surveyor). Meanwhile, though, the townsite investment company had made plans for its "pasture reservoir," meant to be a shallow pond, from which water could seep out to the lands nearby. The company filed for water rights for that pasture reservoir—also in 1908, but in October. The water for the reservoir would be conveyed from Horse Creek through that old ditch built in 1884, plus an extension. Frank Yoder was well acquainted with the townsite company—it had a variety of ranch holdings around the Horse Creek area, and he leased some of them. His development group and the railroad company knew of each other's reservoir plans and negotiated exchanges of money and water to serve their different ventures. In 1908, Clarence Johnston, as state engineer, put a warning stamp on both ventures' water right applications as he approved them; the stamp provided notice that Horse Creek was "largely appropriated" and that they would only be acquiring rights to surplus or waste waters and could not injure earlier appropriators.[26]

That 1908-right water for the townsite company's pasture reservoir was the priority date to which Nelson Sherard, and the state engineer, had tried to limit the company in the failed abandonment effort of 1916–1917. The Hawk Springs reservoir typically could only fill in winter and get the water it needed if the railroad company's winter diversions through the old ditch were limited to its 1908 right. The railroad company's right of October 1908 was later in priority than the May 1908 rights of the Hawk Springs project. If the company could instead claim use of its larger 1884 right, through that old ditch, the Hawk Springs reservoir and its farmers would not get the water they counted on.

The Hawk Springs reservoir was built, and in due course, new farmers on the former prairie lands it served formed themselves into a water users' association, took over the Hawk Springs reservoir and ditch operation, and reorganized themselves as the Horse Creek Conservation District.[27]

Skip forward to the 1930s. The New Deal administration in Washington brought new policies to the West, intended as "relief" to improve the economy and the productivity of the land. A major initiative closed remaining public lands to new settlement and put the land under federal grazing leases managed in conjunction with nearby ranchers (as Mead and Warren had proposed years before). Federal programs also put in some irrigation and stock-watering improvements statewide. Other programs bought out subsistence dry farms labelled as unprofitable, and put those lands, too, under federal grazing leases.[28]

In the worst drought they had ever seen, people who had struggled to get a good water supply to their lands were determined to protect their water rights as best they could to keep their farms going. Horse Creek Conservation District farmers relying on the Hawk Springs reservoir felt that way. Water stored in Hawk Springs reservoir was low, and the farmers blamed the townsite company ranch. The company in turn declared that it could and did take Horse Creek water in winter by virtue of the company's big old 1884 right, easily predating Hawk Springs reservoir. The farmers' district filed an abandonment claim against the 1884 right, saying all the company had used and could use was its 1908 pasture reservoir right.[29]

Once again, the crowds who came to watch the case were so large that the Board of Control hearing moved to the House chambers in the state capitol building. Nelson Sherard, now in his sixties, did not use Hawk Springs reservoir water. But he was a longtime neighbor to the farmers who did, and he could testify about the old ditch to the townsite company's ranch. He was joined by the other witnesses who had seen that ditch left dry in the 1890s and early 1900s. Clint Donahue, the water commissioner, said he had allowed water to go down the ditch of the railroad ranch only under the priority of October 1908. It was not until 1932, a bad drought year, the commissioner said, that the ranch manager had proposed that the water go into that ditch with a much higher priority—1884.[30]

The townsite company and the people to whom it had leased its various ranch lands testified to a different set of facts. Those people included Frank Yoder himself, who had over the past twenty-five years become mayor and then legislator for a larger town just north, built alongside the federal Bureau of Reclamation's North Platte project. By the time of the lawsuit, Yoder was living in Cheyenne while continuing his ranch operations near Horse Creek. He and other witnesses for the townsite company all said that since 1909, the company's ranch had used the big 1880s ditch, always relying on the water right priority date of 1884—the right that would trump the farmers in the conservation district.[31]

After hearing all the testimony, the Board of Control ruled much as it had in the earlier attempted Horse Creek cleanup: the board declared the 1884 right had been abandoned.[32]

The courts, at both the district and Wyoming Supreme Court level, again reversed the board's decision. The court opinions made it clear that the judges never grasped the significance of the October 1908 priority date. They never discussed that right, which the company had applied for and gotten for its ranch reservoir. The Wyoming Supreme Court justices simply noted that water had run through an old ditch built in the 1880s; the ditch had been used initially and then sat unused for years, but it started carrying water again in the winter of 1908–1909. The top court ruled that if anyone had wanted to claim that the 1884 right had been abandoned and the ditch left dry for years, they would have had to say so before any water started running through the ditch again in late 1908. No one had done that. So, failing what the court considered a timely objection, the unused 1884 right could be, and was, essentially revived by its owner in the winter of 1908–1909, the court said. Ruling that the 1884 right was still good, the Wyoming Supreme Court cited precedent that explicitly relied on the old land law idea that "the law abhors a forfeiture."[33]

With this new rule on abandonment, the Wyoming high court joined other western courts that had come to the same conclusion. The rule meant, in practice, that a water right, however long left unused, could be revived and come roaring back to life. Once revived, an old long-unused water right could completely disrupt the pattern of water use that neighbors had built up over the intervening years. Neighbors who wanted to prevent that had to act, filing an abandonment complaint in a "timely" manner—and that meant, *before* the water right started being used again. In the absence of that kind of initiative among their neighbors, landowners with unused water rights could rest easy and not put water to work until they had money, time, inclination, or new technology.[34]

The hitch, of course, as the courts in Wyoming and throughout the West may well have known, is that few abandonment claims are likely to be filed "timely" under that rule. In the small world of Wyoming irrigation, as the twentieth century wore on, the slim margins did not tend to attract many newcomers. Neighbors expected to work lifelong alongside each other and each other's children and grandchildren. It was therefore a major decision to make an enemy of such neighbors via an abandonment claim and its bitter contest of opposing witnesses. It was also a major expense, involving lawyers, engineers, hearings, and sworn testimony. Better to let sleeping water rights lie if they stay asleep and unused. If, however, it becomes clear that someone—perhaps a new owner—is planning to start using water covered by a dormant right, then it can be worth going to the State Engineer's Office to complain of abandonment. But then the complainers must act fast. The Wyoming courts have insisted, since 1939, that the complaint be filed before water is used again under the old right. Sometimes, neighbors rush to the engineer's office in a race against the backhoe next door that is laying new pipe to carry water anew under an old water right.[35]

The court-sanctioned revival of the old 1884 water right on Horse Creek has created tensions that continue to haunt that creek valley. The scarce water in Horse Creek is used enough all year long that the creek is typically managed ("regulated") all year by a water commissioner according to priority date. But since the 1960s, a new factor has been added to Horse Creek: use of groundwater. Post–World War II technology made pumping up groundwater a practical way to irrigate. In response to increasing pumping of groundwater, the State Engineer's Office successfully pushed the state legislature in 1957 to require groundwater wells to have permits from the state engineer and to provide that he could regulate them, like surface water, according to priority. The 1957 law officially acknowledged the potential that different groundwater sources, or groundwater sources and *surface* water streams, could be "so interconnected as to constitute in fact one source of supply," and approved state engineer authority to regulate interconnected sources as one, according to priority date.[36]

The Horse Creek Conservation District pumped groundwater steadily from the 1960s into the 1980s to help supply its Hawk Springs Reservoir. Meanwhile a family that had ranched along Horse Creek above the reservoir since the 1940s bought the old townsite company ranch with its small pasture reservoir in the early 1960s. Proudly considering themselves to be the first in Wyoming to use sprinkler irrigation instead of the old flood method, they too used groundwater, for their sprinklers.[37]

Tapping groundwater on Horse Creek only added to the tension and complexity of water management there. Pumping groundwater to help fill Hawk Springs reservoir prompted plenty of complaints from neighbors and a pumping limitation order from the state engineer in the 1970s. In 1979, the Hawk Spring reservoir users once again brought an abandonment case against the old 1884 water right. The Board of Control ruled that the 1884 right covered only the ranch acreage held by the new owners and declared that therefore, about half the old 1884 right had been abandoned (much as True had decided in 1916). This time, the board's decision was affirmed by the district court.[38]

Then in 2009, the Hawk Springs reservoir users demanded that groundwater and surface water use be regulated in priority together. That would mean that the groundwater wells, all of late date, could not be pumped in irrigation season until most surface rights were satisfied. Even though the need for such regulation had been foreseen in the 1950s statutes, that kind of joint priority regulation was not undertaken anywhere in the state until State Engineer Pat Tyrrell did so in the early 2000s on a creek in central Wyoming. In 2009, the Hawk Springs users' target was, once again, the old 1884 right—what remained of it—on the former townsite company ranch. The ranch typically diverted water from the creek under that 1884 right in winter (because in summer, still older rights took the water available). The ranch irrigated land with the 1884 right—irrigated it in winter. Hawk Springs people argued that the real purpose of running water from that old right on the land was to recharge, in winter, the groundwater wells that the ranch used in summer. And taking the water from the creek for that purpose in winter meant that Hawk Springs reservoir couldn't get the water it relied on taking from the creek in winter.[39]

As the dispute wore on, once again influential people became involved. The owner of the old townsite ranch was a state senator, Curt Meier, who was elected state treasurer in 2018. His lawyer, Harriet Hageman, ran unsuccessfully for governor that year. Meanwhile, Meier was skeptical of fair treatment in his water case because while it was still before the state engineer, the sitting governor, Matt Mead (no relation to Elwood Mead), had a ranch with a little irrigated land served by the Hawk Springs Reservoir. State Engineer Tyrrell had authorized a groundwater-surface water study, held hearings on it, and in 2013 and 2017 issued orders for controls. He did not require regulation of groundwater and surface water on Horse Creek together by priority, concluding that the connections between the two water sources in this valley were complex and the impacts of one on the other might take years to occur. But his orders aimed at finding a workable limit on both winter irrigation use of surface water and summer irrigation use of groundwater. With the help of some wet years, a version of equilibrium apparently was reached. By 2017, groundwater use declined. Upstream uses of pre-1884 water rights shifted so that the 1884 right became more useful in irrigation season. Meier retooled his system on the old townsite ranch so that his sprinklers could tap into ditches carrying 1884 Horse Creek surface water in summer. Still, more friction is likely in future dry years.[40]

––––––

In other decisions, from the 1930s on, the Wyoming Supreme Court has thrown up more barriers against declaring any water right abandoned. That made it harder for water to be available for new, perhaps more successful uses. The court repeatedly overturned abandonment orders—overruling the Board of Control no matter what the board had decided, either for or against abandonment. Through its opinions, Wyoming's high court has set rules for examining abandonment challenges—including an examination of whether failure to use water was "voluntary"—that can be hard to follow as a practical matter.[41]

In only one case, nearly fifty years after the 1939 decision allowing revivals of Wyoming water rights, a prominent member of the Wyoming Supreme Court wrote a majority opinion that refused to apply "the law abhors a forfeiture" to water law. He declared, with considerable judicial vigor (unleashing not one exclamation point, but two), "We cannot call up the abhorrence-of-forfeiture rule in order to rescue [an irrigation company] from an abandonment of a water right in lieu of requiring that the applicable statute pertaining to abandonment be applied and given its plain English-language meaning. We are not the legislature. Indeed, we do abhor forfeitures, but it is the legislature that has established this rule for forfeiting water rights—not the court!!"[42] But that lone pronouncement, never repeated by the court, is the exception that proves the rule. It is very hard to get a water right declared abandoned in Wyoming.

Members of the Board of Control have watched successive court decisions on abandonment with both exasperation and consternation. In the 1960s, the state engineer described the abandonment process dictated by the courts as "so cumbersome and expensive for anyone wishing to force an abandonment that it is seldom utilized." He declared that the result was "many thousands of acres of water rights on the records in Wyoming which have not been utilized over a long period of years, and in some instances have never been used." Though they were rights only on paper, they were nonetheless a cloud on the rights of anyone who came in later, a discouragement of new ideas, because of the ever-present threat of revival. Fifty years later, in 2012, State Engineer Tyrell voiced the same complaint.[43]

The court's reluctance to allow abandonment reveals, to the water supervisors' minds, a profound misunderstanding of water and water management principles. In water, there must be an opportunity to recognize failure and make room for an orderly change to new players with new ideas. Early-date water rights allowed to sit unused on a stream awaiting revival are, by contrast, like time bombs. They may be useful for new people and ideas, as in the case of the senator's recharge plans on Horse Creek, but only by disrupting the local water use patterns, on which all the other water users on a stream have come to depend. Far better, the supervisors believed, to replace a failed old idea with a new idea via a new priority date that might not supply as much water but would not trump the existing users. The new idea would receive water while the people already using water on the stream received water with their earlier-date rights. That approach would recognize the interdependence of water users—the product of idiosyncratic stream hydrology and water user habits—and retain the distinction between rights to water and rights to land that had been vindicated after the struggle over the Little Horse Creek water rights sale decades earlier. The legislation that overturned Justice Potter's decision in that case had highlighted the special nature and the limits of a water right. But as the court began to disfavor abandonment and allow unused rights to persist and be revived, the decisions described a Wyoming water right as a private property right, like a title to land.[44]

Wyoming water users acquiesced, in the 1930s and the decades after. Perhaps they welcomed the court's rejection of a distinction between water rights and land rights when it came to abandonment. Those crowded hearings in the state capitol about Horse Creek in both 1917 and 1934 showed how the idea of abandonment could alarm water users. Small owners as well as wealthy ones were uneasy with it. Many who had come to Wyoming and tried to irrigate, whether on small or big ranches, hoped to make their valleys into blooming Edens, but by the 1930s, the number of people who could stay on the land was getting winnowed down.[45]

Drought and economic stress could make people more desperate to challenge others' rights, but lean times also meant everyone knew the problems of irrigating and raising a crop and how hard it could be to find the cash to fix a damaged reservoir or ditch headgate. Left unrepaired, faulty structures could cut back use of a water right for years. The court rulings making it hard to prove abandonment were to the advantage and possibly the relief of many water users who wanted to see their water rights as their property. The concept of "private property" gives a water right some sanctity, in legal and political parlance. In an irrigator's view, that could be an appropriate return for the blood, sweat, and tears that a family puts into digging ditches and creating irrigated fields in Wyoming's landscape.

Anti-abandonment rulings from the top court gave Wyoming water users the power to stay in the water community, waiting for a better day to put their water to use. But limits to that opportunity remained. Public ownership of water was still represented by the state engineer, the superintendents, and the local commissioners. Their view of good water management, recognizing both user interdependence and the need to replace failure with new ideas, asserted itself with remarkable effect in a different context: water right transfers.

Transfers of a water right to a different location—watering different lands or serving some other completely new use—had been banned in 1909. But the 1909 statute included an exception: a water right could be moved without loss of priority to certain "preferred" uses, like water supply for towns, cities, and important industries like railroads. [46] As Wyoming grew and changed, slowly, after 1909, its towns, cities, and railroads did find they needed larger water supplies. They found room to do so under that exception to the transfer ban. They wanted a secure supply, and of course that came through a water right with an early priority date. The town or railroad would typically want to get an old irrigation right and change it to a new use, often in a new place.

To do that, there were details to be resolved—how much water could be transferred, for use at what times? The Board of Control slowly worked out its own way of answering those questions. Reservoir construction was continuing, and the state engineer in 1920 (Frank Emerson, who extended federal project permits and negotiated the Colorado River Compact), argued that reservoir water should be freely transferable. Building a reservoir, he said, meant creating an asset that its owners should be able to use as needed. But he underlined the unwavering concern in the engineer's office over the very different proposition of transferring rights to water that was not stored in a reservoir but simply diverted as it naturally flowed down a stream. Echoing Mead, he wrote that such a transfer "would make a speculative commodity of the water of the State. The equilibrium of conditions on our streams would be continually changed." Rights to a stream's natural flows that had not been stored—called "direct flow" water rights—had to be governed by the limits imposed by the engineer's office, including strict controls on transfers, to recognize the interdependence of those who use such waters.[47]

For Wyoming towns, acquiring existing rights to naturally flowing water and transferring the rights to municipal use was often their best option. Reservoirs were expensive to build and maintain—and would have a priority tied to the date they were planned and approved, junior to existing users. As towns grew, early-date water rights from the ranches around them seemed the best sources of water supply, particularly as residential areas expanded onto ranchlands that had water rights. In response to town proposals, the Board of Control began to work out ways to allow those transfers in ways that would consider and protect "the equilibrium of conditions on our streams."

A family from near Douglas became prominent in that process. Spencer Bishop, age twenty-two, had come to Wyoming in 1874 from upstate New York to find his fortune, and soon was managing the army hay fields serving Fort Fetterman and hauling freight near what became Judge Carey's ranch. Bishop went back home to New York to get married after he homesteaded a place on LaPrele Creek eight miles from the fort. He began a ranch and freighting operation with the homesteader next door. More and more people came to settle nearby. Spencer's oldest son Loren Clark Bishop (known as L. C.) was born on LaPrele. L. C. dropped out of school in eighth grade to earn money and help the family, and in 1908, he managed to get a job with a survey crew for a LaPrele

Creek irrigation project. The crew was run by Emerson, the future state engineer. The LaPrele irrigation project was, like Wiley's project in the Big Horn basin, launched under the Carey Act, with federal land allocated to it under state supervision. It was the brainchild of a doctor who arrived in Douglas when it was a tent camp in 1886 (and ten years later switched from doctoring to managing a big sheep ranching company). He partnered with a preacher and another sheep rancher-turned-developer to launch the project to water about twelve thousand acres of large ranches and small homesteads. The project took water directly from LaPrele Creek south of town and supplemented it with a reservoir to store high flows of the creek. L. C. became the project engineer and got construction going. There were plenty of financial and water supply troubles, but people settled and irrigated under the LaPrele project, and the settlers organized into an association that took over operations in the 1920s. The LaPrele settlers became quite a community—irrigators' association and women's club meetings, as well as school and church events, were family social occasions.[48]

In 1919, when Emerson became the state engineer, L. C. became superintendent of Water Division I, which encompassed the North Platte River and the southeast section of the state. He had spent some years in the army, including during World War I, and was a crack shot. Family lore had it that the skill sometimes helped him on the job, as when cantankerous irrigators, after seeing his marksmanship, decided not to challenge his closing of their headgates in priority. When a new federal reservoir on the North Platte above Casper was proposed and built in the 1930s, primarily as a jobs project in the 1930s, L. C. opposed the idea floated by the state engineer to give the new dam a 1904 priority date because it had shown up sketched in on government maps of that era. Such a date for a new federal reservoir would upend accustomed water use for rights dated after 1904. Bishop's stalwart stance paid off. He became state engineer in 1939 and held that job till 1956. In 1963, his own son Floyd became state engineer in turn, putting a stamp on Wyoming water management into the 1970s.[49]

L. C. Bishop presided as state engineer in 1940 when the Board of Control considered the proposal for a water rights transfer to the town of Greybull. The town wanted to take over for its supply an 1893 irrigation right to water flowing in a nearby creek. Greybull had been growing as a small industrial center, supported by refineries handling oil from production wells in the Big Horn basin. The town was located where the river, by then called the Greybull, flowing in from the west, joined the much bigger Big Horn River. There were, however, a lot of demands and early water rights on the Greybull River, like those of the Wiley project and the Farmer's Canal. The town, therefore, had turned to the rights of a ranch on Shell Creek, flowing from the Big Horn mountains to the east, in order to add to its water supply. There were also water users on Shell Creek, of course, and the creek is still known today for its feisty water fights. In 1940, Shell water users argued that the town of Greybull should not get a right to more water under the ranch's 1893 water right than the ranch fields had consumed. The town wanted a larger amount of water—the total that the ranch had diverted from the creek to take to the ranch fields. The Shell Creek people argued that the town should be able to get not the amount of water diverted in the past, but only the amount consumed.[50]

The reasoning behind that distinction is this: to get the water to the fields through dirt ditches and to flood entire fields, ranchers typically must divert more water than their hay crop can consume. Some share of that water comes back to the stream, seeping through the ditches and running off the fields. That water is called "return flow." Return flow can, depending on geology and topography, support healthy riparian areas and wildlife. It can also resupply a stream with water that the next users can divert in turn—water they rely upon. Return flow is one of the factors making for interdependence among water users. The people on Shell Creek were worried about seeing all the water that the ranch formerly diverted put into the pipeline planned by the town of Greybull—with no prospect of any water returning to the stream.[51]

The Board of Control headed by L. C. Bishop agreed to focus on past consumptive use. The board determined that the ranch crops had only consumed about 60 percent of the water when the ranch made use of that 1893 right. The board therefore allowed the town of Greybull to put only 60 percent of the original 1893 water right into its pipeline. The local district court upheld that decision, and there was no further legal challenge.[52]

In years that followed, the Board of Control tended to exercise the same caution in other transfers to towns or railroads. The board carefully eyed the hydrology and uses on the stream involved and allowed transfers of sometimes all the water diverted under the original right, and sometimes, a reduced volume equal only to the past consumptive use, depending on the local conditions.[53]

L. C. Bishop's son Floyd became state engineer in the 1960s. He'd been a construction engineer on the Alcan Highway and a bomber pilot over Japan in World War II. He came home to settle into civil engineering for the next dozen years or so, focusing on water as his father had done.[54]

The postwar years stand out as a time of massive economic growth in the United States—particularly, in Wyoming eyes, in downstream states whose growth made them ever thirstier. Federal dam construction enthusiasm in the 1930s had supported growth in those states. Bureau of Reclamation dam building continued in the 1950s, but the big projects affecting Wyoming typically only made existing uses more secure or supplied power and irrigation water for downstream states. Some smaller federal projects supplied new irrigation water. The Bureau's momentum came from Floyd Dominy, a Nebraska farm boy and University of Wyoming graduate who got his start helping put in tiny earth dams to catch the rare rain on the northeast Wyoming plains in the 1930s but whose real talent was in marshaling congressional support for big projects. He shepherded the 1950s' Colorado River Storage Project, with new dams built from Utah to northern Arizona, including a new reservoir crossing the Wyoming border—and encouraging more growth downstream of Wyoming. Meanwhile, Wyoming legislators put some state money into small local reservoirs for irrigators and towns. There was some new demand for water for a few new plants in Wyoming burning Wyoming coal from nearby mines, and the plants required water to make electric power. And in the mid- and late-1950s, major drought struck.[55]

The combination of modest growth and major drought meant competing demands and changing uses for Wyoming water. Water rights transfer proposals multiplied. The State Engineer's Office got the legislature to add steam power plants and industry to the list of the favored few exempted from the water rights transfer ban. Provision for temporary change in the use of water rights for two years, for short-term uses like highway or railroad construction, also made it into the law books. In 1956, the Pacific Power and Light Company, based in Oregon, was granted Board of Control approval to move an 1890s irrigation right on the North Platte River downstream to its new coal-fired power plant between Casper and Douglas. The board allowed the plant to take in its new diversion all the water covered by the old right—not just the water that ranch crops had consumed—and divert it from the river year-round rather than only during the traditional irrigation season.[56]

That changed after Floyd Bishop became state engineer in 1964. He'd had private clients perplexed by the difficulty of putting water to new uses. To handle the pressures for change, he wanted Wyoming water rights cleaned up and put in order. It was he who, soon after coming into office, decried the court-imposed abandonment hurdles as "cumbersome" and "expensive." Barriers to abandonment, in his view, jeopardized newer rights and new development by leaving state water records littered with unused rights that could be revived.[57]

In the case of water for the power plant, Bishop and the rest of the Board of Control got the power company to backtrack, cutting the 1890s right it had moved and some of its other water rights by nearly 30 percent to keep more flows in the North Platte for other users. That mimicked what the board had done in the town water cases. A few years later, when the power company needed more water, Bishop and the board allowed it only summer use and only half the water of the former irrigation water right that the company transferred to its plant.[58]

The biggest challenge, however, came not from industry but from irrigators looking for more water. The droughts of the 1950s were still more severe than the previous ones, and the Wheatland irrigation project, one of the biggest in the state, was desperate for water. Wheatland's lands were some seventy miles north of Cheyenne, on the spot on the Laramie River that John Gordon had pointed out to Judge Carey in the early 1880s as a good cropland site for the company Carey and other stockmen formed to claim and sell land and water to settlers. Early on, there had been a major contest over water right dates between the company and Laramie River irrigators on the west side of the mountains. The Wheatland project in its search for a reliable water supply had built an ambitious reservoir and tunnel system; in the 1950s drought, the project, now owned by its farmers, was looking for more water.[59]

The governor asked all major users on the Laramie River to investigate new water supplies, and eventually, the Wheatland project found a big old ranch to the west across low mountains, known as the Ringsby Ranch, with rolling irrigated hayfields, all possibly for sale. The attraction was the ranch water rights. But could they be transferred to Wheatland? Moving irrigation water from one set of fields to another was clearly prohibited by the transfers ban of 1909. Irrigation was not a preferred use exempted from the ban. But some of the key water rights on the ranch predated the 1909 ban. So the district hoped that water under those rights could be moved, and lawyers indeed persuaded the Board of Control that it had to allow such a transfer.[60]

Bishop and the rest of the Board of Control accordingly allowed the transfer for those pre-1909 rights. Not wholesale, however; they applied the tests the board had developed over the years for transfers to towns and to industry. They required the Wheatland farmers who wanted the water to show exactly what the old ranch had diverted, what its hay crops had consumed, and what return flows had come back to the creek from the hayfields. In the end, the board, as it had done in other cases, limited much of the water the Wheatland project could get to what the old ranch had actually used under its water rights, rather than allowing Wheatland to have the larger amount the ranch had diverted from the creek.[61]

The board's approach had a prominent local critic. University of Wyoming law professor Frank Trelease, nationally known as a water law and policy expert, advocated a more liberal water rights transfer policy. In 1960, he asserted in a report to the legislature that "in general there is no essential difference between the property aspects of land and water." He expanded on that idea in 1966, zeroing in on the impact of the transfers ban a year after the board decision limiting Wheatland's transfer. For over fifty years, Trelease argued, that ban had unfairly and unequally deprived water users of the ability to move their valuable resource to a different use and get paid for it. Within the limits of the transfer ban, there were some who could sell their water rights—if their buyer happened to plan a "preferred use." But those unlucky enough to attract only agricultural buyers or others not covered by the exemption to the transfer ban were stuck and could not sell their water rights without selling the land it served.[62] The result, Trelease wrote, was that agricultural innovation was stifled. The transfers ban, tied to the priority system, meant that

> Wyoming agriculture still lies in the mortmain grip of the pioneer. The pioneer pattern of agriculture is impressed upon the Wyoming land. For the most part, the irrigated land is near the rivers, watered by gravity flow from ditches that follow the contour lines. Much good land lies above the ditches. Today, pumps can put water on those lands and lands can be watered with pipes and sprinkler systems.

The post-1909 rights on Ringsby Ranch that Wheatland could not transfer continued to water only hay land and pasture instead of Wheatland's more valuable crops like sugar beets, dry beans, or barley, Trelease argued. That exemplified the mortmain—"dead hand"—grip on land and water that he deplored. The problem might have been avoided by Mead's old idea of a system of state water leases, regularly reviewed for renewal. Mead had never seriously proposed water leases for Wyoming, though, probably because he knew his audience too well.

Now in 1966, Trelease pushed the legislature to focus on water rights transfers. He endorsed a transfer process much more generous than the Board of Control had adopted, one that would generally allow transfers of water rights, prohibiting only the aspects of a proposed transfer that clearly injured other users.[63]

New transfer legislation was enacted a few years later. But the legislature enshrined in law the board's approach to transfers, not Trelease's. In the early 1970s, the legislature enacted a law officially allowing the permanent transfer of water rights for all purposes—not just for a preferred use—and adopting for all permanent transfers the screening process the board had developed. The temporary water right transfers, initially for such uses as highway construction, that had been allowed by a 1959 statute, included a simple rule of thumb calculation for how much water could be temporarily moved.[64]

The new law for permanent water right transfers explicitly stated a series of more complex restrictions: no amount of water could be transferred that exceeded the amount or the timing of historic diversions, or that increased the amount consumed historically, or that resulted in reduced return flows, or that in any way injured the rights of other water users. The focus on consumption and the patterns it creates has tended to mean that only the previously consumed amount can be transferred. The new statute also said that the board had to weigh any economic loss to the county from which the right was being transferred. Since the early 1970s, the Board of Control has regularly enforced its own transfer restrictions that had thus been put into statute, and its actions have been backed by the state Supreme Court. [65]

———

As the transfer statute has been applied since 1973, however, there has been a curious result, one that Trelease would likely still deplore. A Wyoming water right is hard to lose, due to the Wyoming Supreme Court. If, however, you try to transfer it to a new use, you *are* likely to lose time, money, and water. Given the court's opposition to abandonment and indulgence of water right revivals, a water right holder can often revive an old, unused water right in its original place and use. Other users on a creek may suffer, but they have no recourse.

As the board now applies its transfer rules embodied in statute, however, anyone who wants to put an old right to a new use or on a new location has significant hurdles to overcome. The water right holder must show whether the water right has been used well and recently. The answer to that question, yes or no, leads to a complex tree of yes-no questions that will produce different results. The sequence is easiest for nearby moves to the same kind of use; those proposals may not have to produce a lot of evidence on past use. If a water right is to be moved to a new use or to a distant location for the same use, however, the board sometimes requires detailed evidence on the old right on such issues as consumptive use, return flow, or other impact on fellow water users. The board may end up advising the owner of an old water right that has not been used to invest in putting that water right steadily to the original use, however uneconomical, for three to five years. Even after that investment, the amount of water that can be moved will almost inevitably be less than could be diverted from the stream under the old right. Instead of "Use it or lose it," a new axiom for Wyoming water law could be "A Wyoming water right is hard to lose until you try to move it."[66]

To some degree, that result is in response to the state high court's opposition to abandonment and sanction of revivals. Given the court's stance on abandonment, the supervisors seem to have injected into their transfer reviews a scrutiny of water rights much stricter than they or the court impose in abandonment cases. Further, they sometimes recommend revivals since those have court approval. Notably the board requires—and recommends in revival—consistent use of a water right, if water has been available, for three to five years before a water right transfer can occur. Five, of course, is the number from the abandonment standard, imported into transfer law. All that is required to avoid losing a water right in an abandonment case, however, is using water at some point within the last five years. That is quite different from having to use water consistently, over an irrigation season for instance, every year for three to five years in order to transfer a water right. The scrutiny applied to a water right being transferred is therefore a lot tougher than the scrutiny the courts allow on non-use of a water right that a neighbor claims has been abandoned. Overruled by the courts when it comes to abandonment, the water supervisors have been able to follow their view of the nature of water rights and water user interdependence when it comes to the transfer arena, and they have stood their ground there. There can be considerable impact on other users from moving and giving new life to a water right that has not been steadily used in past years. The board therefore requires that evidence of considerable regular use. The water supervisors have sought to allow change but have continued to impose major restrictions on how much water can move under a water right transfer.[67]

With current technology, board members now scrutinize satellite photographs to verify, quantify, or defeat claims of consistent use. The examination of the evidence is painstaking and often requires further research, testimony, and an inspection of the site by the superintendent, in addition to the photographs. The extent to which the detail required is left to board member discretion, and the resulting length of the process, can frustrate users. A few complaints have surfaced in recent years.[68]

Economists, and many lawyers from Trelease on, do not think much of this Wyoming water transfer process. In the allocation of water, they argue, the goal should be putting the resource to its "highest and best" use, and markets are the way to do that. Restrictions on transferring water to new uses only hobble any markets in water. Markets in land or minerals are far more effective in moving those resources to their "highest and best use." Changing the use of land can be subject to some restrictions, like zoning laws, but real estate markets nonetheless function well. By contrast, the extensive restrictions on changing the use of water in Wyoming, economists have said, mean that the market is unduly limited. Since the ability to transfer property is often regarded as a key feature of ownership, the critics note, restrictions on water right transfers in Wyoming mean that holding a water right doesn't mean owning the water, as a person could own land.[69]

That verdict would, of course, not disturb Elwood Mead or his successors in the Wyoming State Engineer's Office over a century and more. For them, water has always been different, and they have not believed water rights should include all the rights that come with land ownership.

One leading economist in this topic today argues forcefully that Mead took the wrong approach, and as a result, western water law today requires major recasting. Terry Anderson, a senior fellow at the Property and Environment Research Center in Montana and at Stanford University, has explained why. The prerequisite for a market that can put water to its best use is private rights to water, rights that are both well defined and well enforced, Anderson writes. Mead's system, with its concept of state ownership and issuance of state permits, has cast a cloud on private water rights, Anderson argues. Those aspects of Mead's system, even as selectively adopted by other states, have unnecessarily and sadly impeded markets in water in the West, he concludes. He argues that it did not have to happen this way, claiming that in the late-nineteenth century, the prior appropriation system was evolving on its own into a private rights system that could support a smoothly running market. Fears of monopoly in water were largely unfounded, he believes, and when people proposed changing the use of water, nineteenth-century courts made decisions that ensured enough consideration of the impacts that transfers could have on other users. Private rights and the early court decisions could have been the foundation for effective markets for water and the flexible allocation of water for the best economic results, with no need for some broader consideration of general public welfare, Anderson contends. Mead and state administration systems interfered and have long hampered much-needed water markets, he concludes. Adding to the problem, the federal construction of big, often uneconomic irrigation projects only made things worse by ensuring that politics rather than economics have dictated who uses project water where, he argues.[70]

Much of that argument leaves out common nineteenth-century realities and twentieth-century experience, both well demonstrated in Wyoming. Across vast and little-populated terrain, water rights in the nineteenth century were not well defined or enforced. Money and people to do that were scarce. Fear of monopoly had reasonable grounds; the expansive 1880s Wyoming water and land claims, intended by stockmen to keep small settlers out, demonstrate that. Further, the Wyoming court clearly showed itself, on Little Horse Creek, unable to consider third-party impacts of a water-rights transfer. The courts simply did not understand how water worked on the ground. It was because the Wyoming high court demonstrated ignorance of water realities on Little Horse Creek that the state agency and Wyoming water right holders together got the severe "no-transfer" provision put into state law. Leaving water transfer decisions to the courts would have meant no real consideration of the water use needs of anyone or any place other than the interests of two people at either end of a transfer, as Mead had argued. A system of purely private rights, whose nineteenth-century emergence Anderson sees as promising, meant to Mead a riot of "organized selfishness . . . more potent than unorganized consideration for the public interests."[71]

In 1929, an observer looking back from only forty years' distance saw the early years differently than Anderson. Moses Lasky believed that inequities and uncertainty prevalent in the late-nineteenth century were ample reason to adopt administrative systems like Wyoming's. It was appropriate to make a water right "only a grant of a privilege of user by the state" based on a variety of considerations including diligent water use and have such systems displace the private property rights of prior appropriation, Lasky wrote in 1929. The state administration systems should be celebrated as the key step in a welcome trend toward "economic distribution of water," with state agencies in the lead, he wrote. On the verge of his career as a nationally acclaimed trial lawyer, Lasky also chided Wyoming's courts and legislature for "timidity" in failing to champion the power of the Board of Control to declare water rights abandoned.[72]

As the twentieth century wore on, the Board of Control kept tight supervision over water rights transfers, but experience led the water supervisors to soften and finally eliminated the transfer ban in Wyoming. They had it replaced with a set of restrictions that grew out of agency and water user examination and intimate understanding of the impacts of transfers. The board's process ends up defining and enforcing some Wyoming water rights—those rights that come to it to review—as never before, much as economists like Anderson have said is needed for a functioning market. Meanwhile, Wyoming's 1959 temporary transfers law is in regular use, though it can be used only to transfer water rights temporarily to certain new purposes. Anderson and other modern economists are welcoming the idea of temporary water use transfers rather than permanent water rights transfers, as promising steps to a workable water market. They see temporary transfers as the best way to avoid the destruction of agricultural communities that has accompanied some permanent water rights transfers in other states.[73]

Wyoming water users appear to have been reasonably happy with the way both abandonment and transfer laws have progressed. The court's dislike of abandonment makes it hard for the board to clear non-users off the water right records. As a result, the users have considerable control over their water rights and over their own ability to stay in or rejoin the game on their stream. That in turn means that having a right to use water becomes more like owning private property, and users tend to like that. As for the transfer laws, only in very recent years since 2015 or so has the detailed level of scrutiny required in moving water from one field to another seemed to chafe some users, because of the time and expense involved.

With those changes, Mead's concept of a state system morphed over the decades into a system of community management that distributes rights in water among key players. By the 1970s, the state agency and the water users had different sets of rights to water. The state agency represented the public. Once the state agency granted permission, individuals had rights to access, use, and manage water within limits defined by the extent and nature of the use. The state retained significant rights, however. The state could include or exclude individuals from the right to use water; the state could decide when and whether and how much of those rights can be transferred to other places or purposes. Factors such as the politics of federal water projects, aptly criticized by economists, shifted more rights to users as big projects like that on the Shoshone River shredded the state's ability to require user diligence in order to keep a right to water. Court resistance to abandonment did the same. The state nonetheless retained its rights in water, including the power to approve or deny transfers. The understanding that the issue is public waters, rather than purely private rights, is fundamental to the community management system that emerged. Despite the critics, agency and users to this day suspect that a market in private water rights that lacks the restrictions they have created would fail to support the welfare of the people in the state.

Wyoming's longstanding hostility to permanent transfers as a means of speculation in water rights that could disrupt communities was only reaffirmed, in the eyes of both water users and the board, by a highly publicized case from the late 1970s. In that case, an irrigation project that had failed years before tried to sell its never-used water permits, under a plan to move those permits over one hundred miles down the Green River to provide water for a new power plant. The state engineer approved the change, but on appeal, the Board of

Control voted against him and disapproved the transfer. The Wyoming Supreme Court joined them, upholding the flat denial of that transfer.[74] The facts of the case clearly alarmed the board and shocked the Supreme Court, as its opinion made plain. Though the case involved old unused water permits, not unused adjudicated water rights, it demonstrated what could have happened in Wyoming if the court-sanctioned power of users to revive old unused rights had not been restrained by the longstanding Wyoming suspicion of water transfers and speculation, embodied in the Board of Control's cautious process that had been put into statute in 1973.

Wyoming water records remain littered with old rights that might be revived in their old uses and inaccurate information on where and how some water is used. Further, there are major obstacles to taking an old right and doing much new with it with any speed. All that hobbles the ability of a society to use its water resources to respond to new needs. Much of the trouble comes from the court's intervention against abandonment, in disregard of the nature of water, water rights, and water user communities—in disregard of the testimony and arguments of a Nelson Sherard. Blocked by the growing court taboo on abandonment, the Board of Control exercised its expert judgment elsewhere. The water supervisors learned from the needs of power plants, towns, small irrigators, and longstanding and large irrigation projects like Wheatland.

As a board, the supervisors recognized what needed to be done to allow change. They formulated a policy built upon their and the users' understanding of the peculiar nature of water, water rights, and water user communities, and the responsibilities for public welfare entailed in water decisions—in recognition that they dealt with public waters. Court resistance to abandonment, however, has meant the water supervisors have been barred from applying their expertise to the entire arena of what happens to water rights that have lost their usefulness. Economic stagnancy, for lack of turnover, has sometimes resulted. That means ineffective water allocation. The dead hand of the pioneer, the demon that Trelease identified in the 1960s, managed for many years to play a role in Wyoming water rights and Wyoming water use.

5. FACING THE NEW

But the principal defect of the system, the one capable of working the greatest injustice, is inherent in the very theory itself, in its fundamental conception. The prior appropriator, in order to carry out a purpose regarded by the law as beneficial . . . *may* divert and consume, without returning to its natural channel, *the entire water* of a public stream, no matter what may be its size or length, or the natural wants of the country through which it flows.

—JOHN NORTON POMEROY, LEGAL COMMENTATOR, 1893
(ITALICS IN ORIGINAL)[1]

Places: Green River, southwest Wyoming; Salt River, southwest Wyoming; Wind River, west-central Wyoming.
Time: 1965–1990.

Energy production man camp. Michael McClure Collection,
American Heritage Center, University of Wyoming.

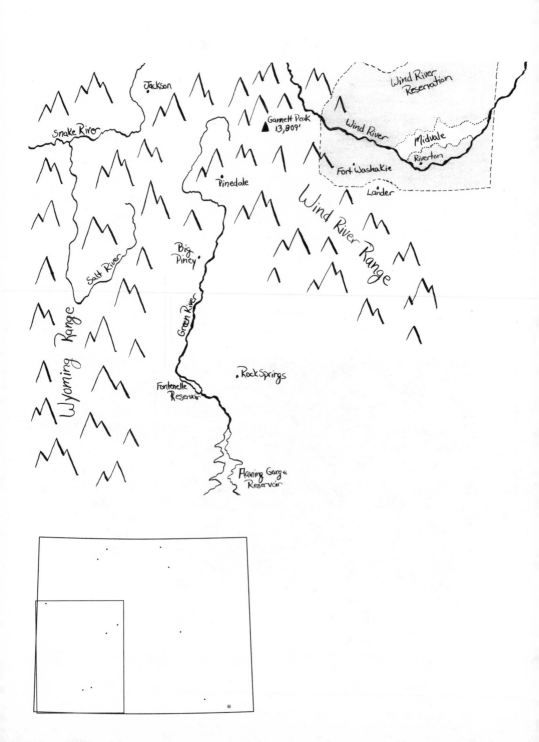

WHEN THE SUPERINTENDENT OF a Wyoming water division envisions the surrounding landscape, what appears is not a mental map of roads, but a web of creeks, ditches and water rights, stretched from the hillsides through the valleys and onto the bench lands nearby.

In this big landscape with its few people, by 1970—over three-quarters of a century since the water management system started—the weave of that web varied across the state. Active "regulation" of a stream by state officials, allocating water to users by water right date, did not occur unless and until some user on that stream formally called for her water. Users often tried to avoid bringing in state staff. Nevertheless, in some places users typically called for water, and active regulation happened regularly, so the web was tightly woven. In those places, users might be many, or the water scarce, or the hydrology complex. Some combination of those factors could prevail, or there might be regular conflicts between neighbors that erupted as water disputes even though the real issues were not about water. So the superintendent, his staff, and the water rights on the books were almost always present. In other places, the web was loosely woven—where users were few, water abundant, hydrology simple, users self-organized into well-run irrigation districts, or some combination of all those elements ruled. There, superintendent and staff rarely appeared, and users managed for themselves, with water rights dates faded into the background, sometimes to the point that no one knew what they were. The web with its spots of loose weave and tight weave reflected the community water management system that had emerged over the decades. The state had greater or lesser presence on the streambank, depending on the location. Water users' ability to manage the water by themselves waned or waxed in response.[2]

That did not mean that a place of loose weave or tight weave was necessarily peaceful. "I carry a rifle, not a shovel, when I go irrigating," said one young rancher in southwest Wyoming in 2020. "I hear you, it's definitely a blood sport," replied an older man who ranched over one hundred miles away. Some fifteen years earlier, in a low-water year for the plains of eastern Wyoming, the water superintendent there commented wryly, "It's going to get very western around here."[3]

Users on any stream could sometimes enjoy access to a "free river," as the state engineer and superintendents call it. Users could take as much water as they needed, well above the amount set in their water rights. That could go on as long as in anyone else's eyes (including the water commissioner's), the water taken wasn't literally wasted and all the users on that river were satisfied enough not to call on the superintendent to restrict everyone to their formal rights. In late spring, when most Wyoming streams, fed by mountain snows, are at their highest, a "free river" and loose weave of water rights could prevail in a good water year in many places across the state.[4]

In the early 1970s, however, a pamphlet called *The River is Free* was distributed by the League of Women Voters of Wyoming. Their idea of a free river was completely different. They were talking about a river free from much human interference, a river whose water was not taken to work in irrigated fields or power plants or city pipes but left to run on its own. They were talking about keeping water flowing for the sake of rivers themselves—on rivers Wyoming people would choose to protect as wild, scenic, or recreational. That had begun to happen nationally.[5] But could the needs of a river be woven into that web of water rights governing Wyoming water management? In prior appropriation and in Mead's system for water use to help create a new society, nothing required that any water be left in a river. Leading nineteenth-century legal commentator John Norton Pomeroy of California highlighted that as a major defect in prior appropriation West-wide—allowing users to dry up a stream, with no concern for "the natural wants of the country through which it flows."[6]

The "wants" of fish had always gotten some attention in Wyoming's water system. Superintendents worried about fish winding up in ditches. As early as the 1920s, the US Reclamation Service built Diversion Dam on the Wind River, serving the Midvale district, with a fish ladder, one of the first in the country, to help fish navigate past the obstruction. Reservoirs could create new, cold flows on the river below—making life there difficult for native fish but attractive for introduced game fish. The reservoirs themselves provided good fishing, and in the 1950s and 1960s, the Bureau of Reclamation finished the last of its big dams and reservoirs (most justified by the power they would create, but also creating good lake fishing), including Flaming Gorge crossing the Wyoming-Utah border on the Green River, Fontenelle further north on the Green River, and Boysen and Yellowtail on the Wind and Big Horn Rivers. Eventually, proposals for new dams began to include in their list of benefits the idea of ensuring some river flows below for fish. But it was not clear, within the bounds of the water rights system, whether rivers might be treated as anything other than conduits to the headgates and pipes people needed for farms and industry.[7]

In the 1960s and '70s, however, there were people in Wyoming who were thinking about water differently than in the past. Several of them came from western Wyoming. That part of the state, traversed by the Rocky Mountain chain heading northwest, is home to river valleys among high mountains, a place where the beauty of the rivers makes people think about rivers' needs. Eastern Wyoming, where the railroad gave birth to the capital city in the 1860s and where grasslands merge into sagebrush steppe that then runs up against the mountains, had hosted many of the early water management debates that went with settlement and so had provided the context for many of the early water management decisions. Western Wyoming had seen somewhat slower settlement, but in the second half of the twentieth century, it was becoming less isolated, playing more of a role in statewide dialogue and water policy.[8]

Western Wyoming hosts the state's highest peak, sitting astride the Continental Divide in the Wind River Mountains at 13,810 feet, while the snowpack on that range feeds both the Green River and the Wind River— sources, respectively, for the Colorado and Missouri-Mississippi river systems. The heights of Yellowstone National Park, in the far northwest corner, feed not only the Missouri-Mississippi system (partly through the Shoshone River running east through the top of the Big Horn basin to join the Big Horn River heading north), but also the Snake River, the largest tributary of the Columbia River. The Snake runs down through high country with long winters and the valley of Jackson Hole, where the Teton range and land fronting it in 1950 were (after decades of local debate and rancor) combined into the Grand Teton National Park, first started in 1929. Skiing and tourism there were only in their infancy in the 1950s—but thirty years later, they gave Teton County residents the highest per-capita income in the state. The Salt River, running north to meet the Snake just as it heads into Idaho, comes through the Star Valley settled by Mormon irrigators in the 1880s who managed to make it into dairy country despite the long winters. On the Green River to the south of Wind River Mountains, fur traders met in rendezvous with Shoshone and other native people in the 1820s–1840s. Early cattlemen came up the river for winter shelter on native hayfields in the foothills and stayed. Further south, running through the Red Desert, the river passes railroad and mining towns, built for the Union Pacific line heading to Utah, that hosted much of the early population of the area.[9] In the drought and depression of the 1930s, when some ranches in the area were reported abandoned, a young man from the mining town of Rock Springs found a way to make a living by planting trout in the streams on the south faces of the Wind River Mountains and taking vacationers up there to fish in the summers. On the north side of those mountains, the Wind River runs through the wide valley where the Eastern Shoshone and then the Northern Arapaho people were forced after the Civil War onto the reservation that was eaten away under settler and government pressure, which in turn spawned two towns, a ranching and mining town in Lander and the irrigation project town in Riverton. [10]

Tom Bell grew up in Lander, the son of a coal miner. A contemporary of Floyd Bishop, he too flew on bombing missions in World War II, losing his right eye, and nearly his life, to shrapnel. After the war, he went into the Red Desert to recover: "It was important to see the beautiful earth," he said. He went on to get a master's degree in wildlife management at the University of Wyoming (he studied pronghorn migration, and started the rodeo club), to work for the Wyoming Game and Fish Department, and then to ranch and teach science and to spur a pioneer museum in Lander.[11] In the 1960s, Bell saw threats to the landscape and wildlife he loved—Wyoming was increasingly being looked on as a potential center for producing energy for the country—a "national sacrifice area," some said, expected soon to host dozens of power plants burning Wyoming coal. There was even a proposal for underground nuclear explosions at the edge of the Red Desert to boost natural gas production. A series of major coal-fired power plants were discussed for the Powder River basin in Wyoming and Montana, and Wyoming US senator Cliff Hansen, a former governor, proposed federal legislation to build an aqueduct to serve industrial development there. Bell created the Wyoming Outdoor Coordinating Council in 1967 to organize and advocate against the developments he considered destructive; in 1969, he created a newsletter that became *High Country News* to get the word out. Both organizations are still thriving fifty years later.[12]

Bell and the council soon turned their attention to the upper stretch of the

Green River—undammed for some fifty miles as it flowed south out of the Wind River range and, in the 1960s, one of the first six rivers nationwide considered for designation by Congress as a "wild and scenic river." When the national act of that name establishing protection for rivers was passed in 1968, the Upper Green River in Wyoming was left out. Wyoming's three-man congressional delegation, supported by a turnout of local residents in the Rock Springs-Green River area, said the state needed to keep open opportunities to develop the river for other uses—and, they said, the state wanted to do its own river planning.[13] At the state level, therefore, Bell took up the cause of the Upper Green. He gathered some people from Upper Green communities and from Casper, the central Wyoming town-turned-oil-industry-hub since the 1920s. They believed, they said, that Wyoming leaders who had opposed federal protection for the Upper Green nonetheless appreciated its value as a free-flowing river. Bell's group argued that there was a growing new industry using Wyoming water—the recreation industry. The group cited 1960s economic studies to show how for recreation, the value of the Upper Green River—its fish and wildlife and its natural setting—would only increase as Americans had more leisure time and better transportation to places once considered remote. Subsidies for water used by agriculture and other industries, however, were likely to shrink the supply of water for recreation despite its increasing value. So government or citizen groups had to take steps to secure and protect key waters for recreational use, Bell's group argued:[14]

> Much is said in Wyoming about bringing "industry" to the State. Few seem to realize that we have a built-in industry which relies only on the God-given assets available to us. However, the value of those assets are dependent on the wise use and management by resource managers and the public. They need only be recognized and protected.
>
> The tourist and recreation industry is recognized as such by our leaders, but often gets no more than lip service in the immediacy of living here and now. Because the industry is so new, it suffers from a time lag in a state still oriented to a rural, early-American West.[15]

Those opening paragraphs of the group's formal brief sound like they came directly from Tom Bell. He was a fan of Wyoming history and rodeos, but in the here and now, he believed in fighting for the beautiful earth. In early 1969, he and his colleagues asked State Engineer Floyd Bishop to grant them a water right to protect some flows in the Upper Green River for fish and wildlife. They filed that brief arguing that this water right would be for an industrial use, with recreation as the industry. The right they asked for would be within the first fifty-one miles of the Green River, as it left the Green River Lakes in the Wind River Range.[16]

Meanwhile, in 1969, Paul Stauffer, a member of the next generation after Bishop's and Bell's, had just come home to the Star Valley in far western Wyoming. Stauffer's father had a drugstore in the Star Valley, and Paul had gone off to get a pharmacy degree. A bright student, he was urged by his professors to become a professor himself, but he realized that if he committed to academia, "I wouldn't be able to live where I wanted to live." Stauffer was an avid fly-fisherman (he titled his recent memoir "Did I Fish Too Much?" The answer, clearly, is "No."). When Stauffer came home to the Star Valley with his wife and baby son, he found that the Salt River had changed. Years of US Department of Agriculture aid to ranchers to switch from flood irrigation to sprinklers and to denude their riverbanks of water-thirsty willows had resulted in lower quality fishing in a wider, shallower river. Some traditional flood irrigators had warned as much but to no avail.

In 1971, a couple of years after Stauffer's return, the river flooded disastrously for the first time in local memory, cutting new channels and eating into irrigated fields. Some landowners responded by charging into the river with bulldozers to straighten the channel and send the river flows barreling downstream. Fish habitat and spawning grounds, as well as downstream landowners, suffered. Stauffer responded by pulling together a group of landowners, fishermen, and government agencies aiming first to get willows back on the banks: the Save the Salt River Committee.[17]

A contemporary of Stauffer's from the Wind River valley, Dick Baldes, was graduating with a master's degree in fisheries. He had grown up in Riverton, the town right by the big bend that the Wind River takes as it heads north. As a boy, Baldes fished, played, and swam in that river as his back yard. But when he came back from the university and his first jobs in wildlife conservation in other states, he found the river devoid of everything but carp and suckers and unsafe to swim in. He took a job with the US Fish and Wildlife Service to work in its office on the Wind River Indian Reservation. His goals included enhancing wildlife on the reservation, which after all the land cessions still encompassed some two million acres stretching up to the crest of the Wind River Mountains, including a vast wilderness that reservation leaders had protected starting in 1934. Another of Baldes's aims was to bring the river back to life.[18]

Baldes was an enrolled member of the Eastern Shoshone tribe. People in Riverton, where early business high hopes had been dashed by the slow development of irrigation, tried a lot of different ventures over succeeding years to keep going—oil drilling, uranium mining, hog or beet processing. Uranium mill tailings were dumped by the side of the river in the 1950s. As the irrigation districts on the north side of the river did manage to grow, however slowly, their diversions left less and less water in the river during summer. Midvale was looking for still more water, and so was the city of Riverton.[19] In the 1970s, as Baldes returned home to work for the US Fish and Wildlife Service and was dismayed by a depleted river, Riverton started eyeing groundwater as a water supply for its small airport and an industrial park. When the city announced plans to drill wells for those sites in 1975, tribal authorities objected, because it appeared that use of that groundwater could affect tribal members' irrigation. The tribal councils cited the original 1868 treaty setting up the reservation to cover much of the Wind River valley "for the absolute and undisturbed use [of the tribes]. . . as their permanent home."[20] Those lands, as the tribes pointed out, were made habitable by their waters, so the groundwater and surface water in the Wind River valley belonged not to Riverton, and not to the state of Wyoming, but to the tribes.[21]

Both the Eastern Shoshone and the Northern Arapaho tribes had been taking strategic steps since at least the 1930s, as entities, to make reservation resources benefit their people. They had achieved a turning point in reservation agriculture, shifting away from the irrigated-farming model imposed by Congress and instead taking on ranching, grazing cattle on tribal uplands while irrigating bottomland hayfields for winter feed, in the pattern that worked best in most of Wyoming. The Shoshone, via a lawsuit against the federal government in the 1930s, got funding to buy back some outsiders' ranch properties in the ceded northern section of the reservation, and the federal government returned all the unclaimed ceded land to the reservation. The Arapaho, at about the same time, got a federal loan to buy a big ranch northwest of Wind River Canyon and start a successful tribal cattle ranch. Both tribes successfully fought off "termination," under the federal policy of the 1940s–1950s that eliminated federal services and obligations to protect tribal lands and sought to relocate native people away from reservations and into cities. By the 1960s, ranching and related irrigated hay lands supported only about 14 percent of reservation population full time. Underemployment and unemployment were significant and apparently growing. Oil companies did develop oil and gas reserves on the reservation and paid royalties to the tribes, which (with ups and downs in revenue) provided per capita payments supporting individuals and revenues for tribal government investment in public infrastructure. The Bureau of Indian Affairs had managed tribal resources, including water and oil and gas, since the 1880s. Federal antipoverty programs of the 1960s, however, encouraged participation by tribal governments independent of the BIA. A 1968 US Senate report concluded that the dominant US policy toward tribes and their governments had been "coercive assimilation," resulting in destruction of Indian communities and individuals, and in the growth of an ineffective bureaucracy at BIA that helped perpetuate tribal poverty. The roots of destructive federal policy were "a continuous desire to exploit, and expropriate, Indian land and physical resources" and intolerance of tribal communities and culture, the Senate report acknowledged. At Wind River, the tribes kept working toward creating economically viable communities and sought to keep their culture intact. A major step for that was the creation in the late 1970s of Wyoming Indian High School, run by the tribes, which was achieved after a ten-year effort by parents who as children had left home for boarding school or suffered disdain and ridicule in white-run local schools that got federal money for enrolling tribal students. As adults, they were determined to give their children "respect, accomplishment, and dignity" instead, and to use the federal funds for their own schools.[22]

By the late 1980s, however, unemployment at Wind River reached nearly 80 percent, and the tribes' lawyer told the US Supreme Court that it appeared that agribusiness, dependent on abundant water, was the tribes' "only certain hope" for a reliable economic future. Since the early 1960s, tribes around the West had begun asserting rights to water resources. They had successes in claiming high priority rights from nineteenth-century treaties, under the rule of the 1908 *Winters* case. When in 1975 the Shoshone and Arapaho tribes cited the 1868

Shoshone treaty to object to Riverton's groundwater drilling plan, Wyoming state government ultimately responded with a lawsuit. In 1977, the legislature endorsed, with nearly unanimous support, a bill to start a "general adjudication" of the Wind-Big Horn River system. Congress had enacted a law twenty-five years earlier providing for such general adjudications. It meant that if a general adjudication were undertaken in a river basin, then a state court rather than a federal court could decide a controversy over what water rights the tribes and the federal government held there. On the Wind River, the state's lawyers saw state court as a significant advantage in their effort to defeat the tribes' claims. The elaborate state court adjudication process formally began two days after the legislature endorsed it in 1977, inaugurating a review of all water rights in the basin, a herculean task that ultimately took three and a half decades.[23]

Bishop's successor in the State Engineer's Office believed, as the lawsuit got underway, that it would eventually "end years of uncertainty and speculation" about how much water the tribes could claim in the Wind-Big Horn system and therefore how much water other users could develop. Baldes, on the other hand, saw the lawsuit as "good for the river." The research and attention that the litigation would bring could, he believed and hoped, make more people learn about the river, care about it, and ultimately take action to let it keep more of its water.[24]

————

Wyoming's state engineers in the 1960s and 1970s, meanwhile, faced pressure from added, conflicting directions. The major drought of the 1950s had left traditional Wyoming agriculture eager for more water, as demonstrated by Wheatland's search for more supplies. Scattered towns around the state, like Casper in central Wyoming, had long been supported by oil and gas production and servicing. Now, increasing electricity demand nationwide, plus a combination of pollution concerns and the dependence on foreign oil eventually highlighted by the "energy crisis" of 1973, led to what galvanized Tom Bell in Lander: demand for Wyoming's little-tapped but massive low-sulfur coal reserves. John Wold, a prominent geologist in Casper who became the state's lone congressman, had successfully invested in federal coal leases in the 1960s in anticipation of a boom in coal that soon materialized. A major power plant tapping Green River water was built in the 1970s near Rock Springs. By the end of the 1970s, Wheatland itself—because of easy access to water on the Laramie River—was the home of a new coal-fired plant that shored up the town's economy. There began to be talk of people speculating in water rights for energy projects.[25]

At the same time, downstream states were continuing to develop at a pace ever faster than Wyoming's. They had their own water ambitions, particularly on the Colorado River, which made people on headwaters in Wyoming's Green River worry about water for their future development, despite the protections of the Colorado River Compact. In 1948, to flesh out those protections and obligations, Wyoming and the other states in the upper reaches of the river had agreed on the Upper Colorado River Compact. That agreement allocated a specific percentage of available water to each upper basin state (Wyoming, Colorado, Utah, and New Mexico), while also putting in place a procedure for ensuring that the upper basin states could meet the 1922 compact requirement that water use in the upper basin not deplete flows below a certain level. Post–World War II population growth and federal investments in dams and infrastructure tapping the Colorado River for water and power, serving both agriculture and cities, soon made for steady economic growth in Arizona and Colorado. Meanwhile, Wyoming couldn't generate much new demand—or corresponding federal investment—for all of its share of the river under the 1948 compact. The poor economics for agriculture in the high desert meant that even a modest federal reservoir built in the 1960s on the middle stretch of the Green couldn't generate the new irrigated agriculture it had been designed to create. Instead, it has served a wildlife refuge, and its projected mine and power company customers use the reservoir only for back-up "insurance" water supplies they have never used. Full Wyoming use of its compact portion of Green River water was much dreamed of but not realized.[26]

Bishop, state engineer from 1963–1974, worried that "economists and federal-type planners" were likely to give priority to downstream, low-elevation states with more attractive climates when they eyed water investments. By the early 1970s, he was frustrated that not only were federal funds going to "social programs of various kinds" instead of water development, but "the trend of environmental opposition to all water development projects is increasingly difficult to counteract." Writing in 1972, after the political upheavals, riots, and assassinations of the 1960s and early '70s, he subscribed to a 1900-era view of farm settlement as a pressure relief valve for US cities. He bemoaned national lack of interest in that idea: "Even though the expenditure of federal funds for water development projects in Wyoming would help to solve some of the social problems of the overpopulated areas of the nation, there seems to be little sympathy for this viewpoint at the federal level."[27]

Bishop argued that Wyoming must pay more attention to Wyoming water matters and invest in the State Engineer's Office, which in 1961 had so few staff and such an outdated filing system that if Elwood Mead had walked in, one staffer wrote, "he would have felt right at home." Bishop felt that the state's water users could not remain comfortably wrapped in isolation. The federal government, he warned, was increasingly activist, not just in water projects to serve downstream economies, but in issues like water pollution and the impact of reduced river flows on native fish, birds, animals, and plants. Since those issues cut across state lines, he could foresee federal intervention in state water management, possibly changing Wyoming's system.[28] As soon as he came into office in 1964, Bishop began warning:

> Too little emphasis has been placed on the administration of water in Wyoming for many years, and if this trend continues the ultimate result can only be federal control of our water, which has historically been a state responsibility. This would be a tragic thing in the eyes of most Wyoming water users.[29]

Wyoming water users, however, were suspicious of change, at any level. They were fearful that new people and new industry would somehow take their water rights away. The water superintendent of the Powder River basin, meanwhile, expected change to take the form of industrial demand for large volumes of water there. Power plants were projected to accompany Wyoming coal production there, and power plants would need water. The water superintendent reported with dismay, however, that almost all irrigators disliked the idea of industrial development. He thought it was because they didn't understand how Wyoming water law would work to allow new industrial users of water to get senior rights on streams only if they paid irrigators for them, and only if existing stream use patterns were protected.[30]

Law professor Trelease, as he argued for a new law to make water rights transfers easier, saw a similar problem. "The mortmain grip of the pioneer" imposed on the Wyoming landscape by the ban on water rights transfers enacted in 1909 had, he believed, stifled innovation and the use of new technology in agriculture and other industry. Further, the transfers ban had left successive generations of Wyoming water right holders with what he called an "heirloom attitude." They feared letting water rights move or change in any way—and that became a factor in the state's poor economy. The Wyoming water user, Trelease wrote,

> feels that water is his most precious asset, his heritage, his birthright. To sell it would be sinful. Laws against sin are much in favor. In part this attitude may come from a misunderstanding, a fear that stability of water rights is at stake, that water will be "taken" from irrigators without compensation, as may be done in some eastern states. In part it seems to stem from desires to preserve the status quo of rural Wyoming, to prevent neighbors from selling out, to prevent the loss of tax revenues for counties and school districts in areas subconsciously feared to be marginal.[31]

Wyoming water users fundamentally did not want to see their familiar system upset. They had reached a comfortable position by the 1960s, sharing power over water management with the state engineer staff. That was particularly the case, of course, on the streams where the web of water rights management was loosely woven, because water right regulation rarely occurred there.

Bishop was not content to sit tight and hope that the national changes he saw coming would bypass Wyoming. At the end of the 1960s, he had his office start drafting a complete recodification of Wyoming water laws to address problems he saw. The rewrite included modernizations, setting new numbers for fees and salaries, and adding oil and gas drilling to the purposes for which temporary transfers of water would be allowed. He wanted to limit time extensions for putting water to use under a permit (the familiar issue that had long dogged the State Engineer's Office). Further addressing "paper," or unused rights lingering on the books, Bishop launched a proposal to give the state engineer authority to initiate abandonment proceedings—just as True had done on Horse Creek in 1916. Bishop regarded the lack of that authority as the most serious gap in water law, and in the water law rewrite, his proposals to get that authority aroused the most opposition from water users.[32]

Eventually, as part of a reordering of Wyoming water laws passed by the legislature in 1973, Bishop got the authority to initiate abandonment actions—but in vain. User hostility to that kind of initiative by state engineers has remained so strong that the authority has only been used once (and unsuccessfully) in the forty-five years since. Instead, the Board of Control has turned to forceful persuasion; when users ask for help in getting their water or moving/changing their water use and superintendents find a mire of overlapping or unused rights, or rights not accurately recorded, they convince users to clean up those problems. Sometime that means persuading users to "voluntarily abandon" their right or some portion of it. In that way, some unused rights have been eliminated, though many still litter the state's records. And the superintendents themselves prefer to preserve water rights, particularly old ones that may predate the interstate compacts made on many rivers, to keep water in use in Wyoming. In such situations, they tend toward encouraging revival of use under a water right and then transfer to other lands, instead of voluntary abandonment.[33]

The final 1973 reorganization of the water laws also included the new procedures for transferring a water right to a new place or a new use, enshrining the board's exhaustive look at transfer impacts on water use patterns rather than the relatively permissive process Trelease had once recommended. Putting the board's process into black and white made sure existing water users, and prospective new ones brought in by the energy crisis, would know that water rights transfers in Wyoming were possible but only with scrutiny.[34]

Wyoming water user resistance to change is still demonstrated, however, in an interesting way. Today, coming up on fifty years since the new transfer law went into effect, it is not widely known. Water users who have never been involved in the transfer process still insist that in Wyoming, water cannot be moved away from the land it irrigates. They believe that the 1909 ban still applies. Mead's original goal of blocking speculation plus the "heirloom attitude" Trelease identified militate against water right sales. People welcome the idea that they and their neighbors lack the power to sell off their water right easily to the highest bidder. They cite it as one of the best features of Wyoming water law.[35]

Soon after the reworked water laws were approved in the early 1970s, change arrived. Big coal mines did open, primarily in the Powder River basin, and some new power plants were built, though not nearly as many as had been forecast a few years earlier. One major power plant built on the Green River, plus a new coal mine to serve it, brought in so many new workers they had to live in tent camps. The legislature enacted a new law requiring that companies proposing big new industrial projects must work with towns and counties to ensure housing and all kinds of services would be beefed up to handle the new development. Soon, "man camps" of trailers to house project construction workers were familiar sights on once quiet roads. Wyoming became by far the biggest coal producing state in the country, and for a time the biggest in the world. Rather than becoming home to legions of power plants, the state shipped off hundred-car trains of coal to many US states.[36]

Wyoming was suddenly a wealthy state. The energy industry—particularly coal, but also oil, gas, and uranium—became the big economic driver for a state whose government one day in 1968 had only eighty dollars in the bank. Having taken a major (and controversial) step in 1969 by imposing a "severance" tax on the extraction of minerals and hiking that tax in subsequent years, the state government began to see big revenues. Legislators long used to thinking small began to dream that someday Wyoming government—if its services remained minimal—could be funded simply by income from a savings account of stockpiled mineral revenue. Provision for a portion of severance taxes to be dedicated to a "permanent mineral trust fund"—earning interest revenue that helped fund state operations—became part of the state constitution in 1975.[37]

For the new power plant on the Green River, some enterprising families had proposed to sell some of their old, unused water permits downriver for the plant—the plan defeated by the Board of Control backed by the Supreme Court. For water users who paid attention, that case confirmed the competence of the Board of Control as a water transfers watchdog. Meanwhile, new proposals for using Wyoming water for energy kept cropping up—including schemes to use water to help turn coal into synthetic fuels or to transport coal to market via pipelines. Those plans in turn spawned a new batch of "temporary filings" cluttering up state engineer records with papers that were not yet even applications for water rights permits, but just sketched ideas for water diversions and reservoirs for industry.[38]

Considerable portions of the new state government wealth went into investment in major water infrastructure, much as Bishop had hoped. At the federal level, there was discouraging talk of cutting federal largesse for water projects. In Wyoming, the state's leadership was determined to step in with their new revenues to build the dams and reservoirs the federal government had not provided.[39]

Ed Herschler started his unprecedented three terms in 1975 as a popular governor, whose slogan was "Growth on Our Terms." He was a World War II decorated Marine sergeant, railroad company lawyer, and legislator from a longstanding ranch family near the coal town of Kemmerer in southwest Wyoming on Fontenelle Creek, which joined the Green River just where the federal dam intended to spur irrigation had ended up creating a wildlife refuge. One of Herschler's flagship projects was creation of a new set of funds for water development fed by state tax revenue from coal, oil, and gas. Towns wanted new water infrastructure for growing populations, and irrigators needed to extend the life of facilities that were often sixty years old or more. Nearly all the population, therefore, could unite behind the idea of using a chunk of the new mineral revenue for water facilities.[40]

The Wyoming Water Development Office, independent of the State Engineer's Office, was created and well funded. The new agency inherited files on all kinds of water projects that had been proposed over the decades and began to accept proposals for both new and rehabilitated water supply projects for irrigated agriculture and for towns. About two-thirds of the costs of each project could be covered by outright grants from state funds and one-third by low-interest state loans. State legislators were soon awash in water development proposals, as well as energy money, and they created a special committee to oversee water development. The Wyoming Water Development Office, envisioned as Wyoming's own Bureau of Reclamation, developed political relationships in Cheyenne much like those in Washington that had kept bureau water projects steadily funded for decades. Mutually beneficial ties between the development agency, its legislative oversight committee, and water using entities ensured that a significant chunk of the new mineral revenues would go solely to water projects dotted around the state to give credit to legislators for bringing state money home. Over the next thirty-five years, some $1.4 billion in mineral revenues went to planning and construction of water infrastructure.[41]

Accordingly, a good number of agricultural water users were shored up rather than undermined by the economic changes they had so feared. Some people transferred some water rights—profitably—to growing towns or industry, but there was no wholesale transformation of Wyoming water use. Rather, longstanding irrigation projects, along with some new reservoirs and groundwater wellfields for cities and towns, got infusions of state mineral tax money. Wyoming was able to update infrastructure and provide new water supply facilities in a way it never could have done without the mineral revenues. The portion of the mineral tax revenues earmarked for water development was soon considered money sacred to water projects.

To get the money for a water project, irrigators had to be in or create some form of public entity that could legally hold title, receive state money, incur debt, and raise revenues to repay a state loan. For those who could meet that requirement, the cost-share funded all kinds of improvements: rehabilitation for private projects built long before, like Wheatland, or Hawk Springs on Horse Creek, or LaPrele near Douglas; new irrigation reservoirs in places like the Greybull River in the Big Horn basin, where additional storage made it possible for the Wiley project and the Farmer's Canal to work in far greater harmony; and upgrades like canal lining, piping, and automation for Bureau of Reclamation irrigators such as those on the Wind River, on the Shoshone east of Cody, and on the North Platte near the Nebraska line. The state of Wyoming proudly helped fund an expansion of the big Bureau of Reclamation Buffalo Bill Dam on the Shoshone, and a major pipeline to give towns along the Shoshone an easy-access water supply for homes and businesses. Cities and towns, which have little fundraising ability under Wyoming law, were able to get state funds for water supply projects. Wyoming was king of coal, exported by rail, and in the eyes of East Coast media, the state was profiting from it as "blue-eyed Arabs." But major new industrial water guzzlers, like synthetic fuels plants or coal slurry pipelines, were not built, and power plants remained few.[42]

Faith in "water development" held sway, a "build it and they will come" spirit reminiscent of the early twentieth century beliefs of Wyoming settlers. The water development agency was tasked to "emphasize projects developing unappropriated water" and "give preference wherever possible to projects developing new storage capacity." In the occasional busts that inevitably accompanied the new energy-minerals-dependent economy, even the need to fund schools could not overcome legislative determination to dedicate and protect money for water projects. While university economists and some legislators questioned water development economic analyses, a legislative majority steadily supported investing revenue from "non-renewable resources" (minerals) into "renewable resources" (water supply).[43]

Rehabilitation projects and some new projects were built. But big water project dreams ran into some limits, imposed by out-of-state reality: downstream states watching supplies for their own thirsty farms and cities, and national legislation aimed at protecting the life and health of rivers and the species dependent on them. When the Wyoming legislature committed coal revenues to a new dam on a creek feeding the North Platte in the mid-1980s, the state of Nebraska went to court to kill the project and get a reexamination of all the Wyoming water use arguably affecting farms in Nebraska near the North Platte. The new coal-fired power plant at Wheatland planned to use water from the Laramie River, tributary to the North Platte. It was constructed—but only after the utility provided for lands and research to support endangered whooping cranes who used the river downstream in Nebraska. It was people nationwide, like Bell, Stauffer, and Baldes, whose concerns led to federal environmental legislation, including the 1973 Endangered Species Act. That act was the legal hammer that local and national opponents used to force settlement terms on the rural power cooperative that built the new Wheatland power plant.[44]

National interest in environmental protection persistently imposed more limits. The federal government began to take on a very different role in water projects than it had played for the first three-quarters of the twentieth century. The federal Bureau of Reclamation went through a personality change, from charge-ahead dam builder to self-described river steward. The bureau required that the expansion to the Buffalo Bill Dam on the Shoshone, nearly half of it paid for with Wyoming's mineral money, provide a set amount of water to keep flowing in the river below rather than be diverted. The bureau had become the manager (either officially or de facto) of big rivers, including the Colorado, the Shoshone, and the North Platte, by virtue of its big dams. It had built a pair of new reservoirs on the North Platte in the 1950s and '60s, a big one in eastern Wyoming on the North Platte, and a smaller one above Casper for flood control and irrigation. Those dams allowed more shifting of water while storing it for a new irrigation season, and a minimum flow downstream through Casper. That flow in turn helped dilute the notorious (and smelly) pollution of the river in central Wyoming, which Wyoming governor Milward Simpson in the mid-1950s had highlighted by pushing to reduce the dumping of untreated sewage and refinery waste into the river. In the 1990s, the bureau's Wyoming manager was able to set a new pattern of shifting stored water around in order to benefit fish, while still providing promised power and irrigation water. In the late 1980s, the bureau realized that under the Endangered Species Act, it had to review the operation of all the dams on the North Platte to determine their impact on endangered species, even if changed operations could mean less water for irrigators. The federal Clean Water Act of the 1970s, meanwhile, meant that Wyoming had to prove "purpose and need" (rather than relying on "build it and they will come") for any proposal to disturb the natural flows of a river and its wetlands. That applied to every dam dreamed of by Wyoming irrigators and legislators.[45]

Wyoming political leaders in the late 1980s and early 1990s displayed dogged determination to build dams despite obstacles like the Clean Water Act. In 1988, the legislature began committing water development funds from coal taxes to a dam proposal in south-central Wyoming, on a tributary of the Little Snake River, that some locals had dreamed of for decades. Two University of Wyoming economists questioned the water development agency's economic analysis. Their critique was immediately disavowed by the university's president. The economists had dared to suggest that the state would do better by just investing the money, rather than building the dam. But as Pat O'Toole, a legislator from the Little Snake area put the prevailing counterargument: "People know that water development is an investment in the future, a safe, secure, investment, much more than Wall Street is ever going to be, because that water is going to flow." The legislature authorized pursuit of the dam project.[46]

That project, however, lost its bid for a Clean Water Act permit for lack of "purpose and need." Federal reviewers noted there were no definite buyers for the nearly two-thirds of the project water yield that the Wyoming water development agency said would go to "industry." Only a much smaller dam managed to get a permit. That dam yielded barely more than a third of the water of the original proposal and served only irrigation, yet the irrigators could not pay even a minimum share of the costs of the severely downsized dam.

Legislators again demonstrated their unflagging dedication to the water development dream, in 1992, by authorizing state water funds to pay the entire cost of the downsized dam—totaling some $30 million—and take it on as a state-owned and run project. It was built, and irrigators in the area eventually came back asking for another dam. They included O'Toole, no longer in the legislature but now the president of the Family Farm Alliance, a group with the motto of "Protecting Water for Western Irrigated Agriculture." The new dam was estimated in 2018 to cost $80 million. Legislative commitment to water development had dimmed a little by 2018, but it persisted. State government was in tight budget times because coal was no longer in its heyday. But after quite a fight, legislators committed nearly $5 million to keep plans for that dam going, while cutting $27 million from public schools.[47]

In the end, purely agricultural projects got less than half the money that small Wyoming cities and towns received from the state mineral revenues pouring into water projects. A large majority of the money for city and town water projects went to new infrastructure—reservoirs (that won federal permits, for towns that could prove they were growing), well fields, major water pipelines, or storage tanks. By contrast, half the money for irrigators who had managed to organize into public entities went to extending the life of old systems, while the other half of the funds gave irrigators new structures, often improving the water-delivery capacity of their existing systems by means that included a few new small dams. Irrigators, organized as an entity or not, continued to use by far the most water in Wyoming. They were a major force in the water policy battles within Wyoming that people like Bell, and Stauffer, and Baldes sparked as they sought more water for live rivers.[48]

———

Tom Bell and his group had asked Floyd Bishop to grant them the right to protect flows in the upper reaches of Wyoming's Green River in 1969, just before the new energy development and new money came into the state. Bishop consulted his agency's lawyer—an assistant attorney general and a popular past governor's son who soon speculated in water rights at likely reservoir sites. Bishop and the lawyer both saw the Green River flow-protection proposal as an interesting idea that, if approved, raised all kinds of questions about how to create and manage a new kind of water right. Water rights, as managed in Wyoming since Mead's time, all had familiar features: they authorized diversion of water out of rivers, their use and non-use could be measured, and they could be abandoned. How would the engineer's office apply its expertise to a right to leave water untouched in a stream? Was there expert evidence on how much water fish needed? How to know if it was "used" or "unused"? Those problems should be thought through before any actions are taken, they concluded. The lawyer suggested that Bishop ask the state Game and Fish Department for draft legislation laying out new rules for a new kind of water right. Bishop instead told Bell's Outdoor Coordinating Council that they should draft such a bill.[49]

At bottom, Bishop wrote to Bell, "the proposed water right is of such potential value to all the people of Wyoming that if it could be granted, it should be vested in the State for the people of Wyoming rather than vesting such a right in a limited group such as you have proposed." Well acquainted with Bell, Bishop signed the letter "with best personal regards." He did not deny the permit, but rather planned to keep it on hold till the legislature might act.[50]

Bishop probably could have acted on his own to approve some version of Bell's application. He and the Board of Control prided themselves on their long history as the experts expected to make judgment calls about water, and they had considerable discretionary authority that, over the years, had been used to recognize new water uses. There was and is no official list in Wyoming statutes of authorized "beneficial uses" of water to which new ideas can be added only by the legislature (though some legislators have sought that power). The State Engineer's Office could simply recognize new uses and had regularly done so. Bishop could have decided to recognize water kept in a stream for fish as a "beneficial use" of water, approved a water right permit, and worked with the Board of Control to figure out how to administer it. But at the core of Bell's proposal was a political question, as Bishop's letter to Bell clearly demonstrated: Would the state government decide to protect water flowing in a river because of the great "potential value to all the people of Wyoming"?[51]

For the next few decades, that question of "instream flow" marked the fault line between Wyoming's water rights traditions and Wyoming traditional users, on the one hand, and people in the state who believed water had an important new role to play.

There were a significant number of such people, eager for the challenge of thinking anew about water, as their state changed almost daily around them. Wyoming's members of congress had emphasized the value of Wyoming's making its own river plans, when the delegation kept the Upper Green River out of the national Wild and Scenic Rivers Act. The statewide League of Women Voters, an active group and a serious force in civic dialogue in Wyoming at the time, took those political leaders at their word. After the publication of *The River is Free* leaflet, League members focused their early 1970s work on pushing for a Wyoming wild and scenic river program. Their effort climaxed in a public seminar held in Casper, the state's oil and energy town. The 1973 legislature responded by creating a Stream Preservation Feasibility Study Committee, which included Bishop, other agency heads, legislators, and two "public" members.[52]

Paul Stauffer, in the Star Valley, was surprised to find himself named to the new committee as one of the two public members. He was more surprised to be elected chairman. He and his committee eventually proposed a bill to require review of any proposed river channelization and a bill setting up a new process for the legislature to designate scenic rivers and recreational rivers, case by case, acting upon recommendations from a proposed new state River Protection Council.[53]

The committee also went further. They had been told that it was not possible to create a water right to protect flows. They nonetheless commissioned an independent report from Trelease, the water law professor, and his report convinced them otherwise. Stauffer saw that as "a signal accomplishment."[54] In line with Trelease's report, the committee proposed that key natural flows of water, in selected rivers to be designated by the legislature, be protected by water rights—rights for "instream flows," to be held by the proposed new River Protection Council "in trust" for the people of Wyoming. Further, they proposed that the governor could act unilaterally to appropriate such rights in any river in the state, to be held in trust. The goal would be to protect or enhance all kinds of values, ranging from geological and historical to fish and wildlife. The state engineer in turn would take on some new roles, studying what features of rivers deserved protection and controlling activities that channelized rivers to the detriment of natural habitat.[55]

"With the advent of the new era of industrial development thrust upon Wyoming by the demands of the energy crisis, the ripeness of this controversy is seldom questioned," the entire Stream Preservation Feasibility Committee observed in its final report.[56] Enclosing the report, Stauffer himself as chairman wrote to the governor:

> The committee is not taking a posture against dams, energy development, or future agricultural needs. We are recommending that the future needs of the people be carefully considered in these areas. The people should also, we believe, have the option to insure that certain streams or portions of streams will remain a part of the Wyoming way of life for the values such streams can provide in their natural state.[57]

The idea of stream preservation, however, alarmed many irrigators, for a variety of reasons. The only instream water flows they might feel easy about protecting would have to be flows released by reservoirs. Protecting natural flows might mean no new water rights in certain streams; it might mean water would be required to keep flowing in a Wyoming river that crossed the border and benefited diversions in downstream states. Fundamentally, the idea of a new kind of water right becoming a phenomenon on Wyoming streams—recognized like any other right by the state engineer's office—meant introducing a new player who might have very different ideas for water into streams where for years primarily agriculture had held water rights and had jointly managed what happened to the water. It meant a new player in statewide community water management. If an energy industry looking for water rights was sometimes unwelcome, a new state committee seeking instream flow rights was always so. (And in fact, energy industry people, from the geologists to the miners and rig hands, were often fishermen, and in turn became advocates for protecting river flows. Access to Wyoming rivers and mountains was a quality-of-life bonus that kept many of them from leaving the state to find higher pay elsewhere.) To irrigators, the potential creation of a state-owned instream flow right could mean a new voice demanding state staff to come regulate, on streams and in communities that had settled into comfortable patterns of use that rarely or never required invoking regulation. It could also mean that a stream might not be a "free river," in a water-user view of those words, for as many days or weeks as it had been in the past—or perhaps never. The prospect of losing the opportunity in "free river" spring runoff to take many times more water than was covered by a water right could be a major concern. At times, it sparked opposition not only to instream flows but also, on occasion, to new reservoirs serving other irrigators.[58]

Stauffer's committee in 1974 optimistically anticipated legislative action on its proposal the next year. But in committee hearings held around the state, Stauffer had seen how polarizing the issues were. "At Pinedale [by the Green River] we received some pretty stern chewing on by some of the hard-nosed ranchers," he remembered years later. "Then when we got to Cody [on the Shoshone], the environmentalists who showed up there insisted that our committee's recommendations were not strong enough." One of Stauffer's Salt River neighbors, who had bulldozed the river and whose job was to represent the Wyoming Farm Bureau, publicly accused Stauffer of trespassing on his ranch—by floating by, down the Salt River in a fishing raft. This despite a hotly contested state Supreme Court ruling over ten years earlier, upholding public rights to float on Wyoming streams through private land. The Wyoming Farm Bureau Federation officially commented on the committee proposal: "When the nation's and world's demands are increasing for red meat, Farm Bureau would not favor any stream preservation concept which might hamper present or future agricultural production or its expansion." Bell's Wyoming Outdoor Council, addressing "You good people who care about rivers," by contrast, supported "strong protection" for Wyoming rivers. Laramie resident Ruth Rudolph, an active League of Women Voters member whose husband was law school dean, argued to protect at least "some small portion of our streams," writing, "Assessing the economic worth of such elusive values as quietude, a natural setting, and an undisturbed life cycle is impossible. Yet, the existence of small geographic areas in their original state is essential to the mental and spiritual health of all men."[59]

State Engineer Bishop had told the league's seminar that detailed studies should be done of every possible scenic river, to catalogue possible development uses, before "locking them up by designation as wild rivers." A few weeks after shelving Bell's instream flow application back in 1969, Bishop had touted the development potential on the Green in advice to the governor. Stauffer recalled that when the committee toured streams, Bishop "would immediately try to figure out where the best place to put a dam might be."[60]

In fact, the final report from Stauffer's committee turned out to be the opening of a twelve-year battle over the idea of preserving water flows in Wyoming streams. It became clear that the legislature, which had long followed rather than led in water law, was incapable of endorsing the idea on its own. People working directly in agriculture were only about 10 percent of the state's population, but they were responsible for most of the water use in Wyoming, and they had strong influence in the legislature. About 25–30 percent of legislators came from agricultural backgrounds or had business connections with agriculture, as did the 1970s governors. However, the legislature and the voters had approved in the late 1960s a constitutional amendment giving voters the power to create law themselves by the initiative process (Judge Carey had first proposed such a provision more than fifty years earlier, when states other than Wyoming first adopted the initiative process, but he had failed). Now, in the face of legislative inaction on instream flows, the people who wanted to see new ideas about water enshrined in law turned to the initiative process and launched a massive public campaign for the idea of protecting flows in streams.[61]

The constitutional requirements set for a successful initiative in Wyoming were steep, but after ten years, the statewide effort led by the Wyoming Wildlife Federation (with a former Democratic congressional candidate at its head) succeeded in getting the signatures of some thirty thousand people—just over 15 percent of the registered Wyoming voters who had voted in the last general election. The petition they signed would put a detailed citizens' proposal for an instream flow law on the ballot. For agricultural people and their allies in the 1986 legislature, the prospect for a popular vote on the issue meant disaster. They rallied to prevent it. They enacted a law they liked better that could, nonetheless, be determined under the constitution to be "substantially similar" to the citizens' instream flow proposal and thus legally forestall a popular vote on instream flow. As Cynthia Lummis, the leader of the legislative counter maneuver (herself a member of a family with irrigated ranches, and a future Republican congresswoman and US senator) told her colleagues, the legislature's version had features that protected agricultural interests from what most disturbed them about instream flow rights.[62]

Thus, in 1986, the Wyoming legislature finally passed an instream flow law, recognizing protection of instream flows as a beneficial use—but for certain purposes only. Further, water rights for those purposes could be held only by the state government. The approved purposes were very limited, and the law imposed an elaborate process for creation of such a water right. Legal protection of the flow of water in-stream, not created by a reservoir, was limited to the minimum flow necessary to keep or improve existing "fisheries," not any flow to upgrade rivers and establish new fisheries. There was no other basis for claiming "beneficial use" for a water right protecting natural flows in a stream. Recreational and aesthetic goals for protecting natural rivers were left out. The state Water Development Commission was required to study the feasibility of a dam on any river where protection of natural flows for fish was proposed by the state Game and Fish Commission; the idea was to see if a dam could be the best way to provide flows, and new storage for water users, on any stream proposed for protection. The state engineer had to hold public hearings before approving a water right permit for instream natural flows in stream segments and was required to deny the permit if exercise of it would reduce Wyoming's potential to develop water under interstate compacts or court decrees. The engineer also had to deny any application for a right to protect flows within a mile of state borders. Other water users couldn't be regulated by the state engineer's office for the sake of an instream flow right unless there was a current or future injury to the fishery. Water users could give or sell their rights to the state to become instream flow rights, after all the usual considerations for a transfer were applied. But the law included a new transfer restriction, since water users couldn't change their right to this new use and still own it. Only the state of Wyoming, no one else, could hold an instream flow right.[63]

The attorney general ruled that that statute was enough like the initiative bill to keep the issue off the ballot, and the legislature's version became law in the summer of 1986. The state Game and Fish Department professionals had been conducting instream flow studies on over one hundred streams since the late 1970s and had computer models of likely fish habitat conditions tied to different flow levels. By 1986, the department had an internal list of sixty-two stream segments as candidates for instream flow protection. Given the proven incendiary nature of the issue, however, the department took a cautious approach to proposing instream flow rights under the new law. Tom Annear, educated in wildlife management and aquatic ecology and later a founder of the international Instream Flow Council, led the department's effort, nominating in the first few years only the stream segments least likely to stoke the political fires further. High-quality streams in canyons with few water users around or expected were among the first candidates for instream flow rights. "Proceed carefully and slowly with first streams to get a feel for the new process and insure first filings are successful," a 1986 internal department memo read. By 1992, only seven instream flow proposals had successfully gotten permits from the state engineer, after traversing what one legal observer called "the administrative quagmire" created by the Wyoming instream flow law. One of the new instream flow rights protected flows for a little under ten miles of the fifty-one-mile stretch of the Green River that Tom Bell's group had proposed for protection in 1969.[64]

Paul Stauffer had withdrawn from the statewide fray in the late 1970s after he was called to local Mormon church leadership (though his brother Alan was in the Wyoming House and voted for the final instream flow bill in 1986). Paul had been disappointed to see the committee's proposed bill to control river channelization die in the 1975 legislature, along with scenic river protection. But federal legislation soon giving the Corps of Engineers power over channelization nationwide "essentially solved regulation of river channelization in Wyoming with a few renegade exceptions, some of which occurred in Star Valley," he said. The State Engineer's Office implemented a process to review and disallow channel changes around irrigation diversions. By 1991, Paul Stauffer's church duties had lessened, and he led the local Trout Unlimited chapter and the Star Valley Conservation District (primarily representing agriculture) in restoring the Salt River, with initiatives including willow planting, to restore prime fishing areas.[65]

———

In 1991, and in 1990, Dick Baldes went through heights of excitement and despair. The instream flow law was no help to the Wind River. There were certainly no fisheries to protect on the Wind River as it nears Riverton. Fish populations there suffer insult and injury, from both silt and irrigation diversions, and Baldes had been irate about it for years. The state's lawsuit against the tribes, which he had hoped would bring people to learn and care about the river and then bring it back to life, almost made a big difference in 1990–1991—and then didn't.

When the Shoshone people visited the Wind River valley seasonally in the mid-nineteenth century, the river's flows had naturally varied greatly through the year. Its source in mountain snowmelt meant that the river carried a surge of water in late spring or early summer and dwindled down after that. Fish (and likewise, people) were adapted to that. The earliest gauge data is from 1912, and it shows that in its low months, the Wind River, measured where it turned north (at Riverton) could run as little as 5 percent of the water it had carried at peak. But by the 1970s, the lows in the Wind River measured at that point could often be as tiny as 1 percent of the high flows. Even in good water years, that wasn't much; in bad water years, it was very, very little. The change was primarily a result of development of irrigation from the river on the lands above Riverton, from the three irrigation projects that shared the 1906 permit priority date. Most of that permit went for watering lands intended for the federally built, slow-growing Midvale project. As with most federal irrigation efforts, the bureau has subsidized the project throughout its life. But farmers on the project still had to struggle to make the lands provide a living. Even the Bureau of Reclamation's official history concluded the project was "more notable for its failed ambitions than its triumphs." Wyoming state engineers, following the precedent on the Shoshone, kept extending the 1906 permit. In 1970, Wyoming's congressional delegation, citing the "high level" and "soundness" of the Midvale project, arranged for hydropower revenues across the Missouri River basin to help pay project debt, and that helped the farmers. As settlement grew on the project, so could the amount of water the project diverted. The permit extensions slowly shrank the flows left in the Wind River. That happened in a way that wasn't noticeable with the permit extensions on the Shoshone River—because the Shoshone had massive flows and a big reservoir on the main stem of the river to store and release those flows all summer that the Wind River didn't have.[66]

The originally healthy fish population of the river was recognized by federal engineers when in 1923 they built the fish ladder into Wind River Diversion Dam, that diverted water into Midvale's main canal about thirty miles above Riverton. Bull Lake, feeding the river via Bull Lake Creek above the diversion dam, had been commonly fished by Shoshone people. The water in lake and creek was controlled by the Bureau of Reclamation for the benefit of Midvale however, starting in the 1930s. Along with another small storage reservoir on the Midvale lands, Bull Lake Dam and Diversion Dam can manage a portion of Wind River water as if in a plumbing system, shifting water between handy storage sites and timing releases to the needs of Midvale's irrigated crops.[67]

As the plumbing system was completed, the needs of the river and its fish got less attention. To supply the Midvale fields in late summer, water stored in Bull Lake is released and the creek leading down to Wind River sees a sudden rush of water in late summer months, a pattern unnatural in stream systems fed by mountain snows. As one of the studies Baldes worked on shows, winter spawning habitat for much prized burbot, in the shallows of Bull Lake, can be dried up by these releases from the lake. Trout in the creek heading down from the lake can in turn have their spawning habitat decimated by the high late-season flows. Fish in the lake, the creek, and the river are isolated by dams both above and below them. Decades after the model Diversion Dam fish ladder was installed, fish could no longer use it. The river below Diversion Dam became less and less of a flowing river as the two smaller irrigation districts near Riverton took water for their fields. Those districts had emerged first from the failed ambitions on Wind River of the early twentieth century. They have first dibs ahead of Midvale on the natural flow of Wind River under the 1906 right, and they have no storage upstream. As the result of another Bureau of Reclamation project, however, they can take more water from the river than their 1906 right provides. As part of its dam-building boom era post–World War II, the bureau condemned reservation land to create a major reservoir, known as Boysen. The dam is located at the head of the Wind River canyon thirty miles downstream and north of Riverton and accommodates a railroad and highway through the canyon to the Big Horn basin, as Wyoming state engineers had envisioned decades earlier. The two small irrigation districts near Riverton can take more water from the Wind River upstream of Riverton via contracts with the Bureau regarding Boysen Reservoir. Under the contracts, the districts pay to store high-runoff water in Boysen during spring in exchange for taking water out of the river upstream in summer. The state engineer can and has approved such water right exchanges around the state and got authority in statutes to do so in the late 1940s.[68]

Diversion Dam and the federal project for Midvale plus the two smaller districts took diversions from the Wind River, and the cumulative effect drastically cut river flows nearing Riverton, nearly dewatering it in summer in dry years. Electroshock counts showed that rainbow and brown trout populations in the river managed to survive below Diversion Dam but were reduced by 90 percent or much more after the diversions for the next two irrigation districts. The quality of what river remained also suffered. Temperatures in the shallow water went up. Further, as Diversion Dam directed river water into Midvale's ditch system all summer, the dam accumulated behind it tons of silt that the river had brought down from the mountains. The irrigation district, to ensure its diversion structure kept functioning, regularly had to flush that silt down the river, often many times in one summer—each flush in just a few minutes multiplying by 1,200 times what environmental engineers call "suspended solids" levels in the river below. The result is brown, muddy flows that are hard on whatever fish and tiny macroinvertebrates try to inhabit the dwindling stretches of river below. Memories of river conditions before mid-century vary; but good fishing and swimming has not been common on the Wind River at Riverton for at least the last forty years.[69]

But 1989 brought a surprise, and the potential for real change. In the early 1980s, young tribal leaders had been able to document oil company theft of oil from leases on the reservation, and the tribes recovered millions of dollars in unpaid royalties that went to two key expenses: tribal member per capita payments, and lawyers to argue the Wind River water case. In 1989, the tribes won a significant part of that case. Wyoming's effort to minimize tribal water rights had backfired. The district court and then the Wyoming Supreme Court confirmed that rights to most of the water in the Wind River belonged to the Eastern Shoshone and Northern Arapaho tribes. The US Supreme Court, split down the middle as Justice Sandra Day O'Connor recused herself, left the award standing.[70]

The Wyoming courts had jurisdiction in the "general adjudication" of basin water rights, but they were required to follow federal law on tribal water rights. Considerable federal law on Indian water claims had developed since the 1908 *Winters* decision in which former Wyoming attorney general Van Orsdel had played a role. As tribes across the West began to assert their water rights in the 1960s and 1970s, the *Winters* case had led to court opinions outlining federal law that tribes have a right to the water necessary for the purpose of the reservation established by the government. The amount of water could be determined by the courts to include future as well as present needs, based on the reservation purpose. It was not a right limited to how much water was used by a certain date or a right that could be lost for non-use. Rather, it was reserved in perpetuity for use for the reservation.[71]

Accordingly, the Shoshone and Arapaho tribes wound up with a confirmed right to five hundred thousand acre-feet of water in the Wind River, nearly all the average annual flow of the river. That water right had the priority date of 1868, the date of the treaty establishing the reservation. That meant the tribes' right to the water was recognized legally, for the first time, 120 years after the treaty, and all the water uses and their water rights established on the river by other people after that time suddenly would have to acknowledge a higher-priority right on the Wind River system. The air was vibrant with the sense of potential transformation of relationships along the river. [72]

The Wyoming high court had allocated the water right to the tribes based partly on their current uses, and partly (as the federal courts had dictated) on elaborate calculations of how much water they might reasonably use *in the future*—for, very specifically, new irrigation projects. Something over half the right was for current uses, and the remainder was for potential future projects. If the tribes could put that "futures" water right to use, non-Indian irrigators could foresee competing with each other for an amount of water currently not covered by tribal rights but likely to dwindle increasingly as tribal uses grew. Ironically, their risks and rivalry now were more intense because of the precedent the State Engineer's Office had reluctantly set in 1915 on the Shoshone for leniency on state water permit deadlines for big federal irrigation projects. That policy had allowed Midvale's actual water use to grow over the years toward its original permit total.

The picture on the ground, however, was complex. The nineteenth-century policy that required individual members of the tribes to take ownership of lands allotted for farms, followed by more than a century of land sales and irrigation, had made everything complicated. Non-Indians were irrigating lands on the reservation water project, and tribal members were irrigating lands under the non-Indian projects. Some non-Indians on former allotments might claim treaty-based rights themselves (a tempting prospect for some, though requiring considerable litigation time to confirm).[73]

Fremont County was not known for treating tribal members with respect, as a federal court found years later in a voting rights case. [74] The new water ruling made it seem that might have to change. Within a couple of years, though, that prospect began to fade. There were rapid new developments. From the start, it was clear that the early-date rights for irrigation on tribal members' lands meant that in a dry year, water would have to go to all those fields first— quite a different sequence from the usual. The year 1989 was a dry year. Tensions exploded within months of the state court decision, before the US Supreme court upheld it. Tribal officials lost patience with early water diversions by non-tribal irrigators under the reservation irrigation project and with fish decimation by federal management of Bull Lake for Midvale. They shut selected irrigation headgates and blocked access to Bull Lake. Allegations of peremptory actions multiplied and raised tempers. The state and the tribes worked out a temporary settlement, which included sharing shortages between the tribes and the non-Indian districts and providing state financial support to the tribes via a cash payment and foregone state taxes on reservation oil and gas production. The US Senate held a hearing where Baldes's issues on the health of the river and the fish were heard, and a leading senator blasted the federal Bureau of Reclamation for favoring non-Indian irrigators at the expense of the fish in Wind River and its tributaries. Tribal-state negotiations continued, aimed at a longer-term settlement on how the river would be managed.[75]

Court decisions rarely answer all the practical questions people encounter after the decisions are issued, so there was much yet to be settled, and each side saw its own promising future in the Wind River rulings. The Wyoming Supreme Court had ruled that the 1868 treaty created the reservation for a "sole agricultural purpose." This was distinct from, and much narrower than, the broad purpose of a reservation to create a "homeland"—which the special master in the Wind River case had found and which the top court in Arizona later endorsed for some tribes there. The Wyoming high court's denial of a "homeland" finding for the reservation on the Wind River was a victory from the state's point of view, and it had meant an award of water rights smaller than the tribes sought. It meant the amount of water that the Wyoming Supreme Court allocated for the tribes' water rights did not include water supplies for fish, wildlife, aesthetics, mineral development, or new groundwater development. The amount of water the Wyoming Supreme Court confirmed for the tribes was based solely on reservation irrigation—existing irrigated lands and reservation acreage considered "practicable" to irrigate in the future.[76]

Further, the court's economic calculation of what was "practicable" to irrigate on the reservation was parsimonious compared to the actual history of irrigation in the West. The calculation used included no allowance for subsidy. It would have rejected affirming water for a project like Midvale. In this the state high court again followed the law federal courts had established on reserved water rights. Those courts, more than one hundred years after some of the treaties, had imposed limits on tribal water awards that had not been imposed on water claimed and used by non-native people. The federal courts seem to have absorbed something of Congress's longtime attitude: generosity with Reclamation projects and parsimony with water for tribes. So Wind River tribal irrigation projects that would have had to be subsidized were not considered "practicable," and water rights were not awarded for them.[77]

On the other hand, the federal courts had also said that water rights *quantified* by what could be used for irrigation did not *actually have to be used* for irrigation. The Wyoming Supreme Court, meanwhile, had not specifically addressed how the futures water rights could be used. So after the 1989 US Supreme Court decision affirming their award, the Shoshone and Arapaho tribes considered other ways to use their futures water rights. They created a water code as the law governing water management on the reservation. The code listed a wide variety of beneficial uses, including aesthetic, spiritual, recreational, and instream flow. Under the code, the tribes saw an opportunity to dedicate some of their futures rights to reviving the Wind River by creating instream flows on the Wind River through Riverton.[78]

The measurements made pre–World War I by the State Engineer's Office showed that the natural low flows of the Wind River just above Riverton in mid-July were about 250 cfs. The tribes proposed to use their water rights partly to reestablish flows at about that level: 250 cfs. The tribes' water office had reviewed an economic study showing that the highest and best use of the water was for instream flow and restoration of fisheries on the river. To achieve the 250 cfs in mid-summer, the tribes dedicated to instream flow in the Wind River somewhat under half of the 1868 water rights allocated to them for "future use." Experts advising the tribal water office calculated that keeping that amount of water in the river would not mean major injury to anyone—it would not force non-Indian districts and ditches on the river to make serious cutbacks in their water diversions. The year 1990, unlike 1989, was a good water year. That year, the tribes arranged to put trout in the river. Originally, some were put there with the help of the US Fish and Wildlife Service, which had Baldes in its Lander office. When Wyoming's US senators quietly scotched that, the National Wildlife Society (with a staffer who had also helped lead the fight for Wyoming's instream flow law) funded fish planting from a private hatchery.[79]

That set the stage for a clear demonstration of the differing aspirations and experience of the peoples on the Wind River, starkly displayed in the gulf that soon revealed itself between Wyoming water law and tribal water law. Under the court decisions, the state engineer oversaw the river. In the summer of 1990, official gauges showed the tribes that the final stretch of the Wind River before it reached Riverton was carrying less water than their 250-cfs instream flow called for. The two irrigation districts near the town were making diversions of water that lowered the flows in that stretch. The tribes asked the state engineer to stop those diversions, to protect their instream flow right with its 1868 date. The state engineer was now Jeff Fassett. Having first been a consultant for the state on the Wind River lawsuit and then deputy state engineer, Fassett was familiar with all the details of the river and the uncomfortable twists and turns of thirteen years of litigation. He favored, generally, making it possible to keep water flowing in rivers. Having now become the state's chief negotiator with the tribes, he hoped for a long-term solution—maybe the launch of multiyear planning and investment for a dam upstream to supply water summer-long for farmers and for fish. In the summer of 1990, he worked with non-Indian irrigators along the Wind River, persuading them to cooperate voluntarily to keep more water in the river. The gauge readings showed that the amount of water the tribes had dedicated to instream flow was there for some stretches, including past Midvale's Diversion Dam. But the tribes were concerned about the river more than ten miles downstream from that dam, and then on toward Riverton, where they estimated that the diversions of the other two irrigation districts were dropping the water levels to sometimes only 80 percent of the tribes' instream flow target.[80]

The State Engineer's Office had no experience regulating water use to keep water flowing in a certain stretch of a heavily used river. Wyoming water law was built around the practice of diverting water, for use away from a stream. The very words *instream flow* sparked strong feelings. The state instream flow law adopted after years of controversy envisioned designation of stretches where flows existed and could be protected; on a hard-working river like the Wind, that might conceivably be done under the state law only by someone handing over old water rights to the state. The state instream flow program was just four years old, and thus far had involved primarily new and very junior instream flow rights that could raise few enforcement issues. Further, in State Engineer's Office tradition, enforcement of even ordinary water rights could be a complicated matter, not undertaken unless a senior water right holder formally requested it. If a request was made, enforcement would then mean closing the diversions of all the junior appropriators (in reverse order, most junior first)—but only if shutting them down would mean water would then be available to fill the senior's right. Fassett said he had followed that procedure—with no formal request—to ensure that water got to the tribes' diversions for irrigation on the reservation. But when the tribal water engineer asked the state engineer to specifically shut down the key diversions just above Riverton for the sake of instream flows in the stretch that concerned the tribes, Fassett saw that as a request for "selective" enforcement, which he said the State Engineer's Office had never done and that he would not do. When the tribes complained to the court in the hot days of mid-summer, the state's lawyers went further, arguing that the tribes' water rights were only for *diverting* water and belittling the tribes' dedication of their water rights to instream flow under the tribal water code as a "self-awarded," "unilateral" action. A change to a different use like instream flow had to be done under state law, the state's lawyers argued.[81]

The Wyoming Supreme Court had said that questions of interpretation and enforcement on the ground of its 1988 award to the tribes should go to the district court. The tribes went to that court to get an order requiring enforcement of their instream flow and removing the Wyoming state engineer from oversight of the river—asking, in fact, that he be cited for contempt of the earlier court decrees. The district judge ruled that the tribes could use their water rights as they chose, including for instream flow, without regard to Wyoming water law. The district judge also found that the state engineer "has had difficulty assuming a neutral role in the administration" of the tribes' rights, and therefore the tribes should take over management of the major stretch of the river within the boundaries of the reservation. Because of the interdependence of water users, the judge did require that the amount the tribes could dedicate to instream flow from a futures water right be limited by the amount of water that would have been consumed by crops (not the larger amount that would be diverted for that purpose) if the water had been used for a futures agricultural project on tribal land. That restriction was one the tribes' 1990 dedication to instream flow had apparently met.[82]

The Wyoming Supreme Court rejected the district judge's ruling. In a tortured decision that displayed five different opinions from its five justices, the Wyoming Supreme Court ultimately ruled that the tribes could not put their futures water right into instream flow. The tribes, the court said, must follow state law if they wanted to put those futures rights to a non-agricultural use. And, of course, the state's limited instream flow law would not allow allocation of a previously unused senior-date water right to instream flow. Further, the split court ruled that the state engineer alone must have charge of the river.[83]

The tribes, uncertain of the likely stance of the US Supreme Court and arguing internally after weary years of litigation, did not appeal. Instream flow for the Wind River died very quickly. Former state engineer Fassett, looking back almost twenty years later, said, "We got off on the wrong foot and found it almost impossible to stop the litigation chain. Clearly the hard-fought litigation left ill will among the parties. It damaged relationships. And it damaged the neighborhood."[84]

The tribes' wariness of a US Supreme Court decision, which would affect tribal water rights nationwide, had some foundation. When the US Supreme Court discussed the Wind River case in 1989, reports from the oral argument showed that the high court might be interested in limiting the tribes' water rights and their use. Perhaps reports of the oral arguments also influenced the Wyoming Supreme Court opinions on the instream flow case in 1992. United States Supreme Court justices often play devil's advocate in questioning lawyers, and in the end, the high court issued no written opinion. Several US Supreme Court justices, however, during the hearing in 1989 had questioned the legitimacy of the basic water award, largely in consideration of the other water users who had come to Wind River since the treaty. Those justices, in the oral argument, asked whether the tribes really needed so much water or might just hold other users hostage as part of a plan to sell them water.[85]

In the Wyoming Supreme Court decision in 1992, Justice Richard Macy pronounced, in one opinion among three that prevailed:

The Tribes do not have the unfettered right to use their quantified amount of future project water for any purpose they desire . . . . The Tribes, like any other appropriator, must comply with Wyoming water law to change the use of their reserved future project water from agricultural purposes to any other beneficial use . . . . Our decision today recognizes only that which has been the traditional wisdom relating to Wyoming water: Water is simply too precious to the well-being of society to permit water right holders un-fettered control over its use.[86]

Wyoming Supreme Court Justice Mike Golden wrote the most eloquent dissent. He compared the recognition and implementation of tribal water rights in Wyoming with its series of court reviews to courts' implementation over several decades of the US Supreme Court's 1954 *Brown v. Board of Education* school desegregation decision, which overturned longtime customs in recognition of the fundamental rights of the people involved. [87]

Regarding the original 1908 US Supreme Court *Winters* decision, Golden said that decision's "broad language does not allow a crabbed interpretation of the proper uses of the reserved water by the Tribes."[88] "The Tribes may call for their water for any use to which water may be beneficially put . . . if the only injury to other users comes about because the Tribes are actually using their water for instream flow, and that same injury would exist if the water was used to irrigate corn, then there is no injury" for the law to remedy, Golden wrote, concluding, "The state engineer is bound to make that water available as requested."[89] "I reject the argument that the reserved water is the property of the state and the state engineer thus must have control. The reserved water rights of the Tribes are not within the boundaries of the state but are within the boundaries of the reservation."[90]

The 1989 decision had put the state engineer in an impossible position, Golden said, requiring him to be an impartial "water master" monitoring water use under the court decision while he was also the state's chief negotiator with the tribes over water issues, and was required by the Wyoming Constitution to protect the waters of the state. It made sense for the district court to remove him as water master. That job should go to the tribes, who would carry the burden of monitoring activity under the 1989 decision and turn to the courts if they perceive a violation.[91]

Harking back to 1905, when the federal government cut back the Wind River reservation to make lands available to non-Indian settlers, Golden concluded:

> If one may mark the turn of the twentieth century by the massive expropriation of Indian lands, then the turn of the twenty-first century is the era when the Indian tribes risk the same fate for their water resources.
>
> Today some members of the court sound a warning to the Tribes that they are determined to complete the agenda initiated over one hundred years ago and are willing to pervert prior decisions to advance that aim. I cannot be a party to deliberate and transparent efforts to eliminate the political and economic base of the Indian peoples under the distorted guise of state water law superiority.[92]

Golden was very much in the minority.

————

So—as of the early 1990s, Wyoming water law and management had seen much controversy but proved itself largely resistant to major change. Despite challenges focusing on the value of flowing rivers and the water needs of the tribes, most Wyoming water users' resolute opposition to change had been successful. The state's community water management system persisted without having either to allow many new players into the community or to recognize an independent separate community on the Wind River reservation. There was some shifting of power among the customary players, the state agency, and the long-time water users. The State Engineer's Office was taking on new significance as a lead agency handling instream flow applications and responding to tribal water uses. That pointed toward an increasing exercise of the state's rights in water, which carried a responsibility for public benefit. In years soon to come, state engineers could act in new ways, from accepting new instream uses in creeks with long-set water use patterns to influencing how private water rights could be managed to accommodate new out-of-state demands, including the requirements of federal environmental law. Yet the most significant statewide change since the 1970s—the major mineral development that transformed Wyoming's economy and government budget—had not brought water users the troubles they feared. Rather, it brought water users a dream-come-true cache of money to improve and repair infrastructure. The dead pioneer hand could still be seen in the landscape. Users could hope to protect their heirloom rights and use water the way they always had.

6. MOVING ON

They are used to drought; have they really thought about a fourteen-year drought?

—GINGER PAIGE, UNIVERSITY OF WYOMING
ASSOCIATE PROFESSOR OF WATER RESOURCES, 2018[1]

Perhaps the best available term is *aridification*, which describes a period of transition to an increasingly water scarce environment. Words such as *drought* and *normal* no longer serve us well, as we are no longer in a waiting game; we are now in a period that demands continued, decisive action on many fronts.

—COLORADO RIVER RESEARCH GROUP, MARCH 2018[2]

Places: Platte River, central and east Wyoming; Wind River, west-central
Wyoming; Green River, southwest Wyoming.
Time: 1990–2018.

Green River and hay meadows. Courtesy of Rita Donham / Wyoming Aero Photo.

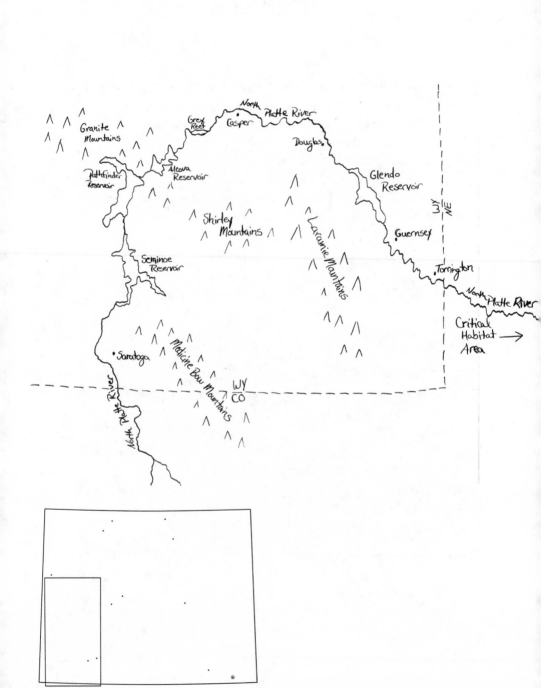

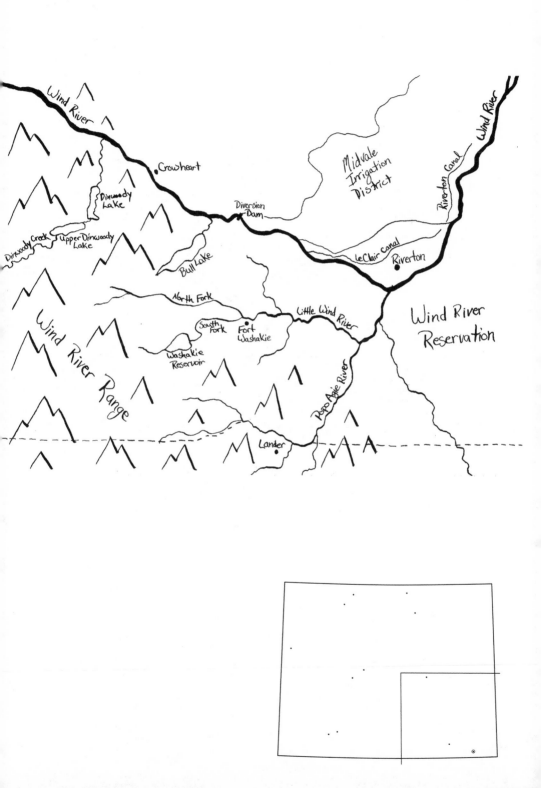

THE DEAD HAND OF the pioneer gripped April Barnes for a long time: she broke free only in 2014.

Born on a farm in Oregon in the 1990s, Barnes traveled to Colorado and then Wyoming, looking for ranch work. She found a spot in western Wyoming working as ranch hand for a woman on Fontenelle Creek. Named for an 1820s French fur trapper, Fontenelle is a Green River tributary that carries snowmelt from the Wyoming Range west of the river. In 1964, the last Bureau of Reclamation attempt at launching one more irrigation project in Wyoming created Fontenelle Reservoir, flooding a stretch of the Green River including its confluence with Fontenelle Creek. The geology was unfavorable, but the Bureau built the dam anyway; the dam nearly collapsed in its first year of operation. The irrigation project never happened. Now Fontenelle's prime feature is the twenty-seven thousand acres of national wildlife refuge, created to make up for the loss of wildlife habitat caused by reservoirs—like Fontenelle—built in the 1950s and '60s.[3]

Ranches up Fontenelle Creek, above the confluence with the Green River, were not flooded out by the reservoir. The lands had been homesteaded and irrigated in the 1870s, with settlers of the usual interesting variety. One, for instance, was a Civil War veteran and lawyer with a wife from a Kentucky abolitionist family; together the two helped keep votes for women in the new state constitution. Another Fontenelle Creek settler was a New York farm boy who spent the Civil War in California mines and married a schoolteacher after he came to Wyoming. His place eventually became the Barnes ranch. Many years later, long after Fontenelle Reservoir drowned out the lowest end of the creek, April, the young ranch hand from Oregon, met Eric Barnes "over the fence" there. Eric's father, when just out of high school in the nearby coal mining town of Kemmerer, had been a ranch hand on the place and slowly bought it, piece by piece. It was a small ranch, with irrigated pasture in the bottom land along the creek and grazing land (for an assortment of wildlife, as well as cattle) on the benches above. Eric's father managed to put his eight children through college from that place. Eric, the youngest, earned his college on a rodeo scholarship and was the only one to come back to the ranch. Eric was a little "wild," April said—but eventually they got married, and together they had a son.[4]

April and Eric did their best to shape their irrigation as Eric's father had taught him, to produce the best forage for their cattle. And they saw plenty of birds and wildlife: sandhill cranes, trumpeter swans, red-tailed hawks, a snow goose once; moose, bobcats, and mountain lions. Eric liked to track the cats but never allowed hunting on his place. The tracks showed that the cats never went after his calves.

To get the irrigation underway every spring, the Barneses had to get right into the roiling, icy flows of Fontenelle Creek. They jimmied telephone-pole-size logs into place to create a dam that would back up the creek enough to send water into the ditches serving the bottomland pasture. It was miserable work, and dangerous; Eric's father was once swept downstream clinging to a log and barely made it out.

In 2014, April heard about the Wyoming office of Trout Unlimited's Wyoming Water Project. Young men and women were working in the Green River basin, as elsewhere in the West, to help ranchers improve irrigation headgate and ditch systems to make them friendlier to both ranchers and trout. On Fontenelle Creek, April and Eric had always seen Kokanee salmon and brown and rainbow trout, and they hated to see them trapped in their ditches. They talked to Trout Unlimited staff, who came to Fontenelle Creek to assess the Barneses' system and see what could be done.[5]

The new irrigation infrastructure proposed by Trout Unlimited offered ranchers in Wyoming the kind of systems many had dreamed of but could never afford. Trout Unlimited had the connections to find money to build new systems in creeks where there was good potential for a healthy trout fishery. As water commissioners for the Wyoming State Engineer's Office can tell you, in Wyoming it used to be that you could always tell the difference between the infrastructure serving irrigation and what was built for the many oilfields scattered around the state. Oilfield pipe would be in good shape; irrigation headgates, ditches, and any occasional pipe looked patched together.

Trout Unlimited's proposed new system for the Barnes ranch was a set of rock structures in the river and simple headgates on the main ditches. The rock structures were specially designed to keep the creek water at the ditch mouth always high enough to feed the ditch. The Barnes could access the water for irrigation when they needed by simply opening their new headgate, with a hand wheel that even their young son could operate, while standing on the bank of the ditch. No more wading in icy water trying to jockey heavy logs into place.

At the same time, the rock structures in the creek accommodated flows and fish moving downstream in a way that recognized and supported the natural dynamics of the creek. The structures were inspired by the work of an Idaho hydrologist who, when working for the US Forest Service, was so disturbed by the destruction of his hometown creeks by clear-cuts and dams that he launched a whole new approach to how people work with rivers.[6]

That was the beginning of a big change in the Barnes' lives, and in their ability to keep their small ranch going.

———

Meanwhile, the role of the state engineer in Wyoming water management had been changing, in response to a variety of pressures.

In the mid-1990s, anxiety gripped Wyoming and reached the State Engineer's Office. It was worry over limited prospects, fear that Wyoming would again see the kind of subsistence incomes and strapped state budgets that characterized Wyoming for most of its history. State income had become dependent on revenues from coal, oil, and natural gas, but as their market prices dropped and oil production declined, it was the end of the halcyon days of the 1970s and early 1980s, when mineral money poured into state coffers, including into water development funds. At the end of the 1990s, state officials forecast a budget deficit, and a blue-ribbon committee dared to recommend that Wyoming adopt an income tax for the first time. State income tax was anathema to Wyoming residents long used to mineral bounty.[7]

In the late 1990s, however, a new kind of mineral production arose that promised the kind of revenue that would save the state from seriously contemplating that step. Natural gas (methane) can be found trapped in coal seams, and a couple of laid-off oil company staffers developed a reliable and cheap way of producing it—getting enough gas from the seventy-foot thick coal seams in the Powder River basin in northeast Wyoming to make a good profit. Someone with just a water-well drilling setup could get down into unmined coal seams to get methane. By 1997, the volume of "coalbed methane" (CBM) being produced in the Powder River basin appeared to be on its way to doubling in the next year. And when final figures came in, it had done better than that.[8]

There were water issues associated with this minerals boom. Coal seams in the Powder River basin are aquifers, and it is water pressure that keeps the methane in the coal; to get methane out of the coal, the water must be pumped out first. From the thick coal seams in the Powder River basin, that can mean a lot of water. In 1997, Jeff Fassett, state engineer during the most contentious years of adjudicating the tribes' rights on the Wind-Big Horn River, was still in office, and he had a decision to make on CBM. Typically, the State Engineer's Office does not supervise oil and gas operation production of water; it's often in small amounts, sometimes used nearby. But as CBM and its water production grew, with potential impact on groundwater supplies, Fassett decided that pumping groundwater in order to produce CBM should come under State Engineer's Office supervision. To do that, he ruled that pumping groundwater for CBM should be considered a beneficial use of water. Some people objected to the idea that simply pumping water up to get at another resource could be considered a beneficial use of the water itself. But the alternative appeared to be letting the state agency that managed oil and gas production be the sole authority in questions of the groundwater impacts of CBM production. Instead, taking CBM water under the State Engineer's office meant that the State Engineer's Office could at some point choose to exercise control over the amount of water produced. The gas developers would have to get permits to use groundwater from the State Engineer's Office, and the traditional water experts could therefore have some say in CBM development.[9]

After making that ruling, however, Fassett and Pat Tyrrell, who succeeded him as state engineer, delayed taking major action for several years. By 2003, CBM alone brought the state and local governments $230 million a year in tax and royalty revenues, equal to the money brought in from all other energy production combined. In 2003, CBM wells also produced what looked like a lot of water. But coalbed water production was scattered among thousands of gas wells. A state project to gather and transport the water in high volumes was studied but proved uneconomical. A large portion of the water that was produced ran down ephemeral creek drainages in the Powder River basin. The key to the boom, and the revenue, was low-cost production—an advantage that would soon evaporate, producers argued, if they were required to reinject the water or manage it for other uses.

But CBM water running in those Powder River basin drainages was not necessarily a good thing. There had never been much irrigation there, and the soils were not friendly to heavy water flows. Water produced from CBM wells varied in quality, but even decent quality water was degraded as it ran through Powder River basin soil or was stored in a small onsite reservoir. The water picked up dissolved solids, including salts. The effect on forage in the ephemeral draws and the cottonwoods they nurtured—resources the ranches depended upon for forage and shade for livestock—could be devastating. Some ranchers in the Powder River basin fought CBM; others were happy to be paid for the water damage and related losses. In an area where few water commissioners had been needed in the past, state engineer staff had to scramble to keep up with the many small reservoirs now being built for CBM water—water which the small storage sites would eventually release downstream. The State Engineer's Office began denying permits if operators drilled before applying for a permit or failed to seek a water storage permit.

In 2007, it became clear, particularly as the industry moved westward in the basin, that more and more wells had produced water for over two years with no gas production at all—the beneficial use, gas production, was nonexistent. In late 2007, ten years after Fassett first declared that CBM water was within state engineer purview, the State Engineer's Office began to use its permitting authority to take a role in managing the industry and its water production. State Engineer Tyrrell, in office since 2001, now required proof of adequate gas production, and he canceled permits that were producing excessive water in proportion to the gas produced. The 1909 statute explaining water rights (and imposing the transfer ban) said "beneficial use" was the basis, measure, and limit of a Wyoming water right; the word "limit" was a guard against waste. That language was still in the law. Pumping out water for little or no gas production was waste.[10]

As it happened, however, 2009 marked the peak of Wyoming's CBM boom. Even during the boom, production of the gas had been only marginally economical. Another new technology, hydraulic fracturing—which began in western Wyoming's Green River basin and spread nationwide—made prolific new gas fields accessible, spurred a gas production boom, brought gas prices down, ultimately helped dampen coal production, and killed CBM production. Tyrrell's action on CBM had come late. As a result, many people remembered only lack of action by the state engineer at a time when water production and management in the face of CBM production was a pressing issue. Wyoming water law had developed with the understanding that water must be cared for and managed well in an arid environment where there was never enough water. Coalbed methane production, however, had presented the unfamiliar situation of too much water, and it had taken years of watching problems develop and struggling to create new policy before the state engineer took decisive action—too late for some places.[11]

But elsewhere, the State Engineer's Office, joining with the Water Development Commission, found it more straightforward to adapt to other challenges and take the lead in guiding water users in adjusting to new developments. As federal policy continued to affect Wyoming water management, it was the state engineer who could best respond on behalf of water users. He was aided further by the water development office, flush with stockpiled energy income (including CBM revenues) and its influence over water users who wanted state financing for their projects. For twenty years, from the mid-1980s to the early 2000s, the engineer and the water development director were a formidable pair in shaping the impact of federal policy on Wyoming water users.

The North Platte River was a prime arena for that work. The North Platte is not Wyoming's biggest river, but it is much in demand as it crosses the state's eastern plains—and even more in demand next-door in Nebraska. With headwaters in Colorado mountains, the river runs north into Wyoming before taking a sharp turn to head east into Nebraska. There it joins the South Platte, also born in Colorado, to form the main Platte River, moving on to join, eventually, the Missouri River near Omaha. The Platte is a workhorse river in every state it traverses. Irrigation from the Platte was early and heavy: by 1900, over four thousand canals tapped the North, South and mainstem Platte, and the river's summer flows weren't enough to serve them all. Colorado's early irrigation projects and cities, including Denver, were founded on the South Platte and its tributaries. The North Platte (along with the Shoshone River) was one of the first places that the new federal Reclamation Service chose to build a big dam. The Reclamation Service (later renamed the Bureau of Reclamation) built Pathfinder Dam west of Casper during 1905–1909. Some Wyoming acreage near the Nebraska line benefited from that project, but most of the water stored in Pathfinder went to irrigation in western Nebraska, where soil and climate was better for growing cash crops. Though the federal North Platte Project, like other federal irrigation projects in the West, eventually had most of its costs subsidized, the original Reclamation Act had envisioned the government's investments in dams to be paid back, and Nebraska fields were the best paying proposition for Pathfinder water.[12]

By the 1970s, forty-nine reservoirs, federal and private, large and small, had been built in the three-state Platte basin, reservoirs capable of storing seven million acre-feet a year to serve farm fields, towns, and cities in Nebraska, Colorado, and Wyoming. Four more federal dams in Wyoming were among those dams built on the river. They helped make it possible to shift water where and when it was needed most, taking full advantage of Pathfinder's storage, boosting the Platte-irrigated acreage in Wyoming, and eventually, in the 1990s, reducing pollution and aiding fish. The State Engineer's Office monitored water use more heavily on the North Platte than elsewhere in the state. Disputes with the sister states (particularly Nebraska) over North Platte basin water through the twentieth century have produced stacks of legal briefs and technical documents, at a cost of tens of millions of dollars. Court decrees and settlements, determining how much water can be used where, are today enforced with satellite photography to help tote up irrigated acres and water consumption.[13]

Before the irrigators came and the dams were built, the Platte was a braided river—overflowing with Colorado and Wyoming mountain snowmelt in the spring and early summer and inching its way in low flows through myriad channels in late summer and fall. In Nebraska, the Central Platte stretch of river has for thousands of years been the crucial, hourglass waist of the migratory flyway for huge flocks of birds moving north toward Canada in early spring and south toward the Gulf of Mexico in the fall. Before the river was dammed and diverted, its fluctuations left it dotted with ephemeral islands and flanked by treeless wet meadows, places where birds could land and rest with good view of any predator.[14]

By 1970, the birds' stopover zone on the Central Platte was estimated to receive only about 30 percent or less of the total flow of the Platte. The river's channel had narrowed in response to the low flows. Several species of birds that relied on the river in Nebraska were at risk of extinction, US Fish and Wildlife Service scientists found, and the birds were determined "threatened" or "endangered" under the 1973 Endangered Species Act (ESA). One culprit appeared to be the radical transformation of the Platte. Dams and agriculture had tamed the old wayward river till it followed a single year-round channel lined with trees. It was no longer such a safe and welcoming resting spot for migrating birds.[15]

In the late 1970s, the lawsuit regarding the proposed coal-fired power plant at Wheatland was a portent of what was to come. A federal court ordered a halt to construction of the dam and reservoir being built for the plant. Federal agencies had issued a loan guarantee and a Clean Water Act permit for the project. The state of Nebraska, joined by the National Wildlife Federation and others, sued the power cooperative building the plant, invoking the Clean Water and Endangered Species Acts. In the settlement that allowed the project to be completed, the power cooperative agreed to create special water releases to the North Platte in Wyoming and a $7.5 million trust fund to maintain Platte habitat in Nebraska. Nebraska and the other plaintiffs had convinced the court that federal agencies approved the power plant without adequately analyzing impacts or consulting with the US Fish and Wildlife Service on the power plant's planned water diversions as they would affect threatened and endangered species habitat in Nebraska.[16]

In the 1980s, it became increasingly clear that the ESA would have a major impact on the North Platte in Wyoming. The big dams on the North Platte in Wyoming, all run by the federal Bureau of Reclamation, faced intense scrutiny under the ESA. The law requires federal agencies to ensure that their operations or actions are unlikely to jeopardize endangered or threatened species, and the agencies must also carry out conservation programs for those species. The Fish and Wildlife Service started pushing for an evaluation of the bureau's big dams on the North Platte in Wyoming in 1980, and proposed projects on the South Platte in Colorado were met with environmental groups' objections because of impacts on the habitat in Nebraska. The Fish and Wildlife Service, the bureau, and the three states on the Platte—Wyoming, Colorado, and Nebraska— worked on ideas for helping habitat and allowing more water development. In 1990, the bureau finally started an official environmental impact assessment of its operations on the river.[17]

Meanwhile. a big non-bureau reservoir built by local irrigators in Nebraska, just past the Wyoming line, saw its federal permit for hydropower come up for renewal in 1987. Hydropower, as a significant use of the public waters, had in the early twentieth century been a focus of Progressive Era thinking like Mead's. The thought was that use of the public water resource, impeding streams and possibly leading to a monopoly in electric power, should require a federal license and regular reviews. Hydropower projects accordingly had federal licenses with fifty-year renewal requirements. After the 1973 enactment of the ESA, any federal proposal to relicense a hydropower plant required consulting with the Fish and Wildlife Service. All that came into play in 1987 when the Nebraska irrigators' hydropower license on the North Platte came up for renewal. Given the decades of a dwindling Platte, the Fish and Wildlife Service was critical of existing operations on the river, concluding that much more water was needed to reach habitat in the central Platte. The agency stated that proper enforcement of the ESA could require that each use, like the Nebraska irrigators' hydropower plant, provide its own contribution to habitat improvement. Alternatively, all the users could develop a basin-wide plan to help the birds' migratory habitat.[18]

One way or another, a federal connection—bureau dams, hydropower permits, federal farm loans, or canal and reservoir rights-of-way over federal lands—meant leverage on every Platte river user in Wyoming, Nebraska, and Colorado to address endangered species issues. A basin-wide approach seemed the best solution; neither individual users on a ditch nor big irrigation districts could effectively work out their own agreements with the federal government. "The ESA is the great convener and the great conciliator. Without the pressure induced by the potential power of the ESA, the states and [their] water users would have had no incentive to resolve the serious environmental issue of the degradation of the Central Platte," recalled a senior lawyer for the Department of Interior.[19]

Top water officials from each state stepped up. For Wyoming, that meant State Engineer Fassett and Water Development Director Mike Purcell. The picture for Wyoming was particularly complicated by the lawsuit Nebraska had filed after the Wyoming legislature, in its enthusiasm for water development with mineral wealth, funded the $45 million dam on a North Platte tributary called Deer Creek. Some major issues on the Platte were therefore in court. The state of Nebraska had first taken Wyoming to the US Supreme Court after the drought years of the 1930s and gotten a high court decree in 1945 allocating most of the river's flows to Nebraska. In 1986, Nebraska sued again, saying that the Deer Creek dam plan and other water uses in Wyoming violated that decree. The US Supreme Court, where such interstate suits are heard, refused to rule out considering environmental damage as part of Nebraska's claim against Wyoming. Together, the Nebraska lawsuit and ESA review led to years of expensive litigation, and ultimately negotiation, to arrive at deals that might serve both people and wildlife on the river.[20]

Fassett and Purcell tried to assess and protect the needs of Wyoming irrigators from the Colorado line to Nebraska. They also had to explain Wyoming water use in the Platte basin to the other negotiators, particularly to negotiators from Nebraska. Meanwhile, the Casper-based manager of the bureau's North Platte Project based, John Lawson, used his detailed understanding of the dams' actual and potential operations to help work through a rethink of river management.

Since 1904, Pathfinder Dam had divided the river into two very different realms of water use. In Wyoming upstream of the dam, the gentle valleys near the Colorado line were largely the world of flood irrigation. Irrigators tapped directly into variable stream flows, resulting in intricate interdependencies among users, their flooded fields, the return flow from those fields, and the entire small water basin. Those irrigators diverted as much as they could, from a "free river" in high water times, and then diverted smaller amounts, in priority order, in the rest of the season (sometimes boosted by small private reservoirs and return flows); all of which Nebraska negotiators found hard to digest, because it was so different from what happened in their stretch of the Platte. Wyoming downstream of Pathfinder as a water-use world was more familiar to the Nebraskans. There were the towns of Casper and Douglas, the federal Kendrick project, and some private irrigation ditches, but the greatest use of water was at the far eastern edge of the state near the Nebraska line, the land of large irrigation districts much like western Nebraska. There, Wyoming irrigators used both river flows and groundwater but, like their neighbors downriver in Nebraska, depended on getting slugs of stored water delivered by the big federal dams upstream when they wanted it. Often the groundwater they pumped for sprinkler irrigation was closely tied to the river. Some of the groundwater resources had been fed by their own years of flood irrigation. Over the line in Nebraska, in fact, farmers tapped a major groundwater "mound" under their fields that their heavy use of Platte water had built up over the years.[21]

The Nebraska lawsuit and the ESA review of the Platte dams brought on about twenty years of back and forth to reach a workable deal and a new beginning. The two tracks of negotiation, closely related even as negotiators kept them carefully separated, wound up in 2001–2006 with a whole slew of new, interlocking management strictures and structures on the river.

In 2001, the "special master" hearing the Nebraska lawsuit for the US Supreme Court made it clear he intended to combine the two tracks, proposing to provide water for the habitat as part of his upcoming opinion deciding the interstate water rights case. That was the last thing either Wyoming or Nebraska wanted. The two states avoided it by announcing on the courthouse steps that they had reached a settlement resolving their conflict as they thought best. Both states adopted new limits on uses of surface and ground water, with some flexibility to substitute new uses for old. In Wyoming, it did mean new monitoring and reporting for the state engineer. But the Wyoming and Nebraska water experts succeeded in ensuring that it was they, rather than distant judges, who nailed down all the tricky details and set up new rules for the river. The US Supreme Court approved the settlement.[22]

Eventually, the river's endangered species issues were handled under a Platte River Recovery Implementation Plan for endangered species, not finalized till 2006. Wyoming, Nebraska, and Colorado agreed under that plan to provide lands and/or cash for habitat and for new or revised water projects intended to deliver more water to the habitat area at key times. Existing dams on the river continued to operate and serve traditional users, but with some new twists to provide habitat water. The increased volume of water pledged was nowhere near what the Fish and Wildlife Service thought was necessary for the species and was more than the states wanted to deliver. Both sides, however, agreed to work in a continued cooperative effort, in stages, to see what water was really needed and what strategies helped the most. The deal thus featured an "adaptive management" approach, recognizing that initial decisions had to be made even as research was ongoing and as all sides debated the scientific conclusions. The water deliveries and habitat improvements decided on were intended to be monitored, to provide more information for future decisions.[23]

To oversee it all, the states created new joint management regimes for the river. There is a Decree Committee for Wyoming and Nebraska, using consensus to resolve water issues before they turn into new grudges and litigation (negotiators noted that their lawyers did not always like this idea). For the endangered species issues, Wyoming, Nebraska, and Colorado joined the federal government in a Governance Committee that includes water user and environmental group representatives. Fassett and Purcell both continued to work on getting the interlocking plans into place after they left state office; Fassett years later took a job as the head of Nebraska's Department of Natural Resources, where he helped administer those plans from the other end of the river.[24]

In the end, people in all three states had to accept the reality that there could be no big new projects built to take water from streams in any part of the Platte River basin for human use. In return, negotiators were able to ensure that many existing water uses in the basin could continue even as the needs of endangered species got attention. The negotiators from all three states ensured that the risks of uses being cut back to help the Nebraska habitat were minimal and would fall where the states, not the Fish and Wildlife Service, chose. Irrigators saw their water uses largely accepted under the ESA in a way they could never have achieved if they'd had to undergo federal scrutiny individually, without state water officials taking on the negotiations.

The choices Fassett and Purcell made for Wyoming took into account the needs of the irrigators at either end of North Platte's course through the state. Pathfinder Dam ultimately played a pivotal role in making possible Wyoming's accommodation of the endangered species issues. Wyoming's water contribution to the habitat came from Pathfinder—but the water development director and the state engineer, working with Lawson at the Bureau of Reclamation, came up with a clever plan for that. They managed to fit this new need, driven as it was by national environmental priorities, relatively painlessly into the operations of the big dam, local irrigation patterns, and the familiar Wyoming water-law system. The proposal was to modify the dam, making it possible in high water years to store more water. Arguably, in some years, the habitat area would get water it would have gotten anyway, but in some years, it was water that the habitat area in Nebraska would never have seen—and federal wildlife officials could control the timing for the water's arrival. To fit into Wyoming water law, the idea was that the modification simply replaced storage space that had been lost to sediment the river had carried and piled up behind Pathfinder Dam over the century since it was built. That new storage space could have the dam's original priority date, 1904, in Pathfinder's case—a senior priority on the river.

The water so stored would be released and shepherded down the Platte when the habitat area most needed it. Ultimately, with the Pathfinder plan, the state engineer, the Board of Control, and the Water Development Commission gave birds a water right with high priority—with Pathfinder's 1904 date—equal to that of the users of the major federal reservoirs and the irrigators they served. The Wyoming legislature and governor approved the plan. After all, water stored under a certain priority date being delivered downstream was familiar procedure. It was much easier to accept than unilateral federal interventions changing the operation and water delivery schedules of the string of big reservoirs on the North Platte. During the litigation, Wyoming dropped its planned Deer Creek Dam, but the dam proposal provided the leverage Wyoming negotiators used to reach their goal. In the final agreement, Wyoming municipalities along the river received their own rights to a slice of the new Pathfinder storage. This amounted to about the same allotment of water they might have received from the canceled Deer Creek Dam, at a cost to the state of Wyoming of about 5 percent of the dam's projected total cost.[25]

Altogether, it was quite a design feat, creating a new overarching structure of interstate and federal cooperation to govern the operations of the river— arguably important for many more Wyoming water users than the Colorado River compact had been, eighty years earlier. It was possible only through the partnership of Fassett and Purcell and their good relations with Lawson, the bureau manager in Wyoming. The years of experience those three had in water management and funding had given them intimate knowledge of federal policy and Wyoming water, water rights, water people, and water politics. They kept coming back to irrigators over the years of work, supplying themselves with local information and concerns throughout the negotiations.

The big Wyoming irrigation districts near the Nebraska line were relieved. Computer modeling based on past water flows showed that the new Pathfinder plan could reduce their water supply somewhat, sometimes. The alternative, however—unilateral federal action changing the operations of the big dams on the Platte—would have affected their water supply significantly. The modeling showed that the Pathfinder plan would most affect the Casper-area federal Kendrick Project (some of whose irrigators tended toward hobby ranching) that had 1930s water rights. The irrigation district there had major problems with selenium levels due to soils inappropriate for irrigation. Mead, in his last years at the Bureau of Reclamation, had approved the Kendrick project in 1935 despite knowledge of the selenium issue. When the twenty-first century plan was proposed for restoring Pathfinder's capacity to take more water under its 1904 priority, the Casper-area irrigation district accepted that it might see less water as a result. The district in return received state aid in reducing selenium contamination and conserving water.[26]

Upstream irrigators near the Colorado line were alarmed by the new Pathfinder plan, however. The modeling showed that the impact to them would be minimal and would occur only when Pathfinder stored water before the irrigation season. But many of them had water rights later than Pathfinder's 1904 right, and on occasion in the past they had had to cut their own water use to help fill Pathfinder. They were wary of seeing any new uses for Pathfinder water, regardless of what the modeling showed, and they didn't completely trust the state engineer or the water development director. They saw an opportunity to block the Pathfinder plan because it required Board of Control action. The board had to approve or disapprove adding new uses—municipal, wildlife, and environmental—to the original 1904 Wyoming water right for Pathfinder. Upper river users officially objected before the board, initially arguing that the silted-in space in Pathfinder had been abandoned and could not be restored. In the end, they dropped that argument but convinced the Bureau of Reclamation and the state to agree to special protective language, which the board adopted. The special clauses provided that the new Pathfinder accounts for migratory birds and North Platte cities could not keep upper river users from tapping the North Platte any time of year.[27]

The birds, or their advocates, similarly wound up not quite happy. As the years went by, the habitat deal did not pan out exactly as promised for them. By 2018, the extra water that the three states managed to supply annually for the habitat amounted to less than two-thirds of what they had promised in 2006. The main problem was failure in Nebraska to build a new reservoir that had been proposed to deliver more water at specific times. This resulted in the Fish and Wildlife Service not having extra water often enough to be able to test its ideas for how to best improve habitat. Rather than insist on a new phase in the recovery program with new goals, however, in 2017–2018, the service agreed with the states to seek an extension of the program and to cut back a little the amount of extra water required. The states promised new approaches to get more water to the Central Platte—including, for Wyoming, a little more water from Pathfinder. That could affect Wyoming irrigators at both ends of the North Platte. The additional releases from Pathfinder for birds could cut down on water available for some eastern Wyoming irrigators. More releases also cut down on water available to the irrigators on the Upper Platte near the Colorado border, who were still uneasy. They harbored major doubts about how and whether the water stored in Pathfinder was really being used in Nebraska for habitat purposes, and they weren't pleased to hear of plans for more releases from Pathfinder.[28]

But an extension of the program—now proposed to last till 2032—would buy all the irrigators more time to keep operating nearly as they always had, for years to come, without federal intervention. And that was what Wyoming irrigators at either end of the North Platte wanted most.

A self-described "flatlander" North Platte irrigator from eastern Wyoming—a member of the Water Development Commission—summed up the dilemma and the benefits and called for irrigator solidarity in 2018. To avoid devastation, irrigators must work together to keep both the joint state-federal habitat program and the Board of Control's protective measure for the Upper Platte in place, he declared. "We need to stand together and say this is how it's going to be: we're going to protect our water rights of Wyoming and to heck with the rest of the world." Framing that as defiance, in *support* of the new North Platte system, was a backhanded compliment to its architects—including the state engineer and the Board of Control.[29]

It was clear that pressure from the outside had forced Wyoming water users to seek protection—and accept restrictions, ultimately enforced by hi-tech monitoring tools—from the state engineer, working in partnership with the water development agency. Some of that was for the sake of a cause—distant wildlife and their habitat—that might not be an irrigator's top goal. All in all, North Platte water users were finding themselves edging into a new water management world.

———

Local interest, rather than outside pressure, meant that aspects of that new world began to be explored at various scales all over the state. Streams reflect what happens in their basins, and people continued to be concerned about the condition of the streams. Over a century, irrigation had inevitably changed the volume and timing of water flows, and accordingly the shape, behavior, and content of creeks and rivers. Helping rivers to function with changed flows, so they could still handle sediment and provide good habitat as well as irrigate fields, attracted increased attention and experimentation from the 1990s into the twenty-first century. Wyoming communities—from Sheridan to Casper to Laramie, and Cheyenne to Evanston—began to see the streams flowing through their towns as quality-of-life assets they could enhance by investment. They put millions of dollars into restoring rivers—hiring expert designers and backhoe operators to create riffles and pools for fish, aesthetics, and better water quality, building streamside parks and walking paths, and trying to get more water to flow through town.[30]

State Engineer Fassett pulled together an ad hoc group in the 1990s to think through how more instream flows might be supported within the narrow confines of the 1986 instream flow laws. The group tried to sketch out ideas for moving and consolidating canal and diversion structures, for instance, to allow water flows to get through a town for the sake of natural beauty, water quality and recreation, before water was diverted to irrigate nearby fields. By the early 2000s, Trout Unlimited had officially opened a Wyoming office of its Western Water Project (the office that later helped the Barneses on Fontenelle) to keep streams flowing for trout. The organization cosponsored studies on Clear Creek through Buffalo and the Popo Agie River through Lander that showed how information on volumes of water diverted from a stream helps indicate aquatic health. The goal was to show what data was already readily available that could aid planning for water conservation and better instream flows. When given a chance, however, the Wyoming legislature frowned on experimentation to help stream flows. The legislature killed years of proposals from Trout Unlimited members to allow instream flows to be one of the water uses (like highway construction) to which private water rights could be temporarily changed. Advocates sought to allow water right holders to at least test out placing their water rights into instream flows, but the proposals went nowhere. The Water Development Commission refused to put money behind any town efforts in support of live rivers as a form of water development. The state engineer and the Board of Control meanwhile continued steady work adjudicating instream flow rights.[31]

In the tradition of Wyoming people and water, water users and administrators found other ways and funding to test how they might get water flowing where people—a town or a fishing group, for instance—wanted to see it flow. Lawson at the Bureau of Reclamation pursued a variety of ways to help Wyoming fish and fishermen. Having changed the timing of releases and storage on the North Platte to help fish spawning, he also worked with the state engineer and the Wyoming Game and Fish Department to help fish on the Shoshone. When state and federal money enlarged the Buffalo Bill Dam there, all three worked out a plan for releasing more water in winter to sustain fish. The plan was adopted in 2004 despite the resistance of the federal Shoshone Project irrigation districts that were used to thinking of the generous flows of the Shoshone River as their own—essentially a perpetual "free river" for the taking. In 2016, the Popo Agie Conservation District began what became a Healthy Rivers Initiative in Lander fostering work between ranchers, conservation groups, and town officials to adjust uses and help the Popo Agie River serve irrigation, drinking water supplies, and recreation with sufficient quantity and quality of water.[32]

Outside of towns, experiments with the value of water flowing in a stream went on as well. Tom Annear, who had become the state Game and Fish Department's water management supervisor, noted that near miracles can happen when there are serious negotiations to "just add water" to a stretch of river. He slowly but steadily implemented the 1986 instream flow law and joined water right holders, including irrigation districts like Wheatland, to keep water in lakes and streams for fish. Paul Hagenstein, who had worked his family's irrigated fields near Pinedale since the 1940s and served as a county commissioner, had initially been fiercely suspicious of introducing instream flow rights into a heavily used creek running through Pinedale. But he later decided to be the first to give an irrigation water right to the state for instream flow, when he could find no more irrigation use for it on his fields or his neighbors'. Above all, he wanted the water right to be of use to his community. Hagenstein's gift was in 2011, twenty-five years after Wyoming's instream flow law was passed. It marked a cultural change to acceptance of the value of instream flow, as Annear at Game and Fish commented: "I went from people calling me all kinds of names . . . to now they say, 'Hi Tom.' . . . From getting death threats to, now, I can't get any attention. It's like with every new idea— first nobody takes it seriously, they laugh at you; then they vilify you and want to kill you; and then, it's common sense." The legislature's refusal to allow users to own an instream flow right themselves, however, continued to hobble landowners interested in helping streams where they lived. In other states water right markets have helped protect or move water for instream flow held by private owners, but that remained impossible in Wyoming.[33]

It became apparent that in water as in energy, conservation is a great untapped source of supply. In the late 1990s the Bureau of Reclamation started offering funding and requiring irrigation districts on federal projects to plan and implement conservation measures like lining dirt canals and automating their control systems. Inter-district rivalries on the bureau's North Platte Project defeated a plan for major incentives there to use less water from the big reservoirs. But the bureau's small Kendrick project district near Casper reduced its selenium contamination problems by using less water. Wyoming water development money went to ditch lining and other improvements, including several for the Midvale district on the Wind River, where using less water may someday help that district learn to accept the tribes' use of their "futures" rights in the Wind River. Towns seeking state water money improved their metering and rate systems and cities like Gillette and Cheyenne plunged into water conservation, including wastewater reuse in Cheyenne.[34]

Conservation can present a conundrum, however, for irrigation on small creeks feeding mainstem rivers. There the conservation goal could be aiding streamflow, recreation, fish, and wildlife. Flood irrigation is common on those creeks, as it is the cheapest way to irrigate. Standard conservation measures could call from changing from flood irrigation to sprinklers or other methods that divert less water from a stream. Yet switching away from flood irrigation can increase how much water plants consume and significantly reduce return flow. Where those return flows support water tables, wetlands, and habitat, conservation may not be the best thing for the environment, nor for the downstream neighbors depending on return flow. The role of return flow in local and regional ecological and economic health generates passionate arguments for valuing return flows and flood irrigation when assessing conservation options. Economists, for their part, urge financial incentives to conserve by diverting less, and incentives have been adopted by some federal agricultural programs. Where the economics are right, moving water from a flood irrigation diversion into a pump for a sprinkler and onto a circular or semi-circular field has become common in some areas in Wyoming. Such moves often involve water right changes that come before the Board of Control. Occasionally, those involve situations that might seem to require the board to view the proposals as a change in use with impacts on return flow that must be considered, though that hasn't happened yet. Wyoming's community water management, formed of state agency and users, has more to do to scrutinize changes in methods of irrigation as a change in type of beneficial use. Perhaps that will help define beneficial water conservation.[35]

Individual ranchers have tailored new ideas for water to local conditions, coming up with changes that would help not only their operations, but fish and wildlife as well. Former state legislator Pat O'Toole, supporter of a new dam in his area, was also a strong advocate for the ecosystem benefits of flood irrigation. He worked to improve water flows for fish in small streams, and he and his wife won a national award for nature conservation work on their ranch. O'Toole's colleague Larry Hicks heading the local conservation district got water development money to help build an elaborate series of projects to create water-cleaning wetlands, and improve stream health and fish passage, as well as rangeland improvements for livestock and wildlife. Comprehensive state studies of watersheds began to help individual ranch projects. A state engineer's office staffer, brought on in the 1990s to help irrigators with federally mandated water conservation planning, ended up in 2002 in the water development office, running "a small water project" program that provided state-funded incentives for water conservation on private lands. To demonstrate public benefit of such a program, it was combined with a new water development program aimed at studying entire watersheds to see if local natural resources could be improved to help stabilize the environment, ranching interests, wildlife habitat and the economy of local communities. The watershed studies could determine the public benefit of small water projects and prioritize them. Though the watershed studies were dropped in priority as water development budgets tightened by 2018-19, the projects that had emerged put considerable focus on improving water quality, keeping livestock out of riparian areas which otherwise could be trampled and degraded, and finding ways to provide water in upland areas to keep livestock better distributed on the range. Those projects typically helped wildlife as well as ranchers. The State Engineer's Office and the Board of Control kept strictly out of water quality issues except where interstate water issues were at stake; the water development office, by contrast, had gotten more involved in water quality problems.[36]

In the southern Big Horn basin, for instance, where sheep men and cattlemen had competed violently for range in the early twentieth century, the range had been beaten and degraded. On Kirby Creek, running west out of the Big Horn Mountains, a combination of drought, floods, and heavy grazing had caused the lower part of the creek to cut drastically down through its banks, in many places by thirty feet or more, causing the water table to drop so that it could not support much riparian vegetation, while grasses were beaten out by greasewood, of little use to browsing livestock or wildlife. In the upper end of the basin, old stock-watering reservoirs in bad shape added to erosion when they broke. The little valley looked nothing like it did in the photos taken by the first settlers. Those settlers had filed water rights with the state engineer from 1897 to 1918, but the ditches they built then sometimes could no longer reach the creek. The basin covered about two hundred square miles, mostly hilly sagebrush-grassland, and some flat valley bottom. Oil drilling, begun in 1914, occupied some creek basin lands. Starting in the late 1990s, neighbors on Kirby Creek ranches joined state and federal staff to see if and how they could change their operations and help turn things around—at least by stopping the continued down-cutting of the creek. Water quality tests showed degraded conditions in the lower stretches of the creek. The Water Development Commission funded two watershed studies to document conditions and suggest management plans, and investment of over $1.5 million in state and federal money came into Kirby Creek for such projects as stream and streambank restoration, small reservoirs, and new pipelines and stock tanks. Landowners thought they began to see results in improving water quality in the lower creek—all through voluntary action. Similar work went on all over the state.[37]

The improvements remained voluntary because there were no regulatory teeth. As a result of the efforts in 1980 of then-Wyoming US senator Malcolm Wallop, the regulatory reach of the federal Clean Water Act specifically excludes the water quality impacts of taking water from a creek for irrigation under state water rights. The act technically could, however, control the quality impacts of water flowing back to a stream from irrigated fields—flows that often carry sediment, and sometimes fertilizer residue as well. But those impacts were, as a practical matter, also left to voluntary efforts because of rural resistance and lack of funding for the regulatory effort that the law could require. In an extreme case, as in the town of Torrington on the North Platte federal irrigation project that extends into Nebraska, fertilizer nitrates that got into town drinking water wells violated federal standards under the Safe Drinking Water Act and pushed the town to install elaborate treatment to remove the nitrates. Management of fertilizer use in the surrounding irrigated fields remained voluntary, and local conservation districts did massive education on the topic to help keep the nitrate problem down.[38]

Conservation districts, post–Dust Bowl creations with local elected boards and funding, worked to keep rural water quality efforts a matter of cooperation rather than regulation. The districts became the prime vehicle for the delivery of state water development money to assess and improve watersheds and build small water projects. The conservation district at Kirby Creek managed to get a key standard for the stream downgraded, therefore easier to meet; Hicks on the Little Snake River, after he became a state senator, pressured the state water quality agency into making sure no Wyoming streams were officially identified as too degraded by irrigation diversions to achieve their best potential use. Repeated sediment releases from an old irrigation diversion dam in need of repair on the Willwood portion of the big federal Shoshone Project, culminating in a release that caused a massive fish kill on the river in 2016, have similarly been dealt with through a cooperative effort aiming to clean up sediment and set operations guidelines for the dam.[39]

In other states, temporary rather than permanent transfers of water rights have been increasingly highlighted as the way to juggle water to serve multiple needs, including fish and wildlife, depending on climate, demand, and economics. Wyoming's temporary water use law from the 1950s has had considerable use in the twenty-first century, but mostly for temporarily shifting groundwater regularly used by agriculture in southeast Wyoming to oil and gas hydraulic fracturing. The temporary change-of-use law could be and was used, though rarely, to move irrigation water to small towns in a drought year. Temporary moves from one irrigated ranch to another were not possible, because the 1950s law allows temporary change only to a different type of use. The legislature's resistance to including instream flow as a suitable new use for temporary changes quashed experimentation on that front. But it seemed possible that frustration over the time and money costs of permanent changes in water use could combine with local interest in live streams, in town and on ranches, to create a kind of pent-up demand that might eventually change how temporary and perhaps even permanent changes in water rights and water use could be handled.[40]

In 2005, the Wyoming legislature adopted and funded an idea that had been gathering support since the 1980s: a Wyoming Wildlife and Natural Resources Trust Fund, to put money into protecting and enhancing habitat for the state's valuable fish and wildlife resources. Fish and wildlife, and their habitat in streams, mountains, and plains, were the assets that at bottom, as many people said, kept people living in Wyoming, foregoing the better pay they could make elsewhere. Those resources also brought in tourist dollars. The director of the new trust fund was a descendant of one of the oldest Green River ranch families, who in his own career had run both the Wyoming Stock Growers Association and a demonstration ranch of The Nature Conservancy. Ranch families, conservation districts, the state wildlife agency, and sportsmen's non-profits brought to the trust projects they had long sought for improved habitat but could never find the funding for. The trust invested nearly $97 million and leveraged in many millions more for eight hundred projects around the state from 2006 through 2017, a quarter of which were for stream restoration or fish passage devices on existing structures in streams. The trust, and private non-profits, funded non-development easements with ranch families to preserve open space as an alternative to zoning. The payments for such easements could often be crucial for the economic survival of ranching operations, particularly when the next generation was ready to take over. Meanwhile, Trout Unlimited, fish-focused, joined forces with the Wildlife Trust Fund, conservation districts, other non-profits, and federal and state agencies to find ranchers like April and Eric Barnes. Their money went to hire local engineers and contractors—some of whom were nearby ranchers with those skills—to build new infrastructure that would help people like the Barneses escape their outdated irrigation system and improve their ranch income, while also making it possible for fish to thrive in the creek that watered their fields.[41]

———————

In the 2000s, Wyoming saw the emergence of a major new actor in both water matters and the state budget: climate change. The phenomenon stalking the twenty-first century was fueled partly from Wyoming, which since the 1970s had responded to the US demand for energy in the form most damaging for the climate long-term—coal. The wealth from taxes on production of coal and other fossil fuels had been enjoyed by everyone in the state, including towns and irrigators tapping the state water development program. It was, accordingly, hard for Wyoming people even to name the reality of climate change. Everyone from oilfield geologists to political leaders, and certainly most members of the state water development commission, initially scoffed at the idea that human activity could affect something as massive as global climate processes. But the US energy market was attuned to the impending reality even if these skeptics were not, and the market's focus was beginning to shift. As natural gas production (much aided by hydraulic fracturing) increased and prices dropped, electric utilities eyeing price and consumer pressure moved away from coal burning to gas and some renewables. Wyoming coal production, which had grown steadily for the previous forty years, peaked in 2008 and dropped by more than 40 percent by 2019, with mines laying off employees, running through bankruptcies and mergers, or closing; even Wyoming coal-fired power plants announced plans to close units. Mineral revenues on which state budgets depended dropped by a similar amount in those years. Governors and legislators began to agonize over the budget and, as in earlier bad times, how to broaden Wyoming's economy and revenue base. Water development funds could not meet demand for rehabilitated or new water facilities. The COVID-19 pandemic in 2020 and the accompanying further drops in oil, gas, and coal production and revenue led to major cuts in the state budget accompanied by stiff voter and legislative resistance to any discussion of new taxes.[42]

In daily life, as the twenty-first century came in, change in old, familiar patterns of temperature and snowfall made itself felt. Scientists at the University of Wyoming, helped by an infusion of multiyear federal grants that started in 2011 to 2013, examined more closely how water moves through Wyoming's landscape. Ginger Paige was one of those scientists, and she expected the answers to be particularly useful in preparing Wyoming people to deal with climate change. Paige originally left her native Connecticut to pursue the challenges of hydrology in the desert Southwest. Then moving to Wyoming in the early 2000s, she joined research faculty and the extension agricultural outreach program that is part of a land grant university's mission. She learned Wyoming politics via the hot debates surrounding the water produced in coalbed methane extraction. The workings of water in the complex geology of Wyoming, now facing the impacts of climate change, have since led her further into terrain made more difficult by politics. Her research embraces water issues from rangeland management to irrigation return flows, including how best to measure water processes and supply scientific data to the process of decision-making on water allocation.[43]

Wyoming's population, small and isolated, faces an information problem, Paige says. Big-picture information like annual total river flows is available, but data for pressing water management needs—to conserve water effectively, or to adapt to climate change—have been lacking. "In Wyoming, we have for years relied on the idea of 'no data is better than data,' because then you didn't know you were in trouble," Paige said. People feared that information on how or whether water is being used could lead to attempts to declare water rights abandoned. In recent years, however, the appreciation and collection of detailed water measurements has increased, with people beginning to understand that the State Engineer's Office needs good water information, not to scrutinize local use of water rights but to manage water in the big interstate water systems like the Platte and the Colorado. Though economic dependence on fossil fuels long kept the words "climate change" out of public discussion, information on climate and water, however, was something Wyoming people had long had an interest in. They knew too well how misunderstanding the realities of climate had shrunk the dreams of agricultural bounty in Wyoming's past.[44]

In the 2000s, Wyoming did not suffer the steady, agonizing nineteen-year drought that dogged the US Southwest. In 2002 and 2003, however, nearly all of Wyoming saw a drought rated "extreme" under official national standards (with peaks to "exceptional" drought in certain times and places in the state). In 2006–2007 and 2012–2013, over half of Wyoming saw that extreme drought again. In the first extreme drought of the early 2000s, Steve Gray, a new PhD botanist put in charge of climate and water data for the state, worked diligently and courageously to find the words that would awaken Wyoming people to what climate change would mean. With snowpack and spring runoff beginning to arrive a month earlier than in the past and a clear scientific consensus on more warming ahead, people would see smaller stream flows in late summer, more reservoir water lost to evaporation, and more of Wyoming's scant precipitation coming as rain instead of snow, Gray explained. Wyoming, so vulnerable to its dependence on water, needed careful new water management thinking to make itself less vulnerable, "regardless of the source of climate change," he said. The water development commission's statewide water plan, published the next year, still studiously avoided the words "climate change." And Gray, and the climate and water data chief who preceded him in Wyoming, moved on to take climate jobs in federal agencies less averse to those two words.[45]

Over the next decade, however, the water development agency sought to recognize less predictable water patterns as it conducted water project planning. In 2018, even a former chairman of the Wyoming Senate committee on agriculture and of the legislative oversight committee for water development investments backed the idea that "climate science" should be included when it came to examining the water needs of a Big Horn basin watershed on his home turf. The development agency found partner states far down the Colorado River willing to commit to steadily funding the winter seeding of clouds with chemicals to cause condensation, in order to get more snow in Wyoming's Green River headwaters, thence to run on down the river. Development staff working on dams and reservoirs used data on the driest years of the past—rather than historical "average" flows—in assessing likely water supply for proposed new reservoirs, and planning staff proposed providing more and better real-time data on water uses and streamflow to support coming decisions.[46]

Paige, meanwhile, has worked with graduate students on issues as local as how much water returns to a stream, and when, from a specific flood-irrigated field. That is the kind of research that can help determine, for people like the Barnes and for the State Engineer's Office and legislators, what methods of minimizing water use can best support both agriculture and a healthy local ecology. Paige and her colleagues are also funded by major federal grants to investigate questions like exactly how snowfall or rainfall in Wyoming can be expected to "partition" itself between surface water flows and groundwater in different locations. The complex geology of "leaky watersheds" in Wyoming can send groundwater off to surprisingly distant basins, she points out, and groundwater can be an important resource for the future. People who use water in Wyoming have considerable knowledge of what they can expect to happen with surface and shallow groundwater in their local watershed, depending on everything from snowpack to their neighbors' irrigation plans, she notes. That knowledge has been gained over the century-plus since agriculture got underway in the state. But what people know and are used to dealing with is what both they and the scientists consider "natural variability" seen for a century. It is not what scientists recognize as the "increased variability" of the past twenty years of climate change—marked by increased length and intensity of drought, variations in precipitation, and warmer temperatures. Climate scientists, who in the early 2000s thought climate change would not have a major impact on the size and severity of droughts, changed their minds about that by 2018—saying there is now a significant threat of droughts lasting ten years or more. Understanding the "partitioning" processes that determine where precipitation goes is important and not well understood in most states, Paige says. Understanding those processes in Wyoming, she says, will be crucial for water users as the climate keeps changing, so that they can manage surface and underground water locally to support their communities. "We don't know if variability is throwing off their local knowledge points. They are used to drought; have they really thought about a fourteen-year drought? If the conditions of 2006 and 2007 lasted fourteen years?"[47]

———

The Wind River mountain range was noticeably affected by climate change. The range is home to the largest concentration of glaciers in the entire Rocky Mountains. Photographs and measurements taken over decades showed that the glaciers in the Wind Rivers lost nearly half their area from 1900 to 2006—and almost half of that decline occurred in thirty-five years at the end of that period, with warming winter and spring temperatures a prime cause. The glaciers have in the past served as reservoirs serving the valleys at the foot of the mountains, releasing water in the late summer season after annual snows have melted. Each glacier's share of streamflow in the creek below varies each year, ranging from nearly 11 percent to half a percent in the last half-century. As the glaciers continue to shrink, scientists warn, glaciers will contribute less to late-season streamflow, at the same time as annual snowpack in the Wind Rivers shrinks and its snowmelt runs off earlier with warmer temperatures. That combination of glacial retreat and declining, early-melt annual snowpack should mean rethinking water use through the longer summer seasons along the rivers below.[48]

Rethinking water use was high priority for the Wind River Water Resource Control Board on the Wind River Indian Reservation. Since the state's lawsuit on Wind River, the tribes' attention to water rights and management has persisted and grown, even as issues of poverty, unemployment, and health care have been top issues on the reservation. In 1991, the Office of the Tribal Water Engineer and the Control Board, both under the joint administration of the Eastern Shoshone and Northern Arapaho, were officially established through adoption of the final tribal water code. The engineer's office has steadily built up its capacity to address water issues, often in conjunction with the Wind River Environmental Quality Commission, whose director ensured that tribal members received relevant training to work on water issues. Even in years when disagreement between the two tribes affected all tribal government and ended funding for water work, the tribal water board retained members from both tribes, and its staff wrote and won outside grants to keep operating.[49]

After devastating local droughts in 2012 and 2013, the Office of the Tribal Water Engineer joined federal scientists in 2015 in a study on adaptation to climate change. The largest reductions in glacier size, and the largest share they contributed to late-season flows, were estimated to be taking place not on the Colorado River side but the Wind River side of the Wind River Range—where the reservation is located. The joint tribal-federal study on the Wind River Reservation investigated what water flows the Wind River basin has seen in the past three hundred years or more. The project combined that with projections of what climate change might bring, so the tribes could assess what might happen in the Wind River basin and how they could best prepare to adapt. Native cottonwood trees, long loved and used by the Eastern Shoshone and Northern Arapaho tribes, are chronicles of water conditions. Slender core samples that can be taken without damaging the trees reveal sequences of growth rings that cover hundreds of years—narrow rings in dry years and wide ones in wet years. Scientists associated with the US Department of Interior's North Central Climate Science Center worked with Tribal Water Engineer Office staff and University of Wyoming extension staff, tribal elders, and tribal high school students to identify good candidate trees appropriate for coring. The university meanwhile helped provide education and resources on Wind River watershed hydrology. Ultimately, the tribal water board, through the drought management plan it builds, hopes to have the data to recognize oncoming drought and manage for it. The board seeks to manage vegetation and minimize drought impact on fish and wildlife, subsistence harvesting, human health, and cultural activity.[50]

Where once the tribes had seemed to have a real chance to think broadly about the Wind River, the court decisions of 1989 and 1992 had in the end only hobbled tribal authority to manage water. The courts had tied tribal water use to agriculture and denied any right to control groundwater despite its interrelation with surface water. Fragmentation of their natural resource base and restrictions on its use continued to beleaguer the people on the reservation. Nonetheless, the joint tribal water board worked to build not only staff capacity in water management but the tribes' presence in water deliberations at the statewide level. In 1989, as the court decree in the Wind River water litigation was pending, the tribes got representation on the state water development commission. State water development grants went to develop safe drinking water supplies on the reservation and rehabilitate the irrigation system; the state funds supplied a match to bring in federal funds for irrigation rehabilitation. The tribal seat on the water development commission was filled from 2006–2014 by Mitch Cottenoir. Cottenoir grew up in Riverton and was his high school student body president; his parents had met at a Bureau of Indian Affairs boarding school in Oregon, where his mother was sent at eight years old. Getting a degree in mining engineering, Cottenoir worked in uranium and in oil and gas, creating his own company. He was pulled onto state commission jobs by two Wyoming governors and became the tribal water engineer in 2009 after serving on the tribal water board for six years. He and the board (which included Dick Baldes after he retired from US Fish and Wildlife) eventually took on a broad portfolio of work.[51]

Inadequate water for agriculture was the water board's focus. Twenty-first century droughts (plus some early-season floods that destroyed irrigation infrastructure) dramatically illustrated the problem. As a lead scientist on the climate study noted, native people in the United States, and in Wyoming, are arguably the people most vulnerable to climate change, and the least empowered to adapt to it. To overcome barriers to climate adaptation, the Wind River board had to work through outside layers of power to get money and simultaneously develop on-reservation capacity to take as much control as possible of land and water.[52]

A major barrier was the dilapidated reservation irrigation system, a relic of decades of neglect by its official manager the federal Bureau of Indian Affairs (BIA). The Wind River irrigation project as originally designed in 1905 was never finished. The design required water storage to make water available throughout the irrigation season. Barely 20 percent of the storage capacity once planned existed in 2018. Reservation fields could be dry starting at the end of June. Since at least 1930, Congress had seen reports highlighting lack of federal funding and consequent lack of water project expertise in BIA. Since the 1960s, the BIA had sought no federal money for operations and maintenance—while never investigating what the system required or the users could afford to pay while keeping their farms and ranches going. BIA decision makers lacked technical expertise to direct any work on the project, but the agency continued to charge user fees. System maintenance was deferred over the decades, so users paid fees for water that might never make it to their fields through the dilapidated system. Without adequate water, crop quality deteriorated or land went unirrigated, and people could not pay irrigation fees. The BIA instituted some reforms, but by 2015, the number of irrigated acres per mile of canal at Wind River was under half the amount considered necessary for any irrigation project to be economically self-sufficient. While the tribes had "historic use" water rights to about 290,000 acre-feet under the Wyoming court decisions, just a little over two-thirds of that amount was being diverted to reservation fields.[53]

Estimates of the costs of catching up with the deferred maintenance on the Wind River project varied from $30 to $90 million. The efforts of the Office of the Tribal Water Engineer initially managed to bring in only about $11 million of that, directing it toward the most pressing problems. Wyoming senator John Barrasso, heading the Indian Affairs Committee, was able to get authorization and eventually funds in 2018 to aid rehabilitation of Wind River and other Indian irrigation projects, so there was hope of larger investment over a series of years ahead. The tribal water board meanwhile worked methodically toward the goal of taking over most of the reservation irrigation system from the BIA, as allowed by federal law. BIA staff on the ground lacked measuring equipment and had to eyeball the amounts of water diverted; discussion showed that an aggressive water user could dominate his part of the system and take whatever water he wanted. Tribal water staff, by contrast, got better training and equipment and worked alongside BIA staff to learn the infrastructure. The tribal board took over most of the system in 2019.[54]

By 2018, Cottenoir's small staff did everything from monitoring water flows in the irrigation system and mediating user disputes to supervising irrigation system rehabilitation contracts and tracking changes in the use of the tribes' 1868 "historic use" water rights. Consistent with the tribal water code, major diversions on the reservation were required to leave sufficient water in reservation streams for fish. Cottenoir's office pulled together federal, state, and non-profit organization money to add fish ladders and fish screens to reservation facilities to support fish movement and populations. In 2018, the new Wind River Inter-Tribal Council, representing both tribes, approved an agricultural management plan that Cottenoir's office had worked on via a series of public meetings on the reservation with university help. Under federal law, having an agricultural management plan in place would allow the tribes to set up a tribal agency that could manage agricultural resources and lands on the reservation instead of the BIA. The plan would also legally set priorities for all federal land management action on reservation agricultural land. Starting in 2014, state water development money had gone into studies to locate sites for reservoirs to serve the reservation irrigation system from the Little Wind River and its tributaries. By 2020, guidance from the tribal water board helped focus those studies to serve key reservation lands, but state water development funds fell far below what building such storage would cost. Meanwhile, Cottenoir planned to share with the state the results of the tribal-federal climate change study, covering the entire Wind River basin. [55]

The Big Wind River, the river's main stem, had been designated in the water rights lawsuit as the source to provide the water for the tribes' futures water rights. In the thirty years since the tribes won confirmation of their water rights, they had not been able to put the futures water to use. Nothing new had happened on the Big Wind in those thirty years. Storage upstream (dreamed of by state administrations for decades before the lawsuit) could change that. The state water development commission began new storage studies in 2014 running through 2020 for the Big Wind River, with the idea of building a reservoir off the main channel to limit environmental and cultural impacts but put some futures water to work for the tribes and supply water to non-tribal districts.[56]

Such a reservoir was expected, by both the state water development commission and the tribal water board, to make the tribes new players on the river because the reservoir would be managed by the tribes. A reservoir might, for instance, provide water to a major proposed new tribal irrigation project, known as Riverton East, on fifteen thousand acres. Water stored for the tribes' Riverton East project would travel down the river past the diversions on the Wind River used by Midvale and the two smaller districts. That meant water going to Riverton East could put sustained water flows where they had not been for years—exactly, in fact, in the stretch of river the tribes had tried to protect for an instream flow back in 1990.[57]

Meanwhile, state officials wanted to cushion the impact of the tribes' futures water use on non-tribal irrigators. That concern had steadily motivated investment of state water development funds in conservation projects for Midvale and the two private small irrigation districts near Riverton. A new source of stored water could help more. It would particularly help the two small districts, which had no storage of their own. The Bureau of Reclamation had, because of the tribes' futures award, made its exchange contract with the two districts only temporary, up for renewal every year.[58]

Climate change crosses human-made jurisdictional boundaries, and handling it requires people to reach across those boundaries. Adapting to climate change on a system as complex as an entire river basin requires genuine, continued flexibility—in infrastructure, in work across institutional boundaries, and in people. It requires cooperation, and cooperation requires trust. In the Wind River basin, with its fraught history of mutual distrust and disrespect among very different peoples, there is little trust. A big water project takes years and countless conversations and negotiations to put together. Perhaps those years, and changing attitudes toward water statewide, might someday foster enough trust among people in the basin to build a new reservoir. Perhaps nothing will come of it. Or instead, people in the basin could make way, without a big new reservoir, for the tribes to put their water rights to use the way they want.[59]

———

The Eastern Shoshone and Northern Arapaho tribes are headquartered on the Wind River, which feeds into the Missouri and Mississippi River system. But the south side of the Wind River Range (where the two tribes have no water rights) provides water to the Colorado River, and the Colorado poses its own challenge for Wyoming in an era of climate change. Nearly 20 percent of the state—about seventeen thousand square miles—is within the watershed of the Colorado River. Wyoming's Green River meets up with the rest of the Colorado River in Utah; Wyoming's Little Snake River feeds a tributary that joins the Green River in Utah before it meets the Colorado River. From Utah, the main stem of the river makes its way down through Nevada, Arizona, and California, whose big economies depend on the river. The growth along the lower river that had prompted concern upriver in the 1920s and made Wyoming and its neighbor headwater states join in the Colorado River Compact only picked up speed in the century that followed the compact. The river with the power to carve the Grand Canyon became so heavily used that it ended in only a dry riverbed in Mexico, where once it had built a big fertile delta on its way to the Pacific.[60]

The Upper Colorado River Basin Compact of 1948 allowed a certain portion of Colorado River basin water to be used in Wyoming, but people in Wyoming have never used all of that. Some had envisioned a major industrial development in southwest Wyoming, based on Green River basin coal, natural gas, and trona (a mineral that is a natural source of soda ash, used in manufacturing glass, chemicals, paper, and other materials). The mines and power plants that did flourish supported the towns of Rock Springs and Green River, but the idea of a major national-scale industrial center there seems to be dimming in the twenty-first century. Fontenelle Reservoir, built in the 1950s for irrigation that never happened, has never been used for industry.[61]

Starting around 2000, some nineteen years of drought dogged the mainstem Colorado River. It was estimated to be one of the worst drought cycles there in 1,200 years. High temperatures from climate change worsened the effect of low precipitation. In 2011, the Bureau of Reclamation, managing the major facilities storing and distributing water for the large economies in the Southwest, issued a dire forecast. The past century had made it clear that the 1922 compact had been based on overly optimistic numbers on how much water the river could supply. With climate change, the forecasts suggested droughts lasting five years or more could be common in the twenty-first century. Looking ahead, the bureau said that unless water use changed, continued population increase would create water demand that would increasingly exceed the limited and likely dwindling water supply. From California to Wyoming, water managers faced a major crisis, and they pushed for planning to meet the threat of long drought.[62]

Few stakeholders, however, think that the answer to the worsening supply-demand imbalance should be new big dams and pipes or some new centralized management for the whole river system. Rather, it has become apparent that the crisis would be met, and might be resolved, by myriad management changes—compromises, experiments, exhausting negotiations and transactions between many water users, other groups interested in water, and water managers—up and down the river. The Colorado River compacts seemed capable of proving themselves remarkably effective and flexible. In the intensifying drought, California came under more pressure from the other states and began serious work to end decades of taking more water than the 1922 compact allocated to it. In 2007, all seven compact states agreed with the federal government on guidelines for dealing with shortages and for management of levels in the key reservoirs that helped balance the needs and obligations of the upper and lower basin states.[63]

Despite the drought, Mexico's water needs (quantified by a 1940s treaty) continued to be addressed, with an agreement on sharing both shortages and surpluses. Native American water rights to Colorado River water, not addressed by the compacts, got more attention, and tribes from Colorado and Utah on south that held river rights expected increasing presence in future negotiations. Environmental concerns continued to be addressed. Efforts to improve water quality in the river (including projects in Wyoming) had started in the 1970s and continued with the support of a key staffer in the Wyoming State Engineer's Office. So had the work, assigned in Wyoming to the same staffer in the engineer's office, to sustain endangered non-game fish on the river with money raised by imposing fees on new water development. The fish had suffered from years of dam building on the river that seriously altered the flow patterns, temperatures, and heavy sediment load of the original wild Colorado River, to which native fish were adapted. The native fish had also suffered from other human activities, from uranium tailings dumps to poisoning for the benefit of introduced game fish. In the face of the 2000s drought, the Bureau of Reclamation continued a planned series of experimental water releases designed to mimic pre-dam conditions and aid habitats in the Grand Canyon. In 2012, in continued drought, negotiators for the United States and Mexico on the Colorado undertook increased joint management efforts on the river, including five years of experimental pulses of water to occasionally renew river flows where they'd been extinguished in the Colorado River delta in Mexico.[64]

So in response to the severe predictions of demand far exceeding supply in a new world of climate change joined with population growth, a whole raft of water use changes, negotiated within compact bounds at a local or regional level, began to emerge and seemed to show the way ahead. Yet in 2017, Brad Udall, a major leader in climate science and water forecasts for the West, warned that despite the major efforts that had been made, people and agencies along the Colorado River were still not doing enough to deal with the risks posed by climate change.[65] In 2018, after further work on the contributing causes of dropping streamflows on the Colorado, Udall and other scholars emphasized that impacts on the river will get worse, and people should give up the idea that what they see now is a "new normal" to which they had already adapted. The researchers warned:

> The best available term is *aridification*, which describes a period of transition to an increasingly water scarce environment—an evolving new baseline around which future extreme events (droughts and floods) will occur. *Aridification*, not drought, is the contingency that should guide the refinement of Colorado River management practices. A very modest starting point is to admit words such as *drought* and *normal* no longer serve us well, as we are no longer in a waiting game; we are now in a period that demands continued, decisive action on many fronts.[66]

In Wyoming in 2005, State Engineer Pat Tyrrell launched a major effort to amass the best data possible on Wyoming's use of water on the Green and Little Snake Rivers, to be armed to face the worst—several years of very low water levels in the big reservoirs on the Colorado River, particularly the reservoir that stores river water coming from the Upper Colorado River basin states of Wyoming, Colorado, Utah, and New Mexico. That reservoir, Lake Powell, was created in the 1960s by completion of a controversial dam that flooded beautiful and remote Glen Canyon. Lake Powell, just above the dividing point created by the Colorado River Compact between upper and lower basin states, was built to provide a "savings account" of water as insurance for the upper basin states that the lower states could receive the volume of water the 1922 compact had allocated to them without triggering a requirement for less water consumption in the upper basin. From Powell, water is released to Lake Mead, further down the river. That reservoir was created by the Hoover Dam, whose construction Elwood Mead had supervised. The lower basin states—California, Arizona, and Nevada—contract for water deliveries out of Lake Mead. In 2005, Lake Powell held only one-third the water it was built to hold. Hydropower and hydropower revenues from Powell, which helped all the upper basin states, were also at risk with low lake levels. Moreover, continued low reservoir levels could mean, under the compacts, that Arizona, Nevada, and California could invoke terms of the compact requiring upstream states like Wyoming to cut back or "curtail" their water consumption to ensure they are not responsible for depleting water flows below the levels that the 1922 compact allocated to lower basin states.[67]

Accordingly, the State Engineer's Office began to invest in substantial measuring devices and data collection on Wyoming water uses on the Green and Little Snake Rivers. Wyoming was potentially in a better position than neighboring upper basin states in case of curtailment; under the rules of the 1948 Upper Colorado River Basin Compact, in a curtailment situation, each state had to cut its water use only in proportion to its past share of upper basin states' water use. Wyoming's uses were already low compared to use in the other states. Nonetheless, if no other preparation were made, the state engineer expected that he might have to respond to curtailment by shutting down water diversions in backward priority. That would mean shutting diversions, starting with the most recent water rights and going back in order possibly to 1922, until Wyoming reduced its water consumption by whatever amount might be required. Many Wyoming irrigation rights on the Green and Little Snake predate the 1922 compact. Those rights were unaffected by the compact and would not be shut down. But those ranches also had 1945-legislated rights to "surplus" water in spring, which helped support the use of their pre-1922 rights, helping to carry water to the fields, often through long unlined ditches. That water, and any additional spring runoff water of a "free river" that they might usually tap, would probably have to be left in streams under curtailment. Meanwhile, towns, industries, and some irrigators had later water rights, likely to be shut down if the state engineer had to respond to a call with priority regulation. The towns and industries, in turn, might then try to buy and transfer pre-1922 irrigation rights, at least temporarily—and such a water market was not welcomed by everyone. State Engineer Tyrrell was determined to be proactive and come up with a plan to handle curtailment but also possibly avoid priority regulation in Wyoming in a severely dry year on the Colorado River.[68]

Projections in 2018 showed that in the lower basin, where the drought had continued the longest, Lake Mead could reach unworkably low levels as early as 2020. The year 2018 was already the third driest year on record. The lower basin states were spurred into further action. Late into 2018, they worked on a Drought Contingency Plan that included requiring themselves to use less water so as to keep more in Lake Mead and adding flexibility that would encourage even further water conservation. It meant that California, for the first time, would be taking actual cuts in compact allocation rather than just slimming down its uses to match the original compact allocation. The plan could be seen as changing an element of the 1922 compact, which had become federal law by act of Congress, so they had to take the new proposal to Congress—as soon as possible. Wyoming and the other upper basin states saw a chance to hop on a now quickly moving train to get what they needed to meet the likely future. The negotiations among all seven states were complex, requiring consideration of the geography, hydrology, and economy of each very different state, much as in the original compact talks in 1922. The complicated conversations were part of a long road headed toward 2026, when the river states and the Department of the Interior expect to come up with new guidelines regarding shortages and reservoir operations.[69]

Wyoming, with a state engineer and lead attorney who had been in office longer than most counterparts in the upper basin, played a major part in what the negotiations achieved for the upper basin states in 2018. First, the upper basin negotiators put provisions into new formal agreements with the lower basin and the Bureau of Reclamation to ensure they would have a say in how federal reservoirs in the upper basin could be used to boost flows to Lake Mead. Second, they put into those agreements a provision that if the upper basin states decided in future to come up with ways to consume less water without being under curtailment, so more water could regularly get to Lake Powell, they could "stamp their names on" the saved water that arrived there. That meant upper basin states would have an account in Lake Powell, which they have not had before, that they could release to avoid curtailment. The states' Upper Colorado River Compact Commission could release the water to meet any compact obligations they might have to the lower states even under those states' new drought-restricted allocations. With such an account, Wyoming need not be forced suddenly into priority regulation: "A program like this lets us control our destiny," Tyrrell told a public meeting in fall 2018, "and we need that."[70]

From 2014 to 2018, what came up the river to Wyoming was not a curtailment demand but an experiment in voluntary cooperation. The idea was to see if water user cooperation rather than water right regulation might help upper basin states conserve water, possibly to store in Lake Powell and release downriver should the need arise. In 2014, the Bureau of Reclamation and several water supply authorities in downstream states, including Colorado and California, put together money for pilot projects testing temporary, voluntary ways to reduce consumption of water in the Colorado River. The first step was just to see if people would agree to be paid for using less water, with no attempt to measure the water or its progress downstream. Tyrrell agreed to make that pilot program available to Wyoming users, though he may have doubted its success among water users traditionally suspicious of the schemes of thirsty downstream states. Trout Unlimited, having worked with a variety of ranchers in Wyoming's Colorado River basin in replacing or moving diversion structures, let those people know that there could be money from this new fund to pay them to stop irrigating in July and not irrigate again till fall.[71]

April and Eric Barnes decided to do just that. It did not change their irrigation pattern much, but it changed their lives. They had already turned off water every year in July. Now they just postponed getting water back onto the soil until October, instead of in late August or September. The big change was in the economics of their ranch. The ranch had never paid out much, and Eric had been under pressure from his brothers and sisters to sell it, which he did not want to do. From 2014 through 2018, money from the pilot project came in, and it made all the difference. The dollars provided a rainy-day fund for things they needed on the ranch and enough money to fix up a house where family members who had left the ranch could come vacation in summer. Pressure to sell the place stopped. By 2016, Wyoming matched Colorado for top participation among all the upper basin states in the pilot conservation program, with eight Green River basin pilot projects including the Barneses'. Altogether, those Wyoming irrigators had foregone use of an estimated five thousand acre-feet that year in return for total payments of a little over $1 million. There was criticism in the basin over drying up Wyoming fields in return for money from downstream states, but participation continued. In 2017, the Barneses and their neighbors—every irrigated ranch on Fontenelle Creek—took up the pilot program offer, testing how ranchers might work together to limit irrigation in an entire small watershed. Eric said he would not welcome permanent limits on irrigation on the Barneses' place—he was concerned about what continual lack of watering in late summer and early fall would do to the soil, the groundwater, the vegetation, and the wildlife. But periodic changes in irrigation patterns, he thought, could work.[72]

For the Green River and Little Snake basins, an important question was where conservation should best occur so that the changing river climate wouldn't undercut prospects for a healthy future for people and wildlife in those Wyoming valleys, where flood irrigation predominated and its return flows were important to riparian areas. Those riparian areas served both ranches and the iconic migrating wildlife herds for which western Wyoming was famous. Perhaps different places in these basins would need different kinds of attention—some forms of water conservation might be effective in some places, and support of traditional flood-irrigated fields and related wetlands would be a good idea in others. The university and non-profit organizations undertook research to explore that.[73]

Experts on the river estimated that the amounts of water that ranchers like the Barneses might forego using in a single year could not help Wyoming forestall curtailment. To accomplish that, water would have to be conserved and flows stored for years in the proposed new accounts for upper basin states in Lake Powell. In 2019, the State Engineer's Office undertook an outreach effort to get every water user in the Green and Little Snake river basins to start thinking about whether it could be feasible to start a Wyoming conservation effort which, if coordinated with the other Upper Basin states into a "Demand Management Program," might save water to be stored in Lake Powell and released by the upper basin states if necessary to forestall curtailment. The Upper Basin states had agreed that any such program would only involve water conservation that was voluntary, temporary, and compensated. How such a program could work and be paid for were among the unanswered questions.[74]

In Wyoming, as in other states, meetings with users to explore the implications of such an idea began. One result was generation of a mass of new data, centering not on diversions as in the past but rather on consumption of water. The 1922 compact, binding Upper Basin states not to deplete river flows, requires those states to focus on details of water consumption, whether facing curtailment or seeking conservation through demand management. The 1920s language thus pushed Wyoming and other Upper Basin states to produce specific local data on consumption, data that economists had argued could support both conservation and water markets. In Wyoming, only individuals seeking to change the use of their water rights had had to come up with such data in the past. Agreements on the North Platte had called for only cumulative basin-wide consumption figures there. Now on the Green and Little Snake, information on how much water is consumed basin-wide could be needed down to the level of individual fields and water rights, and work on generating that data got underway. The 2014–2018 pilot program that had curbed late-summer water use was not expected to be the model for demand management conservation. It had been controversial in the Upper Green River valley, where prominent people voiced "worry about drying up agriculture to wet the whistle of Denver, Salt Lake City, Phoenix and Las Vegas," as one local rancher-legislator put it. There were calls for new high-elevation reservoirs, rather than cuts in water consumption. The Barneses and other Wyoming Green River ranch families who had participated in the pilot did, however, demonstrate that water users on the Green and the Little Snake could consider change. They could be willing to discuss whether to try to stave off curtailment, and possibly contemplate a paid conservation program to consume less water and store it.[75]

––––––––

Two decades into the twenty-first century, Wyoming water users and water supervisors have together taken steps toward something new in their community water management system. There is statewide work on developing reliable water data and local effort to promote good stream flows and water quality. Water users retain their rights to access, use, and manage water locally; some new interests, represented by the state-held instream flow rights, have joined the water user group, and both townspeople and ranchers are interested in water for creek flows or fisheries, as well as for consumption. The major interstate negotiations concluded on the North Platte and the ongoing negotiations on the Colorado to handle climate change have given the State Engineer's Office new experience in working on water across geographic and cultural boundaries. Perhaps that experience can even be applied on the Wind River. Meanwhile, the interstate negotiations have underlined the state's management power over water, particularly when it comes to dealing with the demands of other states or the federal government. The state had long retained its familiar rights to include or exclude people or organizations from water rights and to decide whether and how much those rights can be transferred elsewhere. But because of challenges from outside Wyoming, the State Engineer's Office has had to take a lead role working with users and exercising its right to manage water for public benefit. Under the pressure of changing demands, values, and climate, people can see the importance of state leadership in water, both to protect local water use and to adjust uses to meet new realities.[76]

After the hard days of subsistence agriculture, after the heady days of the energy boom, Wyoming people are now in an age of both inescapable climate change and the decline of a minerals-based economy. Where the economies of Wyoming and of the Wind River reservation go in such a time remains a puzzle, and many are still in denial over the hole in state revenues (as public reports and political discussion attest).[77]

But whatever economic life people in Wyoming work out for themselves in this century, water can help secure the fabric of that life. If the state-and-water-user community management system can continue to welcome to its circle more interests—in river flows, recreation, and water quality—then the capacity for adaptation that the community system has developed over more than a century can come back into play and help people transition, using water infrastructure that minerals funded. The independent reservation community can make its own water use choices. Meeting today's needs, Wyoming's two communities can manage water to support a prosperous rural life in Wyoming, with ranch country, small cities, towns and business, and a thriving landscape.

CONCLUSION
Takeaways

Middle Fork Popo Agie River, Sinks Canyon. J. E. Stimson Collection, State Archives.

WHERE WATER IS SCARCE, water law can inspire something like religious awe. The Prophet Elwood has long been revered in Wyoming.

But water law is not carved into stone tablets. Rather, the fascination and the value of water law lies in how it is shaped by and reflects the people and the places where it develops and the changes both go through. Water connects people who might prefer not to know about each other. Water law reflects what people do with water and land, and it has long responded, and must keep responding, to changing economies and societies.

The story of Wyoming water offers three lessons.

———

First, water law is and must be a living thing, serving people in whatever conditions they face.

For people everywhere, that means water law can and must change over time. For Wyoming people, it means understanding that Wyoming water law has changed over time and must continue to do so. Wyoming water law for many decades responded remarkably well to local needs, changing to accommodate a tough landscape and small-time urban and industrial growth. All the while it remained true to the idea that water should be managed for public benefit, an idea expressed in the very structure of Wyoming water management. Nonetheless, over time, water users began to see their water rights as private property. Harsh physical conditions made some state rules unenforceable, and state high court decisions made it almost taboo for a water right to be lost. For a time, the flexibility to accommodate change seemed to disappear. Yet because water in Wyoming is managed by the community of both users and state water supervisors, experimentation with change could and has continued at various levels. Local people have taken steps to meet water needs they see, and interstate challenges have made clearer the role of the state engineer's office in guarding public benefit from water. Uses, needs, and attitudes to water are shifting, and so is Wyoming's web of people and rules that govern water use.

Wyoming needs to keep moving on and address unfinished business. The state has to defer to the Shoshone and Arapaho tribes as they use all their water rights in order to create their own best future. The big non-tribal irrigation districts on the Wind River near Riverton must work within water use limits that support fish flows in the river and tribal projects including irrigation and recreation. The state should agree that the tribes can market the water covered by their futures water rights if they so choose.

Wyoming also needs to face climate change with attention to both people and ecosystems. That includes incorporating additional science into water investment planning, policy, and public information. The science should include the best regional projections of the possible future climate and local research into the hydrology past and present. In addition, there need to be more ways for people to support river flows where they live as the climate around them changes. Private groups and individuals should be allowed to own instream flow water rights, and temporary change of use to instream flow should be allowed. Engineer and Board of Control supervision will make that work. Meanwhile, when reviewing all kinds of user proposals for changes in water use, the Board of Control should find its way to considering the impact of new proposals on streams, for flows and for water quality, as well as on water users.

All that can happen within the best traditions of Wyoming water law and with respect for the intense effort Wyoming people have made to use and manage water.

———

The second lesson from Wyoming's story is that water law in any location should be guided by one principle: what is at stake, as Mead said, is the *public* waters.

What we deal with are "public waters" because everywhere in the world, water is a resource easy to waste but difficult to keep people from using. Further, water creates interdependence among people and between people and ecosystems. And the welfare of entire societies depends in part on wise use of water.

An understanding of that should inform both the design of effective water law and the analysis of water-law systems. There are many ways of governing natural resources. Private control through private property is only one option; government control is only one option. There are plenty of other possibilities, including a system that distributes some rights to individuals and some rights to government, as Wyoming's system has done. In looking at other systems and planning for the future, we should look for the part that the understanding of public waters has played in creating those systems. In those that function well, it will have had a role, as it did in Wyoming.

———

The third and final lesson from Wyoming water is that anyone who wants to spur changes in water law and management should learn about local water history and aspirations. Water management institutions are a network of rules woven from the experience of a people in a specific place. New challenges create opportunity for reshaping institutions, sometimes moving on from old customs. For people to adopt change, however, they must see the need for it in their place, a link to their traditions, and a path forward.

The combination of climate change and population growth is engaging in water policy new people with new goals. They need to grasp the complicated story of what has gone before and why. Then they can meet the future by combining the best in local knowledge with the latest in science and policy.

NOTES

Introduction

1. *Mead Scrapbooks*, 1913, Elwood Mead Papers, American Heritage Center, University of Wyoming. Elwood Mead was the engineer who drafted Wyoming's water-law system in 1890. His 1913 statement of his view of the only policy suitable for water resources was his consistent view before, during, and after his time in Wyoming.

2. Elinor Ostrom, Nobel Prize winner in economic sciences in 2009, describes natural resources like water (or forests, grazing lands, and ocean fisheries), from which it is difficult to exclude people but which can be easily diminished as a "common pool resource," and for which different societies over thousands of years have created different forms of governance. Ostrom, *Governing the Commons*, 30–33.

3. Eminent writers on modern western water law and its roots include Charles Wilkinson at the University of Colorado, Dan Tarlock at the Chicago-Kent College of Law, and Robert Glennon at the University of Arizona. On basic western water law and its origins, see Wilkinson, *Crossing the Next Meridian*, 21–22, 232–40; and Tarlock, "The Future of Prior Appropriation," 771–76. Instead of states granting water rights for free, Glennon recommends putting a price on water and encouraging water markets, arguing that the water crisis in the United States, born of high use and waste, is resolvable if we "begin to treat water as a valuable, exhaustible public resource"; see Glennon, *Unquenchable: America's Water Crisis*, 23–76, 316–20.

4. Tarlock and Robison, *Law of Water Rights and Resources*, §5:14; Tarlock, "How Well Can Water Law Adapt," 7. Yale property law scholar Henry Smith describes many elements of western water law that smack more of public rights-based governance than private rights; see Smith, "Governing Water," 449, 466–78. The Wyoming Supreme Court in 1992 put it this way: "Water is simply too precious to the well-being of society to permit water right holders unfettered control over its use"; in *General Adjudication of All Rights to Use Water in the Big Horn River System* (Big Horn III), 835 P. 2d 273 (Wyo. 1992) at 280.

5. Tarlock, "The Future of Prior Appropriation," 771–85, notes that original western water law "initially developed as a fair and efficient risk distribution scheme for a regime of many small-scale irrigators in arid and semi-arid areas" (776). He argues that because of agricultural, industrial, and urban growth, supported by federal and state investment

in massive infrastructure in major states like California and Colorado, multi-stakeholder agreements and some forms of water marketing—not traditional water law—tend to govern water in those places in the twenty-first century.

6.   Federal water projects serve less than 15 percent of the lands irrigated in Wyoming, watering about 270,000 acres of the nearly two million acres that were irrigated in Wyoming as of the last full count in 2007. See US Bureau of Reclamation, "Project Information: Shoshone Projects" and "Riverton Unit" (websites), https://www.usbr.gov/projects/index.php?id=422 and https://www.usbr.gov/projects/index.php?id=386; and WWDC, "The Wyoming Framework Plan," 6. The federal North Platte Project serves both Wyoming and Nebraska, and a breakdown of the acreage served in Wyoming is available in WWDC, *Platte River Basin Plan Final Report*, chapter 2, pages 4–5. Hydropower generation at federal water projects helps pay for irrigation operations, but less than 15 percent of that hydropower is used in Wyoming, where hydropower is less than 2 percent of the total energy consumed; WAPA Statistical App., 5, 15; US Energy Information Administration, "Wyoming State Profile and Energy Estimate, Overview" (website), https://www.eia.gov/state/analysis.php?sid=WY. By even the most generous estimation of "urban" population (counting people living in towns of 2,500 or more), Wyoming is 65 percent urban (and has only two cities over fifty thousand people), contrasting to the national average of 80 percent and the 95 or 85 percent urban populations of California or Colorado, respectively. Iowa State University's Community Indicators Program (website), https://www.icip.iastate.edu/tables/population/urban-pct-states, in its 2010 urban population estimates includes as "urban" all people who live in "urban clusters," defined as between 2,500 and fifty thousand people—accounting for Wyoming's "65 percent urban" status.

7.   How to govern natural resources to sustain future human societies is a challenge investigated by many current scholars. A strong line of research now argues that today's troubles stem largely from old assumptions that simple, straight-line cause and effect approaches can accurately assess the impacts of human interaction with nature. The mantra of water management in the early twentieth century could have been, "Build a dam and you will have water when and where you want it; no unexpected consequences need apply," but the expectation of simple, predictable, and controllable outcomes is incongruous in a natural landscape. Research suggests that the challenge for water allocation systems is to maintain dynamic response to both landscape and people. See Folke, Berkes, and Colding, "Ecological Practices and Social Mechanisms"; Gatzweiler, Hagedorn, and Siko, "People, Institutions and Agroecosystems"; Gunderson, Holling, and Light, *Barriers and Bridges*, 8–9; Holling, Gunderson, and Peterson, "Sustainability and Panarchies"; and Holling, Gunderson, and Ludwig, "In Quest of a Theory."

The Setting

1.   Knight et al., *Mountains and Plains*, 3–4, 27–38. Longstanding climatic conditions are set out in *Wyoming Climate Atlas*, by Curtis and Grimes. Wyoming precipitation: Curtis and Grimes, chapter 4, 59–80. Precipitation figures for other states, based

on NOAA National Climatic Data Center figures, are summarized in "Average Annual Precipitation by State," https://www.currentresults.com/Weather/U.S./average-annual-state-precipitation.php. Population estimates are from the historical bible of Wyoming, Larson's *History of Wyoming*; the 1890 US Census Bureau summary of population for each state; and, for Wyoming in 2017, based on US Census reports released May 2018, at http://eadiv.state.wy.us/pop/Place-17EST.htm.

2. Cassity, *Wyoming Will Be Your New Home*, 89, describes Wyoming homesteaders "going against the grain" and in subsequent chapters describes the forces that led to current landownership patterns, pictured in figure 1.3 in Knight et al., *Mountains and Plains*. For acreage requirements for grazing, see Larson, *History of Wyoming*, 173.

3. Larson, 406–7, 431–39, 510–12, describes the history of the state's minerals industry since 1910. Minerals (including oil and gas) represented about 30 percent of the state's gross domestic product by 2017; agriculture contributed about 1.5 percent (commercial hunting and fishing, a lively arena in Wyoming, is counted in "agriculture" in the national statistics categories cited here); and tourism 3.7 percent: see Economically Needed Diversity Options for Wyoming (ENDOW), *Socioeconomic Report, Appendix*, 34, 54, 315. But as a recent commentator on the tourism industry has noted (see Western, "Evolving Wyoming Tourism," 8), tourism is coming up—sales tax from tourism came close on the heels of sales tax from the minerals industry—but the much greater "severance taxes" on minerals removed from the ground, and mineral royalties, have been revenue sources the state has relied on.

4. Jacobs and Brosz, *Wyoming's Water Resources*, 1–4; Wyoming Water Development Commission (WWDC), *Wyoming Framework Water Plan Vol. 1* (see especially sections 3.1 and 5.2). Agricultural commodity and value figures for 2019 (topped by cattle and calves, followed by hay): US Department of Agriculture (USDA), National Agricultural Statistics Service, Wyoming Statistics.

5. Jacobs and Brosz, 3–4, 7; WWDC, "Platte River Basin Plan: Executive Summary," 8–9.

Chapter 1

1. Elwood Mead to Grace Raymond Hebard, March 27, 1930, in "Recollections of Irrigation," 10, Elwood Mead Papers. Mead wrote to Hebard when she was collecting documents in 1930 relating to Wyoming statehood and constitution for the University of Wyoming archives. They were old friends; Hebard had briefly been secretary to the state engineer when Mead held that position in the 1890s.

2. "Theresa A. Jenkins Speech," *Cheyenne Daily Sun*, July 24, 1890, 1, Wyoming State Library's Newspapers Database, https://pluto.wyo.gov/awweb/main.jsp?flag=browse&smd=1&awdid=1.

3. "Judge M. C. Brown Speech," *Cheyenne Daily Sun*, July 24, 1890, 5, Wyoming State Library's Newspapers Database, https://pluto.wyo.gov/awweb/main.jsp?flag=browse&smd=2&awdid=22; Larson, *History of Wyoming*, 259.

4. "Theresa A. Jenkins Speech," 1890.

5. Larson, *History of Wyoming*, 248–49; MaryJo Birt, "To 'Hold a More Brilliant Torch:' Suffragist and Orator Theresa Jenkins," Wyoming State Historical Society (website), WyoHistory.org, published August 29, 2019, https://www.wyohistory.org/encyclopedia/hold-more-brilliant-torch-suffragist-and-orator-theresa-jenkins.

6. "Theresa A. Jenkins Speech," 1890.

7. "Governor Warren Speech," *Cheyenne Daily Sun*, July 24, 1890, 4, Wyoming State Library's Newspapers Database, https://pluto.wyo.gov/awweb/main.jsp?flag=browse&smd=2&awdid=21; Larson, *History of Wyoming*, 263–66. For more on Wyoming and women's suffrage as a territory and state, see Tom Rea, "Right Choice, Wrong Reasons: Wyoming Women Win the Right to Vote," Encyclopedia, Wyoming State Historical Society (website), published November 8, 2014, https://www.wyohistory.org/encyclopedia/right-choice-wrong-reasons-wyoming-women-win-right-vote.

8. Larson, *History of Wyoming*, 253–54; Phil Roberts, "Wyoming Becomes a State: The Constitutional Convention and Statehood Debates of 1889 and 1890—and Their Aftermath," Encyclopedia, Wyoming State Historical Society (website), published November 8, 2014, https://www.wyohistory.org/encyclopedia/wyoming-statehood.

9. Mrs. I. S. Bartlett, "Poem: The True Republic," *Cheyenne Daily Sun*, July 24, 1890, 5.

10. Gillette, *Locating the Iron Trail*, 76–77.

11. Gillette, 76–77.

12. Mead, *Second Annual Report*, 1889, 19.

13. Kluger, *Turning on Water with a Shovel*, 3–13; Mead Scrapbooks, in Elwood Mead Papers; "Average Annual Precipitation by State," Current Results Weather Science (website), https://www.currentresults.com/Weather/U.S./average-annual-state-precipitation.php.

14. Wilson, "Farming and Ranching," 27, 95, 162–64.

15. Wilson, 27, 95, 162–64.

16. Woods, *Wyoming's Big Horn Basin*, 13–24.

17. "The Indians," excerpt from Territorial Governor John A. Campbell's inaugural address to the first Territorial Legislature, October 13, 1869, in "To Make the Desert Bloom," Hoopengarner, 75, app. A, 254–55; "Memorial to President Ulysses S. Grant," *General Laws, Resolutions and Memorials*, 1713–16; "Memorial to the Honorable Commissioner," *General Laws, Resolutions and Memorials 1871*, 143–44; Kruse, "The Wind River Reservation 1865–1910," n.p.

18. "Buffalo, WY," clipping files, Johnson County Public Library and American Heritage Center, University of Wyoming. Water rights records on French Creek for John Fisher, May 10, 1883, in Board of Control (BOC), *Tabulation of Adjudicated Surface Water Rights for Div. II, 1999*. Fisher was predecessor of a key named defendant in Farm Investment Co. v. Carpenter, 9 Wyo. 110, 61 P. 258 (1900), see discussion that follows.

19. BOC, *Tabulation for Div. II*, 365–67; Mead, *Water Right Problems*, n.p.

20. Mead, *Water Right Problems*.

21. Younger sons of British aristocratic families had ranches for health, entertainment,

and occasional profit along the southern Big Horns in the 1880s. The first dude ranch in Buffalo opened in about 1911; "Buffalo, WY," clipping files, Johnson County Public Library. Some cowboys became ranch owners by marrying "girls with checkbooks"— girls whose city families brought them to the dude ranches; Phyllis Hall, interview with author, September 1996.

22. Don Hall, interview with author, September 1996.

23. Nettleton to Mead, January 21, 1888, Elwood Mead—Federal Government Officials 1886, 1888–1890, State Engineer Elwood Mead Records 1886–1892.

24. Mead, "Recollections of Irrigation," 3, Mead Papers; Larson, *History of Wyoming*, 150–58; Gould, *Wyoming: A Political History*, 96–100.

25. Mead, "Recollections of Irrigation," 3–4.

26. Warren Papers; Hansen, "The Congressional Career of Sen. Francis E. Warren"; Larson, *History of Wyoming*, 140–41; F. E. Warren to J. M. Carey, December 11, 1886, box 27, folder 7 in Carey Family Papers.

27. See generally Gould, *Wyoming: A Political History*: frontispiece photo, Francis E. Warren.

28. Warren letters, June 9, 1888, June 15, 1888, Warren Papers. In March 10 1882, (Letterbook 4, p. 110), Warren Papers, Warren commented on measures he got passed in the territorial legislature "without a dissenting vote. This took good financing, 'you bet.'" Larson, *History of Wyoming*, 158–62; Mead, "Recollections of Irrigation," 5, Mead Papers.

29. Mead, "The Ownership of Water," 85–87.

30. Woods, *Horace Plunkett in America*, 52–54, 155–58; Larson, *History of Wyoming*, 304.

31. The territorial statutes of 1887, backed by the stockmen, called for the creation of a territorial engineer, who was to recommend a new water law for the territory; Mead, *Irrigation Institutions*, v.

32. Mead (all), *First Biennial Report*, 32; *Second Biennial Report*, 35; *Irrigation Institutions*, viii, 10–12, 369–73; "Government Aid and Direction," 72–98; "Reclamation and Rural Life," May 5, 1925 speech, Elwood Mead Papers of the Water Resources Collections and Archives, UC Riverside.

33. Kluger, *Turning on Water with a Shovel*, 24; Hays, *Conservation and the Gospel*, 51–53, 69, 74–77. Mead and Pinchot were not allies. When Mead was in Washington a decade later, they occasionally skirmished; Hays, *Conservation and the Gospel*, 243–44.

34. Mead in "Irrigation in Australia," *Independent*, n.d., ca. 1913, clipping file in Mead Papers. Mead described his strong belief about the public nature of water as a major reason he had gone to Australia in 1907 to work for the State of Victoria, noting that there, "laws governing the ownership and use of streams carried into effect the principles for which I had been contending in America for twenty years. The profligate surrender to private ownership of water powers worth untold millions of money has never been a feature of Victorian development." Mead expressed the same views in Mead, *Third Biennial Report*, 57–61; and *Irrigation Institutions*, 374–77. He compared the idea of a state granting term-limited water rights to a city granting term-limited franchises to public utilities; *Irrigation Institutions*, 376.

35.   Mead to State Engineer Clarence Johnston, July 1908, in Johnston, *Ninth Biennial Report*, 76.

36.   Mead, *First Biennial Report*, 22–36; Kluger, *Turning on Water with a Shovel*, 25–26; Conkin, "The Vision of Elwood Mead," 88–97.

37.   Mead, *Second Annual Report*, 96. As territorial engineer, Mead made two annual reports, which thus precede the biennial reports made as state engineer.

38.   Dunbar, *Forging New Rights*, 73–85.

39.   Dunbar, 73–85; Tarlock, "The Future of Prior Appropriation," 779; Tarlock, "How Well Can Water Law," 9–11; Horwitz, *The Transformation of American Law*, 32–43; Pisani, *Water, Land and Law*, 8–11.

40.   Clawson and Held, *The Federal Lands*, 15–27; Carstensen, *The Public Lands: Studies*, xxi–xxvi; Pisani, 11–20.

41.   Pomeroy, *Treatise on the Law*, secs. 12–44; Pisani, 7–23, 63; Wilkinson, *Crossing the Next Meridian*, 17–21, 232–35; Wilson, *America's Public Lands*, 23–33.

42.   Mead, "Recollections of Irrigation," 10–11, Mead Papers; Pomeroy, secs. 12–44; Lasky, "From Prior Appropriation to Economic Distribution," *Rocky Mountain Law Review* 2, 35; Wilkinson, 234–35.

43.   Dunbar, *Forging New Rights*, 73–85; Pisani, *Water, Land and Law*, 11–12, 24–37; Pomeroy, secs. 12–44. As Pomeroy describes, courts went through some gyrations to find a private right to water on the public domain, ultimately holding that use of a stream could lead to a *right* to use it if the "true owner" like the federal government did not intervene.

44.   Lasky, "From Prior Appropriation to Economic Distribution," *Rocky Mountain Law Review* 2, 35.

45.   Dunbar, 60–66, 73–85; Pisani, 24–37; Mead, "The Ownership of Water," 85–87.

46.   Mead to Johnston in Johnston, *Ninth Biennial Report*, 76. Mead cited both Hall and Powell in his 1887 speech to Colorado farmers; "The Ownership of Water," 81–84, referring to Powell's 1879 *Lands of the Arid Region* and Hall's 1886 *Irrigation Development*.

47.   Mead, "The Ownership of Water," 81–82.

48.   Hall to Mead, October 4, 1889, Elwood Mead—Federal Government Officials 1886, 1888–1890, State Engineer Elwood Mead Records 1886–1892. Nettleton expressed his doubts, more mildly, in Nettleton to Mead, July 24, 1888 (same source).

49.   Mead, *Irrigation Institutions*, 207.

50.   Standard elements of prior appropriation law are listed in Tarlock and Robison, *Law of Water Rights and Resources*, §5:44, and described in Wilkinson, *Crossing the Next Meridian*, 235.

51.   Dunbar, *Forging New Rights*, 97–105; Mead, *First Biennial Report*, 68–69.

52.   Mead, *Second Biennial Report*, 34–39; Mead, "The Ownership of Water," 84; Powell's endorsement of tying water rights to the land watered appears in Powell, *Report on the Lands of the Arid Regions*, 43.

53.   Mead, "The Ownership of Water," 82–89. The quotes are on 89, 87, 82, respectively.

54.   Dunbar, *Forging New Rights*, 102–6; Pisani, *Water, Land and Law*, 21–22. *Journals*

and Debates of the Constitutional Convention, statements of George Fox (Laramie) and Charles Burritt (Buffalo), 295, 504–5. Judge Melville Brown of Laramie, chairman of the convention, objected strenuously, but to no avail, to priority based on the date of a water right, *Journals*, 503–4.

55. Mead, *Second Biennial Report*, 37–48 and *Irrigation Institutions*, 82.

56. Mead, 39.

57. Mead, 40.

58. Wyoming State Engineer, *26th Biennial Report 1941–42*, 81.

59. Mead, *Second Biennial Report*, 37–39.

60. Mead, *First Biennial Report*, 56–60; Mead, *Second Biennial Report*, 20–22; Hays, *Conservation and the Gospel*, 265–66.

61. Mead, *Second Biennial Report*, 33–48.

62. Wyo. Constitution art. VIII, sec. 3; Sess. Laws 1890–91, 8 §34; Mead (all), *Second Biennial Report*, 39–41, 51–54; *Third Biennial Report*, 37–46; *Irrigation Institutions*, 268–71.

63. Mead, *First Biennial Report*, 56–69 and *Irrigation Institutions*, 252–59.

64. Mead, *First Biennial Report*, 57, and *Irrigation Institutions*, 247, 256–59.

65. Mead, *First Biennial Report*, 15–21.

66. Mead explained in an 1898 report to the state, for instance, that there were times when the public interest would require, in the case of irrigation, issuance of a new permit to cover land already covered by an earlier state water permit, because the plans based on the earlier permit had not worked out, and the water had not been used. Mead, *Third Biennial Report*, 62–63 (see discussion in chapter 3 of this book). He saw his Wyoming system, creating stability and certainty in water rights in the place of the confusion and litigation that had dominated most of the West, as the necessary groundwork for the economic changes and new water demands ahead. Mead, *Irrigation Institutions*, 261, 377. For the lease and rental fee, see Mead, *Third Biennial Report*, 57–66, quotation at 60; Mead, *Irrigation Institutions*, 374–76. Term-limited permits for resource use were much discussed among Progressive Era policymakers, Hays, *Conservation and the Gospel of Efficiency*, 79–81. Wyoming municipalities could acquire land and irrigation water for municipal use under Wyo. Constitution art. XIII, sec. 5; Mead, *Second Biennial Report*, 38.

67. Wilson, *America's Public Lands*, 43; Nicole Lebsack, "Crook County, Wyoming," Encyclopedia, Wyoming State Historical Society (website), WyoHistory.org, published November 8, 2014, http://www.wyohistory.org/encyclopedia/crook-county-wyoming.

68. Elinor Ostrom and her colleagues have demonstrated through extensive research that resources like grazing lands, classified by them as "common pool resources," have been managed sustainably for centuries governed by common property systems. Resources governed by such systems have historically been called "commons." In economic analysis language, resources that are hard to keep people away from have low "excludability," while resources readily diminished by others' use have high "subtractibility"; common pool resources have both attributes. Typical examples of such resources are grazing lands, flowing water, ocean fish, and forests (many of which are the very resources whose dwindling supplies are a challenge in the twenty-first century). Ostrom, *Governing*

the Commons, 1–28; Ostrom, *Understanding Institutional Diversity,* 24; Dietz et al., "The Drama of the Commons," 3–35.

69.   Frank Benton, identified as a thirty-year Wyoming cattle man, in a speech to the American Cattlemen's Association, 1902, in Dew, "Frustrated Fortunes," 150. Ostrom notes that such sentiment has been expressed by a variety of thinkers since Aristotle. Ostrom, *Governing the Commons,* 2.

70.   Some historians and economists mistakenly cite the deterioration of public range-lands in the United States during the late nineteenth century as an example of Garrett Hardin's famous "Tragedy of the Commons." Hardin's analysis suggested that resources like grazing lands, which he labels a "commons," will inevitably be degraded. He concluded that to avoid that result, such resources must be owned and managed either by private parties or a government. Hardin, "The Tragedy of the Commons," 1243–48. Ostrom and her colleagues have pointed out that what Hardin describes was not a "commons" problem but an "open access" problem, where a common pool resource (though perhaps on paper "owned" by a government) was not managed in any way but was simply left open to whoever might happen to use it. Degradation is indeed the usual result of that situation. Research by Ostrom and colleagues shows that common pool resources can be sustainably used for centuries if they are not left to open access but instead are managed under a common property system finely tuned to local conditions. Dietz et al., "The Drama of the Commons," 11–12.

71.   Cassity, *Wyoming Will Be Your New Home,* 8–19; Board of Control, *Tabulation for Div. II;* Larson, *History of Wyoming,* 133; Gould, *Wyoming: A Political History;* Whitehead, *The Compiled Laws of Wyoming,* 377–79.

72.   Cassity, 12–66; Wilson, *America's Public Lands,* 23–33; Larson, 173–78. The Homestead Act, the Desert Land Act, and other legislation of the period did not stop speculation or land monopolization, or even succeed in giving small settlers the best chance at obtaining public lands, as some backers had intended. Gates, "The Homestead Act in an Incongruous Land System," 315–19, 340; Ganoe, "The Desert Land Act in Operation," 142–57.

73.   Larson, *History of Wyoming,* 190–94; Woods, *Sometimes the Banks Froze,* 42–45; Ann Noble and Albert Sommers, tour and interview with author, June 30, 2018.

74.   Larson, 163–94; Cassity, *Wyoming Will Be Your New Home,* 67–96.

75.   J. David to J. M. Carey, April 13, 1888, box 27, folder 17, Carey Family Papers. "Under the present order of conducting the cattle business, hay is of vital importance and too much of it can not be raised," wrote Carey's ranch foreman in Meeteetse in northwest Wyoming's Big Horn basin. This letter and others to Carey also discussed how best to handle land claims filed by various entry-men to ensure that the lands involved would be held by Carey and Brothers Co. See also Kirk to J. M. Carey, January 24, 1887, box 27, folder 11; J. David to J. M. Carey, March 9, 1888, box 27, folder 17; E. David to J. M. Carey, April 16, 1888, box 27, folder 17.

76.   Total irrigation "enterprises" undertaken in Wyoming, figures most likely provided by the Wyoming State Engineer's Office for decades or five-year periods from before

1860 to 1919, are included in the special 1920 census report on irrigation; US Department of Commerce, *Irrigation and Drainage*, 332.

77.   Blake, Van Devanter, and Caldwell, *Revised Statutes of Wyoming*, 366 (Title 19, Irrigation); 1888 Wyo. Sess. Laws, 55. For Mead comments on the territorial water statutes, see Mead, *Second Biennial Report*, 35–37.

78.   Mead, "Recollections of Irrigation," 5, Mead Papers.

79.   Mead, *First Biennial Report*, 61–62.

80.   Mead, "Recollections of Irrigation," 5–9.

81.   Mead (all), *First Biennial Report*, 66–68; *Second Biennial Report*, 30, 33–35; *Third Biennial Report*, 43–45.

82.   Gould, *Wyoming: A Political History*, 83–113; Mead, "Recollections of Irrigation," 11–13, Mead Papers.

83.   Wyo. Constitution art. I, sec. 31.

84.   Wyo. Const. art. VIII, sec. 1.

85.   Wyo. Const. art. VIII, sec. 3.

86.   Wyo. Const. art. VIII, sec. 2 and 5.

87.   Wyo. Const. art. VIII, sec. 2 and 5; Mead, *Second Biennial Report*, 37–44.

88.   Wyoming Constitutional Convention, *Journal and Debates*, 1893.

89.   Mead, "Recollections of Irrigation," 13, Mead Papers.

90.   Mead, 11.

91.   *Associated Press*, "Appeal of Letter Draws Dr. Elwood Mead to Hyattville: Commissioner of Reclamation who was Territorial Engineer to Attend Pioneers' Picnic," June 19, ca. 1930, Mead Biographical File, American Heritage Center, Laramie, WY.

92.   Mead, "Recollections of Irrigation," 10, Mead Papers.

93.   Mead, *Third Biennial Report*, 149–54.

94.   Cassity, *Wyoming Will Be Your New Home*, 53–87; Larson, *History of Wyoming*, 268–84.

95.   Smith, *The War on Powder River*; Dew, "Frustrated Fortunes," 70; Davis, *Wyoming Range War*, 79–180.

96.   *Report of the Public Lands Commission*, S. Doc. No. 154 at xxi, 58th Congress 3rd Session (February 13, 1905).

97.   S. Doc No. 58–154 at xxi.

98.   Dew, "Frustrated Fortunes," 134–281.

99.   Davis, *Wyoming Range War*, 196–214, 247–54.

100.   Larson, *History of Wyoming*, 284–90.

101.   Davis, *Wyoming Range War*, 200–205; Burritt to Warren, Wyoming Historic Materials, 1892–1951, Brock Papers.

102.   Hays, *Conservation and the Gospel*, 243–44; Mead, *First Biennial Report*, 33–36; Mead, *Second Biennial Report*, 30–36; Dew, "Frustrated Fortunes," 41–82; Cassity, *Wyoming Will Be Your New Home*, 146–59, 266–83.

103.   Mead, *First Biennial Report*, 56–62.

104.   Mead, "Recollections of Irrigation," 13–16, Mead Papers.

105.   Smith, *War on Powder River*, 185; Mead, "Water-right Problems," 32.

106.   Mead, "Water-right Problems," 32.

107.   Larson, *History of Wyoming*, 287; *Sheridan Press*, July 30, 1977, article on Henry Coffeen.

108.   In 1915–1916, for instance, Mead chaired the Central Board of Cost Review for the US Reclamation Service; Gillette, in turn, reported to Mead as chair of the Board of Review for the Northern Division of the Reclamation Service. Kluger, *Turning on Water with a Shovel*, 74; Yale University, *Obituary Record of Graduates*, 128.

109.   Mead, *Third Biennial Report*, 149–50; Board of Control (BOC), "Clear Creek Adjudication," 186–87.

110.   BOC, 186–87.

111.   Mead, *Third Biennial Report*, 151.

112.   Mead, 151.

113.   Mead, 150–51.

114.   Mead, 151.

115.   Faulkner, *American Economic History*, 636–38; BOC, "Clear Creek Adjudication," 182–87.

116.   In what follows, quotes and background come (except where otherwise noted) from the decision and the case file (including briefs for plaintiffs and defendants) deposited in the Wyoming State Archives for Farm Investment Co. v. Carpenter, 9 Wyo. 110, 61 P.258 (1900).

117.   *Farm Investment Co.*, brief of plaintiff at 51.

118.   *Id.* brief of plaintiff at 24–25, 72; see also the Wyoming Supreme Court's summary of the plaintiff's argument, in its decision in *Farm Investment Co.* at 258.

119.   *Farm Investment Co.*, brief of plaintiff at 51.

120.   *Id.* at 26–27.

121.   Moyer v. Preston, 6 Wyo. 308 (1896)

122.   *The Compiled Laws of Wyoming*, sec. 1, reads: "All persons who claim, own, or hold a possessory right, or title, to any land or parcel of land, within the boundary of Wyoming Territory, when those claims are on the bank, margin, or neighborhood, of any stream of water, creek, or river, shall be entitled to the use of the water of said stream, creek, or river, for the purposes of irrigation, and making said claim available, to the full extent of the soil, for agricultural purposes." Sec. 2 then says if the land needing irrigation is "too far removed from said stream," then the landholder is entitled to a right of way for a ditch through others' lands. Whitehead, *The Compiled Laws of Wyoming*, 377.

123.   Pomeroy, *Treatise on the Law; Moyer v. Preston*, 6 Wyo. 308 (1896).

124.   *Moyer*, 6 Wyo. at 319.

125.   Pomeroy, §100, concludes his discussion of the prior appropriation system by saying:

The principal defect of the system, the one capable of working the greatest injustice, is inherent in the very theory itself, in its fundamental conception. This defect is the total absence of any limit to the extent of a prior appropriation,—to the amount of water which may be taken,—except the needs of the purposes for which it is made. The prior appropriator, in order to carry out a purpose regarded by the law as beneficial, of great magnitude,—such, for example, as an extensive system of hydraulic mining, or the irrigation of a large tract of farming lands, or, doubtless, the supply of a municipality,—*may* divert and consume, without returning to its natural channel, *the entire water* of a public stream, no matter what may be its size or length, or the natural wants of the country through which it flows. . . . In this manner the *natural* benefits of a stream to the lands situated upon its bank throughout its entire length *may be* completely destroyed, and the natural rights of all persons who should afterwards settle and purchase lands adjoining the stream *may be* totally ignored, disregarded and abrogated by such a prior appropriation.

Italics in original; idiosyncratic punctuation (",—") is also in original. For a modern critique of western water law, see for example Wilkinson, *Crossing the Next Meridian*, 20–27, 230–47.

126. 1901 Wyo. Sess. Laws 67 §2.

127. *Farm Investment Co.*, 9 Wyo. at 124.

128. Wyoming Constitutional Convention, *Journal and Debates*, 510.

129. *Farm Investment Co.*, 9 Wyo. at 124.

130. *Id.* at 127.

131. *Id.* at 137.

132. *Id.* at 136.

133. *Id.* at 140.

134. *Id.* at 142.

135. *Id.* at 142.

136. *Id.* at 142, citing Kinney, *A Treatise on the Law*, sec. 493.

137. Dunbar, *Forging New Rights*, 113–32.

138. Mead (with Warren's help) became the head of irrigation investigations for the Department of Agriculture's Office of Experiment Stations in early 1899. Kluger, *Turning on Water with a Shovel*, 26–27; Bond, *Fifth Biennial Report*, 68.

139. Report of Edward Gillette, superintendent, Division 2, in Mead, *Third Biennial Report*, 151.

Chapter 2

1. Heritage Book Committee, *Pages from Converse County's Past*, 546.

2. Larson, *History of Wyoming*, 263.

3. Larson, 321, 447–49. Carey's career included territorial delegate pushing for statehood, US senator drafting with Mead the 1894 "Carey Act" fostering state investment in irrigation, switching to Democratic Party after political split with Warren, and election

as governor 1910 seeking progressive ballot and initiative reforms. Larson, 319–25, 330–34. Democrats first used the "grand old man" label for Carey in the 1910 campaign.

4. Kennedy, "Memoirs," 281–88, quotation 285–86. Kennedy was prominent in the statewide Republican Party in the 1910 campaign and felt badly for any Republican nominee who "would be confronted with the irrepressible Judge Carey filled to the brim with venom and highly capable of putting up a campaign which would stir the Republican Party to its very soul"; Kennedy, 285.

5. Larson, *History of Wyoming*, 319–25, 330–34, 447–48; Gould, *Wyoming: A Political History*, ix, 48, 77–82. Fenimore Chatterton, a Republican politician, Wyoming Secretary of State, and acting governor, a little younger than Warren and Carey, wrote bitterly, even fifty years later, of Warren's "Machine" in Chatterton, *Yesterday's Wyoming: The Intimate Memoirs*, 94–103.

6. Larson, 447–48, 125–27, 101–6; Gould, *Wyoming: A Political History*, 52, 75–77; Spring, "Carey Story is a Wyoming Saga," 10; "Joseph M. Carey," Biography, Wyoming State Historical Society, WyomingHistory.org, published November 8, 2014, http://www.wyohistory.org/encyclopedia/joseph-carey; Pexton, "Carey-Bixby Ranch," 4–5 (Pexton is coeditor on Heritage Book Committee, of *Pages from Converse County's Past*, cited throughout this chapter); Larson, *History of Wyoming*, 69. Carey's wife was distantly related to President William Howard Taft. Heritage Book Committee, *Pages from Converse County's Past*, 142.

7. Wilson, "Farming and Ranching," 185–202; Hoopengarner, "To Make the Desert Bloom," 81–89. Quotation, at Hoopengarner 218, is from an 1871 congressional committee report on the organization of "Indian territory" and the irrelevance of the wishes of the native people it affected: "We see nothing about Indian nationality or Indian civilization which should make its preservation a matter of so much anxiety to the Congress or the people of the United States. The fundamental ideal upon which our cosmopolitan republic rests is opposed to the encouragement or perpetuation of distinctive national characteristics and sentiments in our midst. We see no reason why the Indian should constitute an exception."

8. Spring, "Carey Story is a Wyoming Saga," 10; Pexton, "Carey-Bixby Ranch," 4–5; David, *Malcolm Campbell, Sheriff*, 60–70; Flannery, *John Hunton's Diary*, vol. 3, 67.

9. Frink, *Cow Country Cavalcade*, 10: "After 1867, the railroad provided a means of shipping eastward directly from Wyoming. Grazing land could be had for the taking. A man could move in, build a cabin and corrals, and call it a ranch, stocking it with cattle that could range far and wide to fatten on free grass. It was the grass that made cow country"; Clay, *My Life on the Range*, 73, 154. Free grass attracted many, including Scots, who lived in a land where grass for livestock was dearly bought. Clay, 155–56, quotes a prospectus issued by Scottish investors, 1882: "The ranch was secured . . . at a period when a wide selection of pastoral ground still existed. Its advantages are . . . a sufficiency of the most nutritious grasses for 50,000 head of cattle; a low rate of mortality, averaging from one per cent in favorable seasons, to five per cent in severe winters. There is no doubt whatever that the high price of Cattle will continue, as the proportion of Cattle to population

in the United States diminishes rapidly. Even without in any way anticipating the increase to which these conditions point, cattle raising in the Western States of America is at the present time one of the most lucrative enterprises in the world." The houses cattle kings built that hosted adventuring visitors included Frewen's Castle; Woods, *Moreton Frewen's Western Adventures*, 55, discussed later in this chapter. For the Cheyenne Club, see Clay, *My Life on the Range*, 72–78.

10.   Spring, "Carey Story is a Wyoming Saga," 10–11; Frink, *Cow Country Cavalcade*, 38–47.

11.   Larson, *History of Wyoming*, 141. The Republican Central Committee first offered the territorial delegate nomination to Warren, who declined—preferring the governorship, to which he was appointed by President Chester Arthur in early 1885. Larson, 139–41.

12.   David, *Malcolm Campbell, Sheriff*, 70–71. Campbell's reminiscences, dictated to David. Of the ranching industry from the 1870s into the early 1880s Campbell said, "Soon the entire country was to be checkerboarded with ranch limits. The entire country was divided into cattle ranges, the boundary lines of which were defined by certain creeks and rivers." The rule of thumb for recognizing a company's "home range" was ten square miles around water that thirsty cattle could drink. The rule was recognized routinely by men working on the range: "Straying cattle were turned back to their home feeding grounds as a matter of courtesy by the first cowpuncher who found them"; in David, 71.

13.   Clay, *My Life on the Range*, 66–67, 232–47. Clay noted that the WSGA started by protecting its members' interests at cattle markets further east but then "branched off into range protection" against cattle thieves, which got it involved in "sordid" Wyoming politics; 66–67. For non-member views of the WSGA, see Davis, *Wyoming Range War*, 47–49; Smith, *The War on Powder River*, 60. See also Jackson, "Wyoming Stock Growers' Association: Political Power," 571–94; "The Wyoming Stock Growers' Association: Its Years of Temporary Decline," 260–70; "The Administration of Thomas Moonlight," 139–62.

14.   Davis, *Wyoming Range War*, 47–50; Smith, *The War on Powder River*, 27–29, 51–63. "The maverick law of 1884 lit the powder train which led to the Johnson County explosion," says Smith. The disposal of unbranded mavericks was an increasingly controversial problem as herds crowded the range in the profitable mid-1880s. The Wyoming law of 1884 put control of those mavericks solely in the hands of the WSGA. The organization could and did, in effect, ensure that the cattle were auctioned off only to ranch-owner members, never to cowboys, and then kept the cash proceeds of maverick auctions to use for whatever purpose it chose, including range detectives to track down alleged cattle thieves. "Class legislation," said the Buffalo correspondent of the *Laramie Boomerang* in February 1884; Smith, 61. Smith blames much of the blacklist rule and the maverick law on WSGA secretary Thomas Sturgis, who was a close associate of Carey's. Sturgis was described by Clay in 1883 as "the leading man in Wyoming, exercising a wonderful influence" and as secretary of the WSGA, one who "showed great skill in steering the ship" through "sordid" Wyoming politics; Clay, *My Life on the Range*, 65–67. The stock industry in neither Montana nor Colorado adopted the blacklist rule, and though they too had

to deal with mavericks, they did not take the Wyoming approach to that issue. Both states created public commissions to dispose of mavericks found on the range to the highest bidder, cowboy or no; in Montana, key issues such as roundup times, maverick sales, and disposition of maverick sale proceeds were left to local committees. For the wage cut and strike in 1886, "the only cowboy strike in the history of the northern range," see Smith, *The War on Powder River*, 31–33. In a small but telling example of cattle company views of settlers, one settler was allowed to keep a milk cow, but the nearby company branded its calf each year to ensure that the man wouldn't get into the cattle business; Davis, *Wyoming Range War*, 37.

    15.   For post-1887 cattle business, see Clay, *My Life on the Range*, 93–96, 136–44, 247. Appropriate to an era of experimentation in the business world, the larger stockmen in the WSGA and other states in 1887 moved into an attempt at vertical integration, with the American Cattle Trust, to control both beef production and marketing and to escape the grip of eastern meat packers and their low prices. The trust lasted only until 1890. Trust organizers included Warren and Sturgis. Gressley, "The American Cattle Trust," 61–72; Davis, *Wyoming Range War*, 49. On small settlers, see Cassity, *Wyoming Will Be Your New Home*, 82–84. On land and water fraud, see *Annual Report of the Commissioner*, 74–75, 100–101; Davis, *Wyoming Range War*, 36; Larson, *History of Wyoming*, 173–82. In April 1886 reports and a June 5 letter, General Land Office (GLO) agent James A. George named stockmen Frank Wolcott and Thomas Sturgis (prominent associates of Carey's) and well-known lawyer Stephen Downey in fraudulent land filings. George described generally (it was outside federal jurisdiction) the practice of filing oversized water claims under territorial water law: "[They] keep settlers off as effectually as though they had a Chinese wall around the land. They do this by incorporating ditch companys [*sic*] and, by appropriating all the water that the stream affords, debar every settler from using any water in the stream, either above or below." Microfilm selection from US GLO records, "Report of Fraudulent Claim or Entry," and "Supplemental Report," University of Wyoming Microfilm, roll 1, n.p. An inspector in Dakota Territory reported, "The idea prevails to an almost universal extent that, because the government in its generosity has provided for the donation of the public domain to its citizens, a strict compliance with the conditions imposed is not essential. Men who would scorn to commit a dishonest act toward an individual, though he were a total stranger, eagerly listen to every scheme for evading the letter and spirit of the settlement laws, and in a majority of instances I believe avail themselves of them. Our land officers partake of this feeling in many instances, and if they do not corruptly connive at fraudulent entries, modify their instructions and exceed their discretionary powers in examinations of final proof"; *Annual Report of the Commissioner*, 50–51. In Kansas, Nebraska, and Dakota Territories, this inspector reported such land-officer behavior was marked under the timber-culture laws. Carey claimed to have planted, starting in 1883, nearly sixty thousand trees north of Cheyenne, where most of them died. He described his efforts and stated that the law could not be complied with "in the arid region," and federal agents gave him title to the 160 acres involved, under the Timber Culture Act. Larson, *History of Wyoming*, 174. For Carey's lands in Converse

County and fraud charges, see Heritage Book Committee, *Pages from Converse County's Past*, 546; US GLO, "Report of Fraudulent Claim or Entry," and "Supplemental Report," UW microfilm, roll 1, n.p.

16. Larson, *History of Wyoming*, 160 (CY ranch in what became Casper), 304 (irrigation colony created by the Wyoming Development Company and run later by its farmers as the Wheatland Irrigation District). By the mid-1890s, at least, the Wyoming Development Co. under Carey provided "perpetual" water rights with land they sold to project settlers so "every settler owns his water right"; in Carey to J. M. Gordon, August 8, 1895, and text for Wyoming Development Co. circular, 86–87, 378, box 57, Carey Family Papers. On land fraud complaint against Carey, see US GLO, "Supplemental Report" (see full citation in note 15) containing handwritten copy of complaint filed by J. Slichter, W. B. Wood, A. Shaffer, March 18, 1886. The three men noted good soils and timber on the supposed "desert" land and concluded their complaint by saying, "Now we would like if you would come and look at this land as soon as you can conveniently, so that we can get a crop on it yet this season if we get the land, and obliged, yours respectfully." The investigating agent reported that on receipt of the complaint and accompanying survey, he inspected the property in April 1886, and in May 1886 recommended cancellation of the Desert Land Entries Carey claimed.

17. Heritage Book Committee, *Pages from Converse County's Past*, 545–47; railroad arrival in Douglas and Casper, Larson, *History of Wyoming*, 159; John Slichter in BOC, "Testimony in Proof #3321"; US GLO, "Report of Fraudulent Claim or Entry," microfilm, roll 1, n.p. Summary of Slichter testimony in Cheyenne to GLO office, December 1886, in "Brief for Contestee Joseph E. Taylor" (entryman who sold the land to Carey), in Carey Family Papers, box 27, folder 4.

18. *Pages from Converse County's Past*, a book memorializing the creation of the county in 1888, includes interviews and memories written by county settlers, like Jesse Slichter (who died in 1962), as well as pieces on early families and notables written by members of the Wyoming Pioneer Association.

19. Heritage Book Committee, *Pages from Converse County's Past*, 546 (John Slichter family).

20. Heritage Book Committee, 142–43 (Edward David family).

21. Larson, *History of Wyoming*, 182; Carey in BOC, "Testimony in Proof #3322"; Heritage Book Committee, *Pages from Converse County's Past*, 142.

22. Slichter, in BOC, "Testimony in Proof #3321."

23. Larson, *History of Wyoming*, 290–93. In 1894, when the Republican Party in Wyoming was poised for comeback after the 1892 political debacle following the Invasion, Wyoming Republicans kept Warren in the US Senate but ditched Carey in favor of a new face. At the time, US senators were elected by the legislature, but the majority Republican legislature cast not one vote for Carey. Strong constituent views against having two US senators from Cheyenne were reportedly part of the reason. Warren refused to defer to Carey for the Senate seat, writing to a colleague, "Now I am not *against Carey*, but I am *for Warren*"; Larson, 292.

24. Johnston was president of the company in 1894 and was still associated with it in 1918 as a shareholder; Carnes, "The Wyoming Development Company," 49, 106.

25. Larson, *History of Wyoming*, 254–55; Riley, "A Memorial to the Members," 183–84.

26. Slichter, in BOC, "Testimony in Proof #3321"; Carey in BOC, "Testimony in Proof #3322."

27. Carey, BOC. Carey apparently used what he called the "Slichter Ditch" to help distribute water over land already irrigated by an earlier ditch of his own.

28. BOC, *Tabulation of Adjudicated Surface Water Rights, Div. I, 1996*, 312–13.

29. John Slichter appears in a photograph of the "First Jury in the State," in Wyoming State Archives, Bio File–C. W. Horr (John Slichter is misidentified as Jesse in the photo notes); Jesse left the ranch, worked in town, and was county assessor, 1915–18; Charley stayed in ranching and was a two-term county commissioner, Heritage Book Committee, *Pages from Converse County's Past*, 546–47 (John Slichter Family), 544 (Charles Slichter family).

30. David, *Malcolm Campbell, Sheriff*, 5–6, 67–68, 71–72.

31. David, 71–72; Heritage Book Committee, *Pages from Converse County's Past*, 545; Gould, *Wyoming: A Political History*, 36–39. Campbell described a brawl at old Fort Fetterman in 1882 after the roundup; see David, *Malcolm Campbell, Sheriff*, 72–76.

32. Heritage Book Committee, *Pages from Converse County's Past*, 649 (Frank Wolcott), citing Owen Wister diary entry, July 16, 1885. General Land Office, April–June 1886 reports, US GLO microfilm, roll 1 (full citation in note 15).

33. Heritage Book Committee, *Pages from Converse County's Past*, 648–49, 143; David, *Malcolm Campbell, Sheriff*, 130–31, 133–34; Clay, *My Life on the Range*, 138–40, and on Wolcott's plan for the Invasion, 268–69.

34. David, 82–83.

35. David, 68, 135.

36. David, 69.

37. Heritage Book Committee, *Pages from Converse County's Past*, 142–43.

38. David, *Malcolm Campbell, Sheriff*, 69–71.

39. *Buffalo Bulletin* April–May 1956, file 18 in J. R. Smith Papers; "Smith Ranch on Crazy Woman Oldest Place in Johnson County," *Buffalo Bulletin* (Buffalo, WY), August 16, 1956; Davis, *Wyoming Range War*, 16–17. On dry conditions, Mike Whitaker (Water Division II Superintendent), interview with the author, April 2000; on creek diversions, see Division II superintendent Kawulok's report in Christopulos, *Annual Report of . . . 1982*, 109–10. In the 1890s, Mead, aware of Crazy Woman's chronic shortages, had staff investigate a possible trans-basin diversion from Tensleep Creek on the west side of the Big Horn Mountains to Crazy Woman on the east. The idea was determined infeasible. Mead, *Fourth Biennial Report*, 103–4.

40. *Buffalo Bulletin* April–May 1956, file 18 in J. R. Smith Papers; "Smith Ranch on Crazy Woman Oldest Place in Johnson County," *Buffalo Bulletin* (Buffalo, WY), August 16, 1956; Davis, *Wyoming Range War*, 17, 42.

41. David, *Malcolm Campbell, Sheriff*, 70–71.

42.   Woods, *Moreton Frewen's Western Adventures*, vii, 19, 55; Moreton Frewen to Hesse, April 1884, urged "doing all we can to take up the water on the North Fork." Smith cites the letter as preserved in the microfilmed American letters of Horace Plunkett, the third son of Irish nobility, who had established a nearby ranch and started joint operations with Frewen's in 1885. Smith, *War on Powder River*, 96; West, *Horace Plunkett, Cooperation and Politics*, 1–4; Woods, 132.

43.   For Hesse's and Canton's personal histories and invasion role, see Smith, *War on Powder River*, 109, 119; Davis, *Wyoming Range War*, 14–15, 32–33, 50–54. In 1885, to bolster his water claims, Canton managed to summon up a three-man commission—provided for under the territory's early water law—to apportion water in the Little North Fork on the spot, allocating much of that little creek to his company, which boasted plans for irrigating and selling water. Such a commission was provided for in the 1870s "in case the volume of water in said stream, creek, or river, shall not be sufficient to supply the contin-ual wants of the entire country through which it passes, then the county commissioners . . . shall appoint three commissioners, as hereinafter provided, whose duty it shall be to apportion, in a just and equitable proportion, a certain amount of said water, upon cer-tain, or alternate weekly days to different localities, as they may, in their judgment think, best, for the interest of all parties concerned, and with due regard to the legal rights of all." Whitehead, *The Compiled Laws of Wyoming*, 377. A record of the commission Canton invoked is in Johnson County Civil Case #234, Crazy Woman Decree (1889), transcript of petition for apportionment, 1–2.

44.   For 1885 range conditions, see Smith, *War on Powder River*, 95. For Smith's 1886 concerns, see Johnson County Civil Case #234, Crazy Woman Decree (1889), J. R. Smith testimony, 2–4.

45.   Johnson County Civil Case #234, certified transcript of county records, appor-tionment and county surveyor's certificates 1–21; certified transcript of water and ditch claims, 1–25 (showing Smith filed in 1886, his claim of starting ditch in 1879).

46.   Johnson County Civil Case #234, Smith testimony, 4, January 31, 1889. The district courts had the exclusive power to settle water rights questions, under the Water Law of 1886, chap. 61, §9. Evidence was to be taken on the ditches and other water use facilities (and diligence in constructing facilities), chap. 61, §15. The language the Crazy Woman court used regarding irrigated acreage echoed what the 1886 statute required as far as information claimants were to file with the county clerk: "the number of acres of land lying under and being, or proposed to be, irrigated by water from such ditch, canal or reservoir," chap. 61, §10.

47.   Johnson County Civil Case #234, decree of court, July 15, 1889. See also Table No. 3 listing Crazy Woman ditches as in 1889 court decree; in Mead, *Second Annual Report*, 80–81; BOC, *Adjudicated Tabulation of Water Rights, Div. II, 1999*, 447, "Crazy Woman Creek . . . as established by the Decree of the Court of the Second Judicial Dis-trict." Smith was awarded 67 cubic feet per second (cfs), Canton 26 cfs, and Hesse 130 cfs. One irrigator on the creek got rights to water that amounted to one cubic foot of water per second for every hundred acres. Smith did a lot better—his award gave him one cubic

foot of water for every eighteen acres of the 1,200 acres he would irrigate. The standard Mead later imposed statewide, after experimentation with crops, gave people a lot less water per acre than Smith's allotment; Mead's standard was one cubic foot of water for seventy acres. For the controlling position of the Smith right: Zezas Ranch Inc. v. Board of Control, 714 P. 2d 759 (Wyo. 1986), see answer brief of the State Board of Control, June 10, 1985, 10–11.

48. Davis, *Wyoming Range War*, 33, 88–89, 139–41.

49. Davis, 83, 143, 158. Campbell said of Barber, "Perhaps no one on the range was accorded the wealth of affection from the ordinary run of cowboys as well as from the cattle kings as was Dr. Barber"; David, *Malcolm Campbell, Sheriff*, 82.

50. Davis, *Wyoming Range War*, 157–58, 165.

51. Smith, *War on Powder River*, 283. The superintendent of Division II complained in 1906 of the problems of managing Crazy Woman Creek under the court decree: "The irrigation officers cannot follow the orders in said decree and give justice to the appropriators. The decree does not fit the situation and I believe the court was not fully informed of the conditions at the time the decree was granted." He longed for Board of Control review (Johnston, *Eighth Biennial Report*, 68) that did not occur until the early 1980s. Christopulos, *Annual Report of . . . 1982*, 109–10; Division III Superintendent Cooper's report in Christopulos, *Annual Report of . . . 1982*, 118–24.

52. Houston Williams, interview with the author, May 2000.

53. Christopulos, *Annual Report of . . . 1982*, 109–10 and 118–24. *Zezas Ranch Inc.*, 714 P.2d at 760–61, 763–65; Craig Cooper, interview with author, August 2000. The neighbors claimed that much of the large water right on the old Smith ranch had been "abandoned" (see abandonment discussion, chapter 4). The district court and the Wyoming Supreme Court found that original decree language ordering that claimants get the water "necessary and useful" for their acreage, in an amount "not to exceed" the cfs then listed in the order, meant that the water rights had never really been quantified, and that was an appropriate job for the board. The board, in fact, when given this opportunity by the Wyoming Supreme Court, quickly revised all the water rights on Crazy Woman that had been brought into question by the *Zezas* case (those included Smith's, Canton's, and Hesse's) to the statewide one cfs per seventy-acre standard, as much easier to administer.

54. Mike Whitaker (Water Division II superintendent), interview with the author, April 2000; Craig Cooper (Division III superintendent), interview with author, August 2000.

55. Wyoming Session Laws of 1886 provided that water commissioners had the duty to divide water according to priority, chap. 61, §4; but (perhaps since the commissioners were to be paid, by the counties, only per day of actual work) the laws enacted that same session went on to say, in a section captioned "Commissioners to perform duties only when necessary," that "said water commissioners shall not begin their work until they shall be called on by two or more owners or managers, or persons controlling ditches in the several districts, by application in writing"; 1886 Wyo. Sess. Laws 61 §8. That language survived Mead's overhaul of Wyoming water laws in 1890, being re-enacted in 1890 Wyo. Sess. Laws 8 §45. It was repealed in 1901, perhaps due to misplaced confidence among

Mead's successors in the capacity of far-flung water commissioners to be on every stream that needed management, whether called upon or no; 1901 Wyo. Sess. Laws 102 §2. Craig Cooper, interview with author, 2000; Wyoming Statutes Annotated, 41–3–606; substitution of local arrangements for strict enforcement of priority rights is common in western states, Tarlock "The Future of Prior Appropriation," 778–80.

56. Wyo. Constitution art. I, sec 31.

57. Heritage Book Committee, *Pages from Converse County's Past*, 563–65 (Addison Spaugh, foreman of Converse County Cattle Co.), 545–47 (John and Sarah Slichter Family) 135–37 (George H. and Lea Cross family). Olson, *Ranch on the Laramie*, 14–19. On Carey and Kendrick, see Trenholm, *Wyoming Bluebook*, 3:4, 250; Larson, *History of Wyoming*, 387–93; Cynde Georgen, "John B. Kendrick: Cowboy, Cattle King, Governor and U. S. Senator," Encyclopedia, WyoHistory.org, Wyoming State Historical Society, published November 8, 2014, https://www.wyohistory.org/encyclopedia/john-kendrick. Kendrick's acquisition of some of his ranch land is in "Over A Century of History," Trail End State Historic Site (website), trailend.org/kendrick-ranches.html. Starting in 1908, Kendrick built a nearly 14,000-square-foot mansion on a hilltop overlooking the town of Sheridan. It featured a ballroom on the top floor and, reflecting Kendrick's fascination with technology, a specially designed wooden staircase with features that looked like nuts and bolts. Kendrick called the house Trail End, and it is now open to the public as a state historic site. Cynde Georgen, "Trail End Historic Site," Encyclopedia, WyoHistory.org, Wyoming State Historical Society, published November 8, 2014, https://www.wyohistory.org/encyclopedia/trail-end-state-historic-site.

Chapter 3

1. John H. Gordon to his cousin David, December 19, 1881, John H. Gordon Biographical File.

2. Gordon, "Personal History" (unpublished manuscript), 6, 19, 23. Gordon Biographical File.

3. Mead, *Third Biennial Report*, 18.

4. Mead, *First Biennial Report*, 24.

5. In the early 1900s, the State Engineer's Office had inadequate funding and staff for a rapidly growing workload as more lands were settled. The four division superintendents were falling behind in their inspections and adjudications of territorial claims and state permits, and they sometimes paid meeting expenses out of their own pockets; Bond, *Sixth Biennial Report*, 54–58, 66–70. They, and the engineer, regularly took on survey and engineering work of their own to make a living. Nonetheless, the new agency provided a communications network for on-site water users, who had as yet few organizations of their own. The superintendents and their part-time "commissioners" monitoring and regulating streams lived in their divisions, pursued their own livelihoods, and communicated regularly with water users. The superintendents and the state engineer met twice a year as the Board of Control to confirm adjudications and rule on proposals from users

to change a water use or argue that a neighbor had forfeited a water right; 1890–91 Wyo. Sess. Laws, 8 § 25–26, 34; Mead, *Third Biennial Report*, 37–45; Johnston, *Seventh Biennial Report*, 19–20. This made conflict resolution low-cost for users (who did not have to attend). Mead, *Irrigation Institutions*, 247. For the five board members, however, even getting together was expensive; it took days to cross the state by horseback, wagon, or coach on rough roads, and train service reached only a few places; Mead, *Fourth Biennial Report*, 72; Johnston, *Seventh Biennial Report*, 30–35.

Centralized state administration looked elegant on paper, but it was patchy in action. That provided opportunities for users to change how the system worked—a classic example of what economist Douglass North calls an opportunity for entrepreneurs to change the institutions within which they work; North, *Institutions, Institutional Change*, 83–89.

6.   John H. Gordon, "Personal History," (unpublished manuscript), 25–29. Gordon identified the purchasers of his Laramie River ranch in about 1884 as Teschemacher and de Billier; at 28. After the disastrous year of 1886–87, H. E. Teschemacher and Frederick de Billier were ruined, and later they joined the Invasion of Johnson County; Clay, *My Life on the Range*, 77; Davis, *Wyoming Range War*, 143. Gordon's Scottish friend was Andrew Gilchrist, later a prominent ranchman based in Cheyenne involved in establishment of the irrigation colony at Wheatland; *Progressive Men of the State*, 355–57.

7.   Gordon, "Personal History," 25–29.

8.   Mead, *First Biennial* Report, 73–74.

9.   Horse Creek figures in Mead, *Third Biennial Report*, 46. Gordon quote in Gordon, "Personal History," 29.

10.   Johnston v. Little Horse Creek civil case (1895). Regarding George Baxter's biography, see Davis, *Wyoming Range War*, 78, 135, 139, 206, 230; Gould, *Wyoming: A Political History*, 93–96; Mercer, *The Banditti of the Plains*, xlvii; Larson, *History of Wyoming*, 249; Riley, "Memorial to the Members," 179–81. Baxter grew up in Tennessee and married "a wealthy Tennessee girl." He helped raise $100,000 for the legal defense of the invaders, an amount "dwarfing the expenditures of Johnson County" for prosecution; Davis, *Wyoming Range War*, 206.

11.   *Johnston v. Little Horse Creek* civil case (1895)

12.   Johnston v. Little Horse Creek Irrigating Co., 79 P. 22 (Wyo. 1904), 223. Johnstons listed in the case file are James R. Johnston, George D. Johnston, Lizzie D. Johnston, and Harry Homer Johnston. The Johnstons are described in glowing terms in *Progressive Men*, 502–4, probably funded by fees paid by the subject of each profile. James Johnston was one of the founding brothers who went west in 1849, and the quote about the "pluck," etc., common to him and his son is on 504.

13.   *Johnston* 79 P. 22 (Wyo. 1904); Indenture of October 30, 1894 between Springvale Ditch Co. and Little Horse Creek Irrigating Co., Records of the Laramie County District Court, *Johnston v. Little Horse Creek*, Docket # 6–233, box 2, Wyoming State Archives (explaining the week-by-week rotation the companies had arranged). Mead described his interpretation of these facts in his *Irrigation Institutions*, at 262–65 (written while the case was pending before the Supreme Court).

14. Mead, *Third Biennial Report*, 45–46, 52–53; *Johnston* civil case (1895).

15. Mead, *Irrigation Institutions*, 264.

16. *Johnston* 79 P. 22 (Wyo. 1904).

17. In a July 30, 1908, letter to the State Engineer's Office, Mead described the effect of Potter's decision as "mischievous," saying, "Not only did that decision render meaningless and practically inoperative some of the most important features of the State's water law, but, if carried to its logical conclusion, it would throw Wyoming back into the ruck of the arid States of America, whose water laws belong to the lower Silurian period"; Mead's letter to State Engineer's Office, in Johnston, *Ninth Biennial Report*, 76.

18. Mead, *Irrigation Institutions*, 262–65; *Johnston* 79 P. 22 (Wyo. 1904) 233–35.

19. 1905 Wyo. Sess. Laws 97.

20. Johnston, *Eighth Biennial Report*, 81–99. J. A. Johnston left the State Engineer's Office the year Mead left, 1898. He became a director of the Stock Growers National Bank in Cheyenne, part of the financial empire of John Clay, cattle industry banker and key partner in a commission firm handling the marketing of Wyoming cattle. By 1907, J. A. Johnston was head of Clay's commission house office in Denver; *Wyoming Industrial Journal*, no. 4 (September 1, 1899): 19; and *Cheyenne Daily Leader*, no. 109 (December 23, 1905): 10 (available in the Wyoming State Library's Newspapers Database, http://newspapers.wyo.gov/); Woods, *Commissionman, Banker, and Rancher*, 215. Mead had J. A.'s son Clarence work summers for the State Engineer's Office, serve as assistant state engineer, and then join him in Washington working on irrigation and drainage issues for the US Department of Agriculture. Clarence T. Johnston Biographical File, American Heritage Center, University of Wyoming. J. A. and Clarence kept up family ties, as in a 1907 family Christmas party with both attending; *Cheyenne Daily Leader* no. 85 (December 27, 1907): 5.

21. Johnston, *Eighth Biennial Report*, 84; Cooper, *History of Water Law*, 40; Mullen, *Wyoming Compiled Statutes*, 724 (beneficial use, and transfer ban), 725–26 (exception to transfer ban for list of preferred uses); Wyoming Statutes Annotated, 41-3-612 (rotation among water users).

22. Compare beneficial use language in Mullen, 724, Wyoming Compiled Statutes 1910, and Wyoming Stat § 41-3-101 (2011).

23. Johnston, *Tenth Biennial Report*, 17–29; quotations are from 28, 29.

24. Mead, *Second Biennial Report*, 34–46.

25. Johnston, *Eighth Biennial Report*, 70.

26. "Report of Superintendent Edward Gillette, Division No. 2," in Mead, *Third Biennial Report*, 150–51.

27. *Johnston* 79 P. 22 (Wyo. 1904) 229, 232–33. In contrast to the 1904 court, Wyoming people by 1910 had essentially recognized a special kind of property right in water, illustrating Demsetz's principle that property rights emerge when resource values rise. Demsetz, "Toward a Theory," 347–59. Property rights outline the relations between people—privileges and obligations—over a resource; Meinzen-Dick and Nkonya, "Understanding Legal Pluralism," 14. Political scientists, economists, and lawyers all agree that there is no single "property right." Property owners hold a "bundle" of rights, and bundles

can differ. Ostrom and Schlager have usefully identified basic rights to look for in classifying the property relations that different people may have in connection with a resource. They identified the right to access a resource; the right to withdraw portions of it; the right to manage how and when it is used; the right to exclude others from it; and the right to "alienate" or transfer a resource by sale or lease to another's control. Schlager and Ostrom proposed that someone who holds all five of those rights, including alienation, is an "owner," while those holding less than all five are not owners but should be identified with different terms. Schlager and Ostrom, "Property-Rights Regimes," 251–54. Wyoming water users and the State Engineer's Office had understood that there were a variety of possible property rights. They had decided that for water, the right of alienation—in the case of water, the right to transfer the resource to be used in another place—should not be put into private hands.

Mead had prepared water users to think that way in his reports in the 1890s. The average settler, Mead told them all in 1894, "has usually regarded an appropriation of water in much the same light as he regards acquiring title to land, and looks on nothing less than absolute ownership as adequate and proper. We [at the Wyoming Board of Control] have never been able to accept that view." Mead, *Second Biennial Report*, 39–40. Eighty pages later he returned to that theme: "The difficulty is to draw the line between adequate protection to the appropriator and preservation of the rights of the public. To do this involves to many a new conception of property rights, few being able to conceive of any interest in water short of absolute ownership, which the Board is not disposed to sanction." Mead, *Second Biennial Report*, 124.

28. Mead, "The Growth of Property," 12.

29. George Baxter Biographical File; Riley, "A Memorial to Members," 181; Gordon, "Personal History," 30–31; *Denver Sunday Post*, May 28, 1899, reprinted in Read, *Johnny Gordon Had a Dream*, 2006.

30. Lou Blakesley quotation in Johnston, *Tenth Biennial Report*, 56.

31. Big Horn Power Co. v. State of Wyoming, 23 Wyo. 271, 148 P.1110 (1915). The constitution provided that permits could be denied. While outright permit denials were rare, many water permits were issued only after promoters revised their projects to make them more practical in the view of the engineer's office; see Parshall, *Eleventh Biennial Report*, 10.

32. 1890–91 Wyo. Sess. Laws 8 §34, 36.

33. Johnston, *Seventh Biennial Report*, 12–13; see also Johnston, *Ninth Biennial Report*.

34. Mead, *Third Biennial Report*, 63–66: Mead to A. Gilchrist, April 20, 1892, box 1, folder 10, Carey Family Papers.

35. Mead, 63–66.

36. Davis, *A Vast Amount*, 1993; Davis, *Goodbye, Judge Lynch*, 2005; US census figures for Big Horn County (which in 1900 encompassed most of the Big Horn basin) can be found in Historical Decennial Census Population for Wyoming Counties, Cities and Towns, http://eadiv.state.wy.us/demog_data/cntycity_hist.htm; and Land Area and Decennial Housing Units for Wyoming Counties, http://eadiv.state.wy.us/demog_data/cntyhus_hist.htm.

37.   Bonner, *William F. Cody's Wyoming*, 2007

38.   Blakesley, in Johnston, *Tenth Biennial Report*, 56.

39.   Mead, *Fourth Biennial Report*, 93.

40.   Elwood Mead to J. A. Van Orsdel, November 14, 1902, Mead-Van Orsdel correspondence file; "Andrew Gilchrist," in *Progressive Men of Wyoming*, 355–57.

41.   The 1894 Carey Act, as it was known nationwide, attempted to give the support and supervision of state governments to big irrigation projects, which had typically failed for lack of adequate financing and staying power. The Carey Act allocated federal lands to states to set aside for private irrigation projects where the state certified that sufficient water was available. The state was to enter a construction contract with the private company. The contract would include state approval of the amount the company could charge settlers for water rights on lands covered by the company's irrigation system. Carey Act of 1894, 28 Stat 422, 43 U.S.C. § 641 (1976). In Wyoming, the company initially filed for water rights for the whole tract to be irrigated, and the company had to make project settlers owners of shares in the canal, with the price of the shares set by the state and paid for over time by the settler. Mead, *Fourth Biennial Report*, 21. The settlers' water rights were adjudicated in each settler's name and obtained the priority date of the initial water right filing by the company. See, for example, Solon Wiley's Carey Act project, discussed in the text. Big Horn Development Co. Permit 233E, filed December 1896 (creating that priority date), in Wyoming State Engineer's Office files, Cheyenne (hereafter cited as SEO Permit 233E).

42.   Lindsay, *The Big Horn Basin*, 1930; Cook, *Wiley's Dream of Empire*, 1990.

43.   Lindsay, 1930; Curtis and Grimes, *Wyoming Climate Atlas*, 101, figures 6.4 and 6.5, net annual precipitation and annual precipitation, 1895–2003; Bond, *Sixth Biennial Report*, 47–53.

44.   Van Orsdel, *Biennial Report . . .1901–1902*, 31–37, 82–88; SEO Permit 233E, filed December 19, 1896 (creating that priority date), and signed by Elwood Mead, February 10, 1897. Mead in 1902 described his action six years earlier regarding the Wiley project: "When Mr. Wiley took up the building of a canal along the line of the original Hay project [prominent Cheyenne banker Henry Hay was Gilchrist's partner in the original bench land project], we discussed the effect of an assignment of the original Hay permit [to Wiley], but I could not approve the recognition of a priority for his project to date from that of the original Hay filing [1893], because it would work an injustice to other appropriators who had dug ditches and irrigated farms along the river between the time the Hay project was approved and abandoned. In this discussion, I told Mr. Wiley that his project was a new enterprise, having had no real connection with that of Mr. Hay and that I would only issue a permit on an application of the regular form, and only recognize a priority dating from the receipt of that application. He complied with this requirement, and his application as recorded and approved only gives him, in my opinion, a priority dating from its receipt in the State Engineer's Office." Mead to Van Orsdel, November 14, 1902. Hay, a delegate to the state constitutional convention, helped found the Stock Grower's National Bank in Cheyenne (in which Gilchrist had major shares)

and became the bank's president in 1894. He was also state treasurer 1895–1899, and again briefly in 1903. "Hon. Henry G. Hay," in *Progressive Men of Wyoming*, 231–32; Trenholm, *Wyoming Blue Book*, 2:137.

45. Van Orsdel, *Biennial Report . . .1901–1902*, 31–37 (includes the question submitted by State Engineer Fred Bond); Mead to Van Orsdel, November 14, 1902, December 2, 1902.

46. Van Orsdel, *Biennial Report . . .1901–1902*, 31–37, 82–88.

47. "Biography, Josiah Alexander Van Orsdel," Judges of DC District Courts, Historical Society of the District of Columbia (website), http://dccircuithistoricalsociety.org/Biographies/biosalpha.html; Peters, "Joseph M. Carey," 15–16; Trenholm, *Wyoming Blue Book*, 2:172, 203.

48. Van Orsdel, *Biennial Report . . .1901–02*, 31–37.

49. Van Orsdel, 31–37.

50. Van Orsdel, 82–88.

51. Van Orsdel, 82–88. Van Orsdel argued that the construction deadlines in a state water project were imposed under state water laws, "the general statute" dating from 1890, but that the state had subsequently by statute implemented the Carey Act (currently Wyoming Statutes Annotated, § 36–7-101) and set terms for the contract, with no construction deadline. That "special statute," enacted later than the general, must prevail. Van Orsdel noted the state Carey Act statute had also made the water for the project "appurtenant" to the land (appurtenant in a water context means any water rights go with the land in a land sale, unless the water is separately conveyed elsewhere). He also argued, in the June 30, 1902, opinion, that deadlines for *putting water to use* (as opposed to completing construction) certainly could not apply to the developer company in a Carey Act project, since only individual settlers could accomplish getting water on the land. In general, he said, whether an individual had gotten water on the land could only be determined by the Board of Control in an adjudication proceeding, which should consider whether the settler had used "reasonable diligence" to get water onto the land. With this reasoning, Van Orsdel paved the way for permits to acquire considerable practical and legal status in Wyoming water law, which ultimately encouraged numbers of users to believe they need not always bother to get their permits adjudicated.

52. Van Orsdel to Mead, August 19, 1902; quotations respectively from Mead to Van Orsdel, November 14, 1902; Van Orsdel to Mead, November 24, 1902. Mead-Van Orsdel correspondence file.

53. Farmer's Canal was represented by Gibson Clark, a former Confederate soldier from Virginia who was a clerk in the trading post at Fort Laramie after the war, eventually studied law, and, a Democrat, was elected to the Wyoming Supreme Court in 1892 (the year of the Invasion of Johnson County that saw other Democrats elected in reaction). Clark left the court to serve as US district attorney for Wyoming under the Democratic administration of President Grover Cleveland, and then went back into private practice in 1898 when his term expired. Trenholm, *Wyoming Blue Book*, 2:201. Farmer's Canal 1902–1904: Final Decree, June 14, 1904; SEO Permit 233E (full citation in note

41), endorsements for extension of time to show beneficial use of the water (the series of signed notes on the permit, by successive state engineers, initially to extend the time for getting water on the land, and eventually to cut out the project lands which saw no water, runs from 1902–69). The Farmer's Canal and the Bench Canal, with others on what came to be known as the Greybull River, joined together in 1920 to form a district that could build reservoirs on the river, and did so in 1938, 1972, and 2005—with a Federal Emergency Administration of Public Works grant and loan for the first reservoir and a Wyoming Water Development Commission grant and loan for the last one. "About Us: A Brief History of Greybull Valley Irrigation District," Greybull Valley Irrigation District (website), https://greybullvalleyid.com/.

54. Reclamation Act of 1902, Pub.L. 57–161, passed June 17. For the roles of Warren, Mondell and Mead, and Mead's opinion of Mondell, Kluger, *Turning on Water*, 33–34. Key characters in passage of the Reclamation Act testified to their support for a national role in reclamation via irrigation in the West in 1905, as witnesses called by the government in a dispute between Kansas and Colorado over water in the Arkansas River, decided by the US Supreme Court in State of Kansas v. State of Colorado, 206 U.S. 46 (1907). Testimony on the passage of the Reclamation Act and its value from Wyoming players are in vol. 2 of the transcript for *Kansas v. Colorado*, in *The Making of Modern Law*, hereafter cited as "*Kansas* Transcript of Record," at 1179–90 (Mondell), 1289–95 (Warren), 1323–1467 (Mead), 1055–59 (Van Orsdel), and 1047–55 (Clark). For enthusiastic newspaper coverage in Wyoming after President Theodore Roosevelt signed the Reclamation Act, see a report with photographs covering almost a full page: "The Irrigation Bill: How the Wyoming Members of Congress Worked for the Passage of the Measure and What It Means for the West," *Cheyenne Daily Leader*, June 19, 1902. See also Lilley and Gould, "The Western Irrigation Movement," 1966.

55. Deloria and Lytle, *American Indians, American Justice*, 1–12; Hoxie, *A Final Promise*, 1–188.

56. Shurts, *Indian Reserved Water Rights*, 5 ("coercive and vicious"), 119–25; Lone Wolf v. Hitchcock, 187 U.S. 553 (1903); O'Gara, "Home From School," September 1, 2017, WyoFile (website), http://www.wyofile.com/home-from-school/; Trenholm, *Wyoming Blue Book*, 2: 472; Mondell, Frank, "Autobiography, vol. 1," 227, unpublished manuscript, in Mondell Papers.

57. Hoopengarner, "To Make the Desert Bloom," 83–96; "When the Tribes Sold the Hot Springs," Encyclopedia, Wyoming State Historical Society (website), published December 3, 2018, https://www.wyohistory.org/encyclopedia/when-tribes-sold-hot-springs; Mondell, "Autobiography," vol. 1, 221–26, and vol. 2, 308–14. Quotations "preparing for cession" 1897, vol. 1, 226, "fair and fertile region," vol. 2, 309, "grandiose plan" vol. 2, 310.

58. Hoopengarner, "To Make the Desert Bloom," 83–96; Wilson, "Farming and Ranching," 202–18; Kruse, "The Wind River Reservation," n.p.

59. Wilson, 218–23; Shay, "Promises to a Viable Homeland," 558–99.

60. Hoopengarner, "To Make the Desert Bloom," 97–155. Johnston's quote on

promising agricultural prospects is in *Eighth Biennial Report*, 9. He notes Wyoming statutory authority from 1899 for his office to make surveys "to demonstrate the feasibility of various irrigation projects" at 10, and his discussion of his work to survey the project and launch it is 30–60. Chatterton, *Yesterday's Wyoming: The Intimate Memoirs*, 55, 59, 99–103.

61.   Permit #7300, August 7, 1906, State Engineer Permit Files, submitted and signed by Fenimore Chatterton as attorney for the Wyoming Central Irrigation Co., while Chatterton's autobiography, 98, and state records show Chatterton was Wyoming Secretary of State until January 1907. Trenholm, *Wyoming Blue Book*, 2: 29–130. The entry notes that Chatterton remained the attorney for Wyoming Central Irrigation Co. until 1914. Chatterton in *Yesterday's Wyoming*, 124, said regarding the irrigation project on the ceded portion of the Wind River reservation: "I personally had the water rights for 300,000 acres of land—now officially known as the Riverton Project."

62.   The Chicago investor who founded and headed the Wyoming Central Irrigation Co. was Joy Morton, founder of Morton Salt; Ballowe, *A Man of Salt and Trees*, 165–66. The quotation on the attitude of the settlers is from an unpublished, unfinished biography of Joy Morton, written by his son Sterling Morton toward the end of Joy's life; see "Other Business Ventures," 18, Morton Family Papers, box 50, folder 8, Chicago History Museum.

63.   *Cheyenne State Leader*, October 20, 1910, 1, 8, quotation, 8; Hoopengarner, 97–155.

64.   Hoopengarner, 166–97; Mondell, "Autobiography," vol. 3, 657–63. These pages contain details on Mondell's work to get money for irrigation surveys and construction on the ceded portion of the reservation—the portion opened to white settlers, and particularly the area that became the Riverton Project (now home of Midvale Irrigation District) run by the Bureau of Reclamation (in 1920, known as the Reclamation Service). Mondell was successful over several years in ensuring that money for the Riverton Project came from the "Indian appropriations bills," which Congress used to provide funding for Indian reservations. Mondell commented that with $200,000 allocated to the Riverton Project in the Indian Appropriation Act of 1920, "Thus at last the project was underway with an Indian Bill appropriation but under Reclamation Service control, a hybrid but nevertheless an established project," 662. Mondell crossed out these pages and this detail, to shorten his autobiography as serialized in Wyoming newspapers in 1935–36, but the pages still remain, completely readable, in the manuscript at the American Heritage Center, where the newspaper serial version is also preserved, in Collection 01050, box 23, folders 3 and 4.

65.   Hoxie, *A Final Promise*, 158–61, describes how lease and sale of allotments became possible under federal law.

66.   Winters v. United States, 207 U.S. 564 (1908). Shurts (2000) provides an excellent, detailed description of the context and implications of the Winters case in the early twentieth century.

67.   Shurts, *Indian Reserved Water Rights*, 6. *Winters*, 207 U.S. 564 effectively allowed an "inchoate, unquantified, flexible reservation of water"; Shurts, 8–9. Westerners worked around the prior appropriation doctrine "to reserve water and watersheds for coordinated

and comprehensive economic development," 17–34. Pages 103–18 describe the competing interests in Montana's Milk River valley leading up to and after the *Winters* case.

68.   Van Orsdel was a witness in Kansas v. Colorado, 206 U.S. 46, (1907), a dispute between Kansas and Colorado over the Arkansas River. That dispute posed the question: Between downstream and upstream states on the same river, states with different rates of development, who had the best right to water? And did the Reclamation Act mean the federal government was the prime decision maker on what water went where in the West? Federal lawyers called on witnesses from Wyoming (both an upstream and downstream state, depending on the river involved) to bolster the case for a US government role. See "*Kansas* Transcript of Record," vol. 2 at pages cited in endnote 54 for Wyoming witnesses; Van Orsdel's testimony, as attorney general of Wyoming, is at vol. 2, 1055–59 of the transcript. The high court rejected federal intervention, and favored an "equitable apportionment" of water, to sort out the competing needs of Kansas and Colorado. But in 1906 Van Orsdel's appointment as an assistant attorney general in Washington had him working with the lawyers who had made him a witness for the United States in *Kansas*, 206 U.S. 46. One of the cases he soon went to work on was *Winters*, 207 U.S. 564, then before the US Supreme Court. For *Winters*, the solicitor general's brief and also the supplemental brief, the latter signed by Van Orsdel as apparently the lead author, are found in the Transcript of Record (file date November 20, 1906, term year 1907), in the Supreme Court records and briefs database *The Making of Modern Law: U.S. Supreme Court Records and Briefs, 1832–1978*, 180–213, 227–68, hereafter cited as "*Winters* Transcript of Record." The solicitor general's brief highlighted the Montana treaty and its intent: the Indians were to be farmers, and to farm there they needed water. The solicitor general argued further that Montana water law had a hint of "riparian" rights in it, for land that was not in the public domain (where prior appropriation ruled). So, the solicitor general argued, the water for the reservation had no cap or deadline; it should be available whenever and in whatever amount reasonably needed for farming. "*Winters* Transcript of Record," 180–213. (Shurts, in *Indian Reserved Water Rights*, 43–50, ably explains the role riparian rights played in water law discussion at the time.) Van Orsdel filed the supplemental brief, initially focused on narrow jurisdictional questions raised by the other side, but then going to the merits. He launched an argument different from the solicitor general's, saying that the United States, by setting aside land for an Indian reservation, had protected from appropriation all the water needed for that land. Van Orsdel mentioned the treaty intent argument but emphasized this alternate argument, an argument much like his argument in the Carey Act situation in Wyoming in 1902. He summarized his argument as, "The United States by setting apart and holding this land as an Indian reservation thereby reserved it from the public domain and exempted from subsequent adverse appropriation, under either the public-land laws of the United States or the laws and customs of Montana, the uninterrupted flow of all water necessary to its beneficial use," in "*Winters* Transcript of Record," 243.

In its final *Winters* decision, the US Supreme Court homed in on both the solicitor general's argument on the intent of the treaty and Van Orsdel's argument on the authority of the federal government to reserve land and water from the reach of state water law. On that point the high court said, "The power of the government to reserve the waters and exempt them from appropriation under the state laws is not denied, and could not be"; *Winters* 576–77. In its *Winters* ruling on that point, the high court cited a comment it had made in 1899, cited also by Van Orsdel for the same point; "*Winters* Transcript of Record," 245. The court had suggested in 1899 that when the United States owned land bordering on a river, state water law could not deprive the government of the right "to the continued flow of its waters, so far, at least, as may be necessary for the beneficial uses of the government property"; United States v. Rio Grande Dam and Irrigation Co., 174 U.S. 690 (1899) at 702–3. In that case, the United States had opposed the plans of a private company with New Mexico state permits to construct a dam on the Rio Grande. The court ruled that the state's right to impose the law of prior appropriation on its rivers was limited by the superior power of the US government to ensure navigable rivers remained navigable. The court noted in passing that the only other ground for limiting state authority to set the law for use of rivers would be that states cannot destroy the right of the United States, as owner of property by a stream, to the continued flow of water in that stream, at least as necessary for the purposes of the US property. Though this statement was made only in passing in 1899, the court then quoted it with approval in *Kansas*, 206 U.S. 46 and ultimately in *Winters* 207 U.S. 564. Van Orsdel did not cite the 1899 Supreme Court decision in his 1902 opinion on the water rights of Carey Act lands in Wyoming, but he would have been aware of it then. The principle of tribal water rights dating from treaty dates has since been known as the principle of "reserved water rights," echoing the language of Van Orsdel and the top court.

69. Shurts, *Indian Reserved Water Rights*, 202–5; United States v. Hampleman, Case No. 763 (D.Wyo. 1916) (Hampleman was the water commissioner on Owl Creek, and a water user himself, upstream of the Duncans. The action he took as water commissioner appeared to be intended to ensure water reached other settlers, with 1880s rights, at the mouth of the creek on lands that had not been part of the reservation created in 1868.); Wadsworth, "Wind River Indian Reservation Annual Report, Narrative," 1912, box 1, file 107, RG 75, National Archives; "Report of Superintendent of Water Division No. 3," in Parshall, *Eleventh Biennial Report*, 30–31.

70. Shoshone and Arapaho tribes with H. E. Wadsworth, Clerk Mayer, and Inspector Norris, "Minutes of Council Meeting," May 28–29, 1912, box 6, file 215, pp. 1–7, 9–10, 13–14, 20, Shoshone and Arapaho Business Council Proceedings, RG 75, National Archives (hereafter cited as "Minutes of Council Meeting").

71. "Minutes of Council Meeting," 6–13. Comments of Joe Lajeunesse, Shoshone and president of the Business Council 6–7; Calvin Littleshield, Arapaho 7–8; Big Plume (Yellow Calf), Arapaho 8–11; Rabbit Tail, Shoshone, 12; and Cook Tinzona, Shoshone, 13.

72. Big Plume, Arapaho, "Minutes of Council Meeting," 10–11. Big Plume was often known outside his own people as Yellow Calf, his childhood name, and he appears in a 1908 photo of a Wind River delegation to Washington. Kruse, "Wind River Indian Reservation," n.p. He was called Yellow Calf by the reservation agency translator and

transcriber. By contrast to Big Plume's statement, Dick Washakie, Shoshoni, said that the secretary if petitioned would only say the same as his officers, and, referring to Wadsworth, "My friend here has told the truth by telling us to sell a part of our land to improve the other part with the proceeds of the sale. We old people here, we are unable to work, we would like to sell our dead Indian lands and have all the cash right down so that we can make our living on it. What I have heard today I believe I understand everything and I believe it best to do what they want us to do"; 16.

73. Big Plume (Yellow Calf), in "Minutes of Council Meeting," 8.

74. Tisse Guina, in "Minutes of Council Meeting," 20.

75. Inspector Norris, in "Minutes of Council meeting," 2, 6, 11–12, and quotation, 20.

76. Wadsworth, "Wind River Indian Reservation Annual Report, Narrative," 12, 16. The Wind River superintendent worried in 1913 whether farming on remaining allotments was proceeding quickly enough, at 12: "Should the State of Wyoming be found to have control of the waters within the reservation it will be a serious proposition to save the water rights if the State carries out its intention of adjudicating the water rights Dec. 31, 1916." For the agency's response once Judge Riner's decision in U.S. v Hampleman (D.Wyo. 1916) (the Owl Creek case) was issued in late June 1916, see "letter of July 26, 1916," from the Assistant Commissioner of Indian Affairs in Washington to the Wind River agency enclosing a copy of Riner's decision; and "letter of July 31, 1916," from Wind River Special Agent Calvin H. Asbury (holding a temporary post pending appointment of a new superintendent) to W. T. Judkins, Riverton, enclosing a copy of the decree. Asbury: "I would appreciate it if you would give a little attention to the manner in which this decree is being obeyed; and also please explain to the Indians the substance of this decision, and assure them that their right to this water must not be interfered with by Mr. Hampleman or his employees"; in "Correspondence July 26 and 31, 1916," in General Correspondence. US District Judge Riner's decree of June 26, 1916, is in *Hampleman*, Case No. 763 (D.Wyo. 1916), found in Civil Case Files box 120, file 753, folder 2. Riner's decree is not accompanied by a memorandum explaining his reasoning, in either file. For Riner's earlier membership in Wyoming's 1889 Constitutional Convention, see Trenholm, *Wyoming Blue Book*, 2:485. US District Judge Kennedy's decree and memorandum of opinion of October 11, 1926, in United States v. Parkins, can be found in Civil Case Files box 257. Parkins was a non-Indian purchaser of allotments in 1912 who by 1920 failed to pay operation and maintenance costs for the reservation irrigation system, was accordingly denied water for those allotments, and then took water without permission from a different point on the system. Kennedy granted the United States a permanent restraining order against Parkins's unpermitted diversion, ruling that the United States had the right to the water, for the benefit of the Indians, based on the 1868 treaty. United States v. Parkins, 18 F.2d 642 (D.Wyo. 1926), Judge's Memorandum, 3–4.

77. 51 Cong. Rec. H12949 (daily ed. July 29, 1914) (statement of Rep. Frank Mondell); Shay, "Promises to a Viable Homeland," 556–58.

78. Wilson, "Farming and Ranching," 22–31, 216, 222–24.

79. Tom Rea, "Buffalo Bill and the Pony Express: Fame, Truth and Inventing the

West." Encyclopedia, Wyoming State Historical Society, WyoHistory.org, published September 5, 2015, http://www.wyohistory.org/essays/buffalo-bill-and-pony-express-fame-truth-and-inventing-west. Beck, "Autobiography" (unpublished manuscript); Gould, *Wyoming: A Political History*, 93, 185.

80.   Mead, "The Cody Canal," 12–14. The piece includes rosy descriptions of the Cody area the canal was to serve, the magnanimity of Buffalo Bill, and how the Carey Act helped make possible such a project, one that Mead said would not be feasible with only private capital.

81.   Bonner, in "Elwood Mead, Buffalo Bill," 44, concludes: "The financial disarray of the Shoshone Irrigation Company (Beck and Cody's company organized to build the Cody Canal) can be traced directly to its directors' reliance on Elwood Mead's assessment of the economic potential of their project." In footnote 7, Bonner concludes: "When we take into account the construction difficulties he either did not foresee or drastically minimized, Elwood Mead was a disaster as a consulting engineer."

82.   "Permit #2111," State Engineer Permit Files. Mead signed the permit May 22, 1899; he left for Washington later that year.

83.   Bonner, *William F. Cody's Wyoming*, 2007.

84.   Bonner, "Elwood Mead, Buffalo Bill," 43–44 (see his footnote 7), 46.

85.   Correspondence File, December 17, 27, 1915, of Wyoming State Engineer's Office, Wyoming State Archives.

86.   True and Kirby, *Allen Tupper True*, 3–28, 236–39, 464.

87.   "Permit #2111," State Engineer Permit Files, containing True's signature 1915 on time extension for the permit, and a brief explanation.

88.   Charles E. Robinson family, Garland Division (told by his son Woodrow W. Robinson), Oral History, n.d., oral history files, Homesteader Museum, Powell, WY.

89.   Bill Sedwick, Oral History, ca. 1977, oral history files, Homesteader Museum, Powell, WY.

90.   Koelling, *First National Bank of Powell*, 19.

91.   Koelling, 21. For more on the history of the Shoshone Project and the town of Powell, see Churchill, *Dams, Ditches and Water* and Churchill, *People Working Together*.

92.   US Reclamation Service, "Cost of Water Per Acre," 7; Lampen, *A Report of an Economic Investigation*, 53–65, 78, 83–86; Mead, "Shoshone Project: Testimony of Farmers, 1915," Sumner Merring, 3–9, Robert Allan, 7–9, memoranda 32, in Mead Papers.

93.   Mead, "Shoshone Project: Testimony of Farmers, 1915," Albert Shoemaker, 12–13, in Mead Papers.

94.   "Permit #2111," State Engineer Permit Files, contains the date of all the permit expiration notices and subsequent extensions signed by successive state engineers; Wahl, *Markets for Federal Water*, 33–34; Government Accountability Office (GAO), *Bureau of Reclamation*, app. 3, p. 6. Permit 2111 was finally adjudicated in the Big Horn River Adjudication which began in 1977 and was completed in 2014 (see chaps. 5 and 6).

95.   Emerson, *Sixteenth Biennial Report*, 51–58; Burritt, *Report on Water Rights*, 1935.

96.   "Permit #7300" and notations thereon, State Engineer Permit Files. The lawyer for

the Reclamation Service assigned to determine the status of land and water rights for the federal Riverton project in 1918 reported that he had consulted with the Wyoming State Engineer (James True), who said that "he would grant such reasonable extension as the United States might request. He expressed also a desire to cooperate to the fullest extent." US Bureau of Reclamation (BOR), *Riverton Project History*, 1918, 43 (the report of District Counsel E. E. Roddis, in Denver). For use of Indian funds on the project, see Mondell, "Autobiography," vol. 3, 662, in Mondell Papers and *Riverton Project History*, vol. 1, 8–9. The "Indian appropriations bill" for fiscal year (FY) 1918 and 1919 appropriated a total of $105,000 for initial surveys and then for plans and construction of the "Riverton Project." Details include the Wind River Diversion Dam, the Wyoming Canal, Pilot Butte and Bull Lake reservoirs, etc., in *Project History*, vol. 1, 10–12. The FY 1919 appropriation of $100,000 for the project is in May 25, 1918, Public Law 159, 65th Cong. The language of the bill, discussing how the costs of construction will be reimbursed to the government (citing terms of the 1905 cession agreement), distinguishes between Indian-owned and settler-owned land for reimbursement terms. The idea that there were Indian-owned lands within the project boundaries (apparently pre-cession allotments obtained by some tribal members on the north side of the Wind River) may have been used as justification for funding this non-Indian, Reclamation Service project via an Indian appropriations bill. Vol. 1 of the *Riverton Project History*, however, notes that of project lands (i.e., on the ceded portion of the reservation), only 1.3 percent were Indian owned; 81 percent were still "public land" open to settlement; and 17 percent were already patented to settlers. Of all those, lands that were "improved" under the few ditches built by the failed company were selling for ten times the price of the bare lands with no ditches (unimproved land under those initial ditches was selling for four times the price of those bare lands with no ditches.). US BOR, *Riverton Project History*, 1918, 3–6.

97. Holt v. City of Cheyenne, 22 Wyo. 212, 137 P. 867 (1914).

98. In 1914, the Wyoming Supreme Court upheld the state engineer's view that in actual distribution of the water, waste of water had to be avoided so that as many water users as possible could be served. In Parshall v. Cowper, 143 P. 302 (1914), the court upheld a state engineer order to restrict an irrigator to 0.38 cfs rather than the 1.8 cfs that the Board of Control had awarded in an adjudication some twenty years earlier. The state engineer had argued that despite what the adjudicated water right said, the irrigator's ditch in 1912 was not capable of carrying more than 0.38 cfs, and the court concluded the irrigator's water use could be restricted because the law on water commissioner duties (numbered 41-3-603 in modern statutes) provided that water commissioners controlling distribution could prevent waste. *Parshall v. Cowper* (1914) at 303–4.

99. Johnston, *Seventh Biennial Report*, 20; True, *Thirteenth Biennial Report*, 86–87.

100. Johnston, 11.

101. The concept of "relation back," tying a water right priority date to first steps taken to obtain the water, predated Mead's system, and the crucial "first step" was the permit application date under his system. The Wyoming Supreme Court upheld application of "relation back," for Wyoming territorial rights and state-permitted rights, in a number of

cases including Moyer v. Preston, 44 P.485 (1896) at 848, Van Tassel Real Estate & Live Stock Co. v. City of Cheyenne, 54 P. 2d 906 (1936) at 913. For an overall discussion of the doctrine of "relation back" see MacDonnell, "The Development of Wyoming Water Law," 96–97.

102.   For homesteading rates in Wyoming, see Larson, 173–78, 362, 414–16. For the "good faith" test for receiving land ownership following on a federal permit, see Spaulding, *A Treatise on Public Land*, secs. 76, 80, 103, 106.

103.   True, *Thirteenth Biennial Report*, 86–87; 1917 Wyo. Sess. Laws 119. When Wiley relied on unused permits for his project back in 1902, Van Orsdel had noted that a penalty for missing permit deadlines was lacking in the Wyoming water statutes. The 1917 law put such a penalty in place.

104.   James True to Ron Nebeker, September 30, 1918. Correspondence Files, State Engineer General Correspondence 1913–20.

105.   Emerson, *Fifteenth Biennial Report*, 62–63. Correspondence files from True's era show the State Engineer's Office sending notices warning of default, but apparently that practice was followed less faithfully under other administrations. In 1977, the Wyoming Supreme Court in dealing with an unadjudicated permit from 1910 ruled that failure of the State Engineer's Office to send such notices in subsequent years meant that while the water continued to be used, no forfeiture of the water rights under the permit could occur. Snake River Land Co. v. State Board of Control, 560 P. 2d 733 (Wyo. 1977), at 735–39. The court's opinion in the 1977 Snake River Land case shows that for the 1910 permit, the State Engineer's Office in the 1970s followed what had become common practice by then for water rights held only under permit; the permit holders were able to turn in affidavits showing use since the ditch was completed in 1911, and the water rights were finally adjudicated (in part), in 2016. *Snake River Land Co.* at 735; "Permit 10160," John Love, State Engineer Permit Files.

106.   Neighbors might watch to see if water diverted under a permit went to the lands that were originally planned; this had clearly been the case for six decades or so given the facts presented in Green River Development Co. v. FMC Corp., 660 P.2d. 339 (Wyo. 1983). Neighbors might not draw attention to water rights left unused for years, but they would be the first to know and could be ready to object if that water right were suddenly revived and the water diverted from the stream. This was demonstrated in Lonesome Fox Corp. 1981, in BOC, *Order Record Book 27*, at 19. To this day, however, a state permit remains an absolute requirement for a new water right. It is rare in Wyoming to see anyone attempt to claim a water right based merely on years of usage with no state water permit—and any such claims are uniformly thrown out. The legal concept of gaining a right based simply on years of usage is called "adverse possession," or sometimes "prescription," a concept from land law, which at one time had some sway in the water law of other states; see Trelease, *Cases and Materials*, 3rd ed., 203–5; Lewis v. State Board of Control, 699 P.2d 822 (Wyo. 1985).

107.   In 1920, when users faced a drought year on top of a depressed agricultural market, juries in the Big Horn basin refused to convict on charges of water theft, even when

unchallenged facts showed illegal use of water. On some creeks, the superintendent in the basin reported, neighbors decided to turn a blind eye to water rights sitting unused, and no one brought charges that the rights had been "abandoned" and should no longer be recognized; Emerson, *Fifteenth Biennial Report*, 81–82. Modern superintendents report the same problem: "Elected sheriffs & prosecutors don't even like to bring charges against voting irrigators, let alone persuade rural juries to convict," Jade Henderson (former superintendent Division IV, retired 2014), personal communication with the author, October 8, 2018.

108.   The federal program embodied the kind of settlement policy and investment yearned for not only in the Big Horn basin but also in most of the rest of Wyoming. Starting in 1911, the legislature, disappointed in its hopes for population and agricultural growth, invested in a program to recruit farmers and capital to the state; Larson, *History of Wyoming*, 363–64.

109.   Johnston, *Eighth Biennial Report*, 7, 28; Emerson, *Sixteenth Biennial Report*, 54–55.

110.   In interstate water disputes, the US Supreme Court, in *Kansas v. Colorado* in 1907 (206 U.S. 46) and *Wyoming v. Colorado* in 1922 (259 U.S. 419) upset the theories of both those who thought the headwaters "state of origin" had the primary water rights, and those who thought priority dates of actual water use should work across state lines so the earliest development would have the primary rights. Willis Van DeVanter, Warren's former political chieftain on the US Supreme Court of 1922, wrote the unanimous *Wyoming v. Colorado* opinion, much to regional dismay. The high court rulings (and the time and money spent on litigation) gave Emerson and other negotiators from the Upper Colorado River states reason to want to develop their own interstate agreements for water allocation, rather than seek court decisions. Tyler, *Silver Fox of the Rockies*, 15–20, 163–201, 218; Mackey, *Protecting Wyoming's Share*, 41–70; Larson, *History of Wyoming*, 460–62.

111.   US Department of Commerce, *Irrigation and Drainage*, 332, 339.

112.   Mead, "The Growth of Property," 3.

113.   Mead testified in the Supreme Court case of *Kansas v. Colorado*, as did Van Orsdel, Carey, Warren, and other prominent figures in Wyoming (see note 54 above). In his 1905 testimony, Mead said that he admired Italian water law, as instituted in the nineteenth century, because it allocated water to the most benefit to the public (sometimes, the best economic use); rights to water there were allocated not based on time of the application for a right, but on the basis of the opinion of representatives of a variety of interests in government, from agriculture to the military, as to the relative benefits to the public of the proposed different uses. *Kansas v. Colorado*, vol. 1 transcript in *The Making of Modern Law*, 1368. Mead's comments suggest that he might have preferred such a system for Wyoming and the West, but given the established popular custom of priority appropriation, he embodied the concept from Italy as best he could by including the concept of "the public interests" in the constitutional provision: "Priority of appropriation for beneficial uses shall give the better right. No appropriation shall be denied except when such denial is demanded by the public interests"; Wyo. Const. art. VIII, § 3, and in his attention to public control of "public waters" in the structure of the State Engineer's

Office. Mead's support of a short-term abandonment rule, and his emphasis on permit deadlines and the issuance of new permits to replace old unused ones, as seen in the Gray Bull River case, also sought to allow flexibility in water law to accommodate new uses, possibly better serving the public interests.

114.   Mead discussed the proposals he put before Congress that failed, in a speech in Cheyenne in June 1925 covered by the *Wyoming State Tribune* and *Cheyenne State Leader*; found in *Mead Scrapbooks*, Mead Papers. As of 1922, just before Mead began as head of Reclamation, the agency had a lackluster record; many of the twenty-four projects built in the West appeared failed, because farmers had either defaulted on payments or abandoned projects. Less than 10 percent of $135 million spent on reclamation had been repaid, only seven percent of irrigated acres in the West were attributable to federal reclamation; Tyler, *Silver Fox of the Rockies*, 149.

115.   Mead to Grace Raymond Hebard, April 25, 1929, box 1, file 1, Mead Papers. Hebard was head of the Department of Political Economy and Sociology at the University of Wyoming and had been Mead's deputy when he became state engineer years before in Cheyenne. Hebard's work with the State Engineer's Office and her career at the university are described in Mike Mackey's "Grace Raymond Hebard: Shaping Wyoming's Past," Wyoming Historical Society (website), WyoHistory.org, published November 29, 2014, http://www.wyohistory.org/essays/grace-raymond-hebard.

116.   Lampen, *A Report of an Economic Investigation*, 1–2.

117.   Anna Christofferson interview, Powell *Tribune*, 1979, Marse and Anna Christofferson Family, oral history files, Homesteader Museum, Powell, WY.

118.   Betty Christofferson interview, 1996, Marse and Anna Christofferson Family, oral history files, 15–16, Homesteader Museum, Powell, WY.

Chapter 4

1.   Trelease and Lee, "Priority and Progress," 70.

2.   Johnson, *Trails, Rails and Travails*; see 3–4 for "Horse Creek and Bear Creek Valleys," quoting Nelson Sherard article in the *Torrington Telegram* in 1946; and 97–98, for "Nelson H. Sherard and Adella Hubbs Sherard," by Donald Sherard (their son).

3.   For the townsite company activity, see endnote 19, below. For Horse Creek flows, Hinckley Consulting and AMEC, *Horse Creek Groundwater/Surface Water*, 3–4.

4.   Johnson, *Trails, Rails and Travails*, 3–4, 97–98.

5.   On the trial of Tom Horn, see Larson, *History of Wyoming*, 372–74. LaGrange area and other ranchers on the jury are listed in Davis, *The Trial of Tom Horn*, 73–74. Nels Sherard's interest in the trial is in "Reminiscences of Donald N. Sherard," September 21, 2007, as told to his son Stephen Sherard, Wheatland, WY, typed manuscript on file with the author.

6.   Map accompanying "Permit 5937 Sherard Ditch," and two small reservoirs, Permit 494R, March 22, 1904, signed by surveyor John H. Gordon, State Engineer Permit Files; Johnson, *Trails, Rails and Travails*, 98.

7. Horse Creek Conservation Dist. v. Lincoln Land Co., 54 Wyo. 320, 92 P. 2d 572 (1939), Case File Docket 2093, case file: Abstract of Record on Appeal, Appellant (Lincoln Land. Co.), July 8, 1938; 9, Testimony of Nelson H. Sherard (for respondent Horse Creek).

8. *Horse Creek Conservation Dist.* 54 Wyo. 320, 92 P.2d 572 docket 2093, case file, at Abstract of Record on Appeal, Appellant (Lincoln Land. Co.), July 8, 1938, 18–19, Testimony of Hugh Stemler (for respondent Horse Creek); 15, Testimony of Earl L. Chamberlain (for respondent Horse Creek); 21–22, Testimony of Otis N. Lovercheck (for respondent Horse Creek).

9. Hugh Stemler, Otis Lovercheck, and Earl Chamberlain families in Johnson, *Trails, Rails and Travails*, 104–6, 81, 63–64.

10. *Horse Creek v. Lincoln Land* case file, Abstract of Record on Appeal, Commissioner Charles C. Donahue at 25–33, Superintendent L. C. Bishop at 33–34; Horse Creek v. Lincoln Land case file, Brief of Respondent Horse Creek Conservation Dist., September 20, 1938, 13–15. Water commissioner C. C. ("Clint") Donahue came to the area with his parents at age twenty-one in 1904, homesteaded on his own, and then bought an 1883-origin ranch on Little Horse Creek. He married a neighboring young woman homesteader and installed what his neighbors considered an "ingenious" water system taking spring water to house, fishpond, and garden near Meriden. He worked as water commissioner for forty-five years, starting in 1916. His brother Dan, whose wife refused to come to Wyoming without her piano, ran a Meriden store that had a much-appreciated dance hall attached. Bastian, *History of Laramie County*, 241, 225.

11. Hinckley Consulting and AMEC, *Horse Creek Groundwater/Surface Water*, figure 2–14.

12. *Horse Creek Readjustment of Rights*, BOC, Petitions Granted Files: Letter of Frank Kittle, Superintendent of Division I, October 4, 1917; Wyoming State Board of Control, Board order January 7, 1918.

13. Utt v. Frey, 39 P 807 (Cal. 1895); CF&I Steel Corp. v. Purgatoire River Water Conservancy Dist., 515 P.2d 456 (Colo. 1973); Trelease, *Cases and Materials*, 188–205.

14. "Abandonment" and "forfeiture" are not completely interchangeable terms. Abandonment as a legal term generally implies the idea of intent; a statute setting a strict term of period of non-use after which a right will be lost is considered a provision for "forfeiture" of rights, which does not require a showing of intent. Wyoming since 1888 has had what is technically a forfeiture statute for water rights, even though what it concerns is commonly called "abandonment" in Wyoming. MacDonnell, *Treatise on Wyoming Water Law*, 154n830. Both terms have appeared in Wyoming statutes relating to loss of water rights since at least 1913; see 1913 Wyo. Sess. Laws 106 §1; 1973 Wyo. Sess. Laws 176, § 1; and Wyo. Stat. Ann. §41-3-401(a). The confusion inherent in the use of both terms has played out throughout Wyoming water law, with the state's high court tending to resist forfeiture in various ways, including by requiring a show of intent not to use water, and the State Engineer's Office tending to seek strict action of the statute without discussion of intent. More Wyoming Supreme Court cases on abandonment are discussed later in this chapter.

15.   1888 Wyo. Sess. Laws 55, § 14. The Wyoming Territorial Legislature of 1888 declared the fatal period of non-use to be only two years, and Mead's water laws adopted that standard; Mead, *Second Biennial Report*, 26. In 1905 water administrators opposed changing the abandonment period to five years, but some water users seeking that change were also legislators. Since 1905, five years has been the standard; 1905 Wyo. Sess. Laws 39; Wyo. Rev. Stat. 41-3-401; Johnston, *Eighth Biennial Report*, 95–96; Trelease, *Cases and Materials*, 194. On Board of Control power in abandonments, see Van Ordsel, December 12, 1904, opinion, *Biennial Report . . .1903–1904*; Parshall, *Eleventh Biennial Report*, 25, 33–34; 1913 Wyo. Sess. Laws 106.

16.   Lands in Goshen County were acquired by the Lincoln Land Co., the townsite and land arm of the Burlington Railroad, and its subsidiary Goshen Hole Irrigation. Such companies bought lands on possible rail routes and sold them off when and if the railroad arrived; investors or contractors for townsite companies sometimes were also Burlington staff; Overton, *Burlington West: A Colonization History*, 182–83, 473. Some said that the company bought thousands of acres from the British-owned Union Cattle Co. (whose cowboy, Goshen Hale, gave his name to the county) and that Union Cattle Co. originally had acquired its lands by having its cowboys file fake homestead claims to get land titles then conveyed to the company. Downing with Smith, "Recollections of a Goshen County Homesteader," 53–72, 68. Downing did not arrive in Goshen County until 1910, so some of his "recollections" are likely hearsay. For relevant land acquisition by Goshen Hole Irrigation around LaGrange, see the following records located in Goshen County Archives, Torrington, WY: Affidavits of Henry S. Bush, December 6, 1920 and W. J. Turner, October 2, 1974 in Goshen County Clerk Misc. Documents Book 53, p. 658 and Book 373, p. 498; Desert Land entry conveyed to Goshen Hole Irrigation (covering parts of future 66 Ranch and its pasture reservoir), W. F. DeNyse of New York to Edw. Page of Massachusetts, 1886; Page to Goshen Hole Irrigation Co., 1887: Goshen County Clerk Deed Book 7, pp. 116, 349; Desert Land patent from US to W. F. DeNyse, 1891, Goshen County Clerk Patent Book 2, p. 100. Federal agents tried to track the Goshen dummy entry fraud; Carey's colleague Sturgis pointed that out and warned of potential similar investigations into the irrigation colony project on the Laramie River in which he and Carey were invested, Thos. Sturgis to Carey, January 19, 1887, Correspondence 'S,' box 27, folder 12, Carey Family Papers. For Lincoln Land's water rights activity and the territorial court decree, see Horse Creek Decree, First Judicial District, Wyoming Territory, June 12, 1889, summarized in Board of Control, *Tabulations of Adjudicated Surface Water Rights, Div. 1, 1996*, 30. Horse Creek Ditch #1 held by Goshen Hole Irrigation Co. was awarded a water right dated September 18, 1884.

17.   *Horse Creek Readjustment of Rights*, the district court decision, and Letter of Frank Kittle, Superintendent Division I, October 4, 1917, are filed with Board order January 7, 1918, in BOC, Petitions Granted Files. True convinced the 1917 legislature to appropriate $4,000 ($88,000 in 2020 dollars) for an investigation and "readjustment" of rights on those creeks. 1917 Wyo. Sess. Laws 125, §27. (That sum matched what went that year to the popular dream of studying more big dams for irrigation for central Wyoming. 1917

Wyo. Sess. Laws. 125, §26.) The field survey was done by Elmer K. Nelson, Report on the survey of Crow and Horse Creeks and their tributaries, submitted to James B. True, Wyoming State Engineer, November 14, 1917; *Horse Creek Readjustment of Rights*, BOC, Petitions Granted Files.

18.   Tour of Horse Creek area by author with a hydrographer-commissioner staff to the Board of Control, November 2011.

19.   "Horse Creek Readjustment," BOC files—Surveyor's notes on owners of original rights; Hearing Transcript, November 16, 1917, 7, 14; Order of the BOC, January 7, 1918; BOC, *Tabulation of Adjudicated Surface Water Rights, Div. I, 1996*, 29–31 reprints the order of priorities set on Horse Creek by the territorial court in 1889: priorities #8 and #9 belonged to F. E. Warren's Warren Land and Livestock Co., while priority #35 belonged to the Maple Grove Land and Livestock Co. represented by John W. Lacey, former law partner of Justice Van Devanter, and the water right that Sherard and others continued to contest into the 1930s was priority #52. BOC, *Tabulation of Adjudicated Surface Water Rights, Div. I, 1996*, 617–19 reprints the order of priorities set on Crow Creek by the territorial court in 1888 where George Baxter held priority #43. Former US Senator and Governor J. M. Carey also had water rights from Horse Creek but the surveyor found his ditches "in good condition," so they were not included in the board's declaration of abandonment.

20.   *Horse Creek Readjustment*, BOC Petitions Granted Files: List of recipients of Kittle letter October 4, 1917; Elmer K. Nelson, Report on the survey of Crow and Horse Creeks and their tributaries, submitted to James B. True, Wyoming State Engineer, 11–24–1917; Board order January 7, 1918.

21.   The attorney general appealed the district court order to the Wyoming Supreme Court on behalf of the state engineer, but he later withdrew the case. Comments of J. W. Lacey on Horse Creek Readjustment, "Transcript of Board of Control hearing," November 26, 1917, Board of Control Petitions Granted files, 14, 37; draft brief to court, Wyoming Attorney General, n.d., 2, 5, 7. In the matter of *Horse Creek Readjustment of Rights*, "Laramie County District Court Order, May 27, 1918," case file 11–478, Wyoming State Archives; Wyo. Const. art VIII, §2; Wyoming Supreme Court docket file 3–972, 973, Wyoming State Archives; "Judge Lacey, Teapot Dome Lawyer, Dies," *Laramie Republican Boomerang*, February 11, 1936; Trenholm, *Wyoming Blue Book*, 2:466.

22.   Nationally acclaimed trial lawyer Moses Lasky in 1929 described the growth of private property thinking in water law as creating a dichotomy between lawyers and "the layman," arguing that "legalistic thought obscured the layman's natural inclination" as lawyers and courts insisted on a private property content to water rights. He cited early Utah, Colorado, and Wyoming territorial statutes as evidence of laymen's belief in public benefit rather than private property as the concern of water law. Lasky, "From Prior Appropriation," no. 3, 166–69; Wiel, *Water Rights in the Western States*, sec. 567; West Group, *West's Encyclopedia of American Law*, 1998.

23.   For drought history charts, see Curtis and Grimes, *Wyoming Climate Atlas*, 101. In Larson's *History of Wyoming*, 411–46, the chapter entitled "Depression Years, 1920–1939,"

begins: "The title of this chapter may surprise some readers who have accepted the standard American-history textbook interpretation of the 1920's as a decade of prosperity after
a short depression in 1920–21. Although textbooks offer a picture of over-all prosperity in
the nation, they usually concede that agriculture and coal mining did not share in that
prosperity. Agriculture and coal mining were important in Wyoming, as was oil, another
industry that suffered a severe setback in Wyoming in the twenties. Because of deflation
and the 1919 drought, the livestock business suffered extraordinary distress." Cassity, *Wyoming Will Be Your New Home*, 205–44, describes the forces that affected would-be independent small farmers and ranchers in Wyoming in the 1920s into the 1930s and began
the process that cut back the number of small self-sufficient holdings. The Board of Control in the 1920s showed itself perfectly willing to consider each case on its merits, finding
abandonment in some situations but not in others. "Report of the Secretary of the Board
of Control" in Emerson, *Fifteenth Biennial Report*, at 72; "Report of the Secretary of the
Board of Control" in Emerson, *Eighteenth Biennial Report*, 36. But the engineer and the
superintendents saw plenty of reasons for cleanup efforts using the abandonment tool.
In the thirty years since Mead's water system was established, energetic land and water
development had led to confusion and contradiction in water rights statewide, just as on
Horse Creek. Water superintendents across the state reported instances of irrigators using
more water than they were ever entitled to, tacking their old priority date onto it in an
"expansion" of their original water rights. Irrigators thought that was innocuous; administrators thought it illegal and unfair to later settlers. Superintendents themselves meantime
had also fallen prey to the temptation not to inspect personally every water right they had
to adjudicate, because they had simply too many to review. Bond, *Fifth Biennial Report*,
76–79; "Report of William Gilcrest, Superintendent of Water Div. No. 1" (Southeast
Wyoming, including Horse Creek) in Johnston, *Seventh Biennial Report*, 20; Report of
C. E. Howell, Superintendent, Water Division No. 4 (Southwest Wyoming) in Parshall,
Eleventh Biennial Report, 32; True, *Thirteenth Biennial Report*, 86–87; Emerson, *Fifteenth
Biennial Report*, 62–63. As a result, administrators had issued certificates to water rights
which were likely to overstate, based on irrigator testimony, how much water that irrigator had succeeded in using. In the Horse Creek case, the unused rights True sought to cut
back were of all kinds—rights his own office had approved, as well as rights confirmed
in territorial court. *Horse Creek Readjustment*, BOC: Order, January 7, 1918. The BOC
opposed revival of unused rights, and instead required an unused old priority to be lost
and a new right to be issued in order to start use again, as demonstrated in 1922: Gottlieb
Fluckiger in BOC, *Order Record Book 6*, 157, *5 Minute Record Book*, 413. The water right
holder followed board policy and voluntarily abandoned a water right for forty acres, for
lack of use, and applied for new 1922 water right for the same acreage.

 24. "Yoder Family" as told by Oscar Yoder, son of B. F. (Frank) Yoder, 1966–67, in
Johnson, *Trails, Rails and Travails*, 755–56; Cooper, *History of Wyoming Water*, 32–35.

 25. Hinckley Consulting and AMEC, *Horse Creek Groundwater/Surface Water*, figure
2–14.

 26. Hinckley and AMEC, figure 2–14; the townsite investment company's permits

were numbered 1980E, Enlarged Horse Creek No. 1, Lincoln Land Co., and 1415R, 66 Pastures Reservoir, Lincoln Land. Co., both dated October 28, 1908, in Board of Control, *Tabulation of Adjudicated Water Rights, Division 1*, 33; Memoranda of agreements between Lincoln Land Co. and Hawk Springs Development Co., May 24, 1912 and September 1, 1921 (enclosures in letter, Kara Brighton to Randy Tullis, Supt. of Water Div. I, May 15, 2007); stamp on permits signed 1908 by State Engineer Clarence Johnston include: permits 1307R, 1892E, 1980E, and 1415R, in State Engineer Permit Files.

27.   Notes on assignments (undated), entered on applications for Permits 1307R, Hawk Springs Reservoir, and 1892E, Enlarged Lowe Cattle Co. No. 1, supply ditch for Hawk Springs Reservoir, in State Engineer Permit Files.

28.   Cassity, *Wyoming Will Be Your New Home*, 251–91. The Taylor Grazing Act of 1934 closed the public domain, ending its disposal into private hands and establishing the federal grazing lease system still operating today.

29.   *Horse Creek Conservation Dist.* 54 Wyo. 320, 92 P.2d 572.

30.   *Horse Creek Conservation Dist.* 54 Wyo. 320, 92 P.2d 572 at 9, 15, 18–19, 21–22, 27, Abstract of Record on Appeal: Testimony of Sherard at 9, Testimony of Stemler at 18–19, Testimony of Chamberlain at 15, Testimony of Lovercheck, 21–22, Testimony of Donahue, 27. The year 1933 was prime time for abandonment claims—the Wyoming Supreme Court itself took note of the weather, in a different abandonment case, recording that "commencing with about 1930 or 1931, a period of drouth settled over southeastern Wyoming, which became more and more severe from time to time, reaching its height in 1933 and lasting at least through a part of 1934," Van Tassel v. Cheyenne, 54 P.2d 906 (1936).

31.   *Horse Creek Conservation Dist.* 54 Wyo. 320, 92 P.2d 572, Abstract of Record on Appeal 40–42, 43–45; *Horse Creek Conservation Dist.* case file, at brief of appellant, 11, 34–38; "Yoder Family," Goshen County History Committee, *Wind Pudding and Rabbit Tracks*, 755–57.

32.   Board of Control, "Order, Apr. 20, 1934," *Order Record Book 7*, 695.

33.   *Horse Creek Conservation Dist.* 54 Wyo. 320, 334–37, 92 P.2d 572; *Horse Creek v. Lincoln Land* case file, Abstract of Record on Appeal, 49–52, summary of Findings of Fact, Conclusions of Law and Decree, Laramie County District Court, April 22, 1938; *Horse Creek v. Lincoln Land* case file, Brief of Appellant Lincoln Land; *Horse Creek Conservation Dist.* at 334–37. *Horse Creek Conservation Dist.* at 335–36, citing Zezi v. Lightfoot, 68 P.2d 50 (Idaho 1937). In a water right abandonment case two years earlier, the Wyoming court also specifically cited legal commentators on water who discussed the general rule that "forfeitures are not favored in law," in the water context. Ramsay v. Gottsche, 51 Wyo. 516, 69 P.2d, at 529 (1937). For the general policy against forfeitures, see *West's Encyclopedia of American Law*, 1998. The leading treatise in western water law in the 1930s notes that as a result of that rule, abandonment was not easy to find even in the customary nineteenth century water law in the West. Once the abandonment concept was put into statute, the treatise noted, statutes that specifically used the term "forfeit" were essentially a legislative response to court's reluctance to order forfeiture. (Kinney, *Treatise on Irrigation*, §1118, cited in Ramsay v. Gottsche, at 529.) Legislatures were attempting to overcome

that reluctance, to impose a strict standard requiring forfeiture. Wyoming's legislature had used that term for the loss of water rights since 1888 (1888 Wyo. Sess. Laws 55 §14). The Wyoming high court in the 1930s, responding in its turn to the legislature, sought to mute the effect of that harsh word *forfeit* by putting up barriers to abandonment—as in its endorsement of water right revivals in 1939. Property law scholar Carol Rose has noted that courts and legislatures, in a never-ending attempt to reach a balance between clarity and equity, often do engage in moving the law back and forth between "crystalline" and "mud" rules affecting property, particularly when the issue is forfeiture; Rose, *Property and Persuasion*,199–225. The Wyoming Supreme Court appears to have triumphed there, even after State Engineer Bishop later got a "crystalline" rule on abandonment into statute with the language "fails, either intentionally or unintentionally, to use the water" adopted by the 1973 Legislature (Wyo. Stat. Ann. 41-3-401(a)). The court has opted for "mud" for an abandonment rule that looks straightforward in the statute book but in practice rarely allows a loss of water rights for non-use.

34.   For similar decisions in other states, see Hall v. Lincoln, 50 Pac. 1047 (Colo. 1897); Platte Valley Irrigation Co. v. Central Trust Co., 75 Pac. 391 (Colo.); Carrington v. Crandall, 147 P.2d 1009 (Idaho 1944); Trelease, *Cases and Materials*, 196–97. In Wyoming as elsewhere, the neighbor claiming abandonment has to carry the burden of proof on many issues in a hearing before the board or in court, *Ramsay*, 51 Wyo. 516, 69 P.2d, 535; *Hall*, 50 Pac. 1047 (Colo. 1897). One exception: the burden of proving unavailability of water, as a defense, rests on the one charged with non-use. In the Matter of Johnson Ranches, 605 P.2d 367 (Wyo. 1980).

35.   A variety of cases fleshed out how revival could work in Wyoming—and its limits. Sturgeon v. Brooks, 281 P.2d 675 (Wyo. 1955) held that a right to fill a reservoir with a damaged dam that has not held water for years can be revived by dam repair, and use of the water—so that an abandonment petition filed two years after water use restarted would fail; Ward v. Yoder, 355 P.2d (Wyo. 1960), held that a landowner purchaser of land with an old unused right who aimed to clean out its grassed-in ditch could be stopped by an abandonment claim filed before the cleaning was accomplished or water put to use; Wheatland Irrigation District v. Pioneer Canal Co., 464 P.2d 533 (Wyo. 1970) held that the Board of Control could not give the owner of a damaged reservoir a grace period to get the dam repaired in order to avoid an abandonment ruling. In a talk at the Wyoming Game and Fish Commission, former Wyoming Superintendent of Water Division III Craig Cooper described how neighbors noticing truckloads of pipes being delivered could successfully derail a city's plans to revive an old right and put it to municipal use. Fishery and Wildlife Managers Educational Seminar, January 29, 2003, audio tapes on file with author.

36.   Hinckley Consulting and AMEC, *Horse Creek Groundwater/Surface Water*, 1–2–4, 2–13, 3–4–6; Wyoming Statutes 41-3-916 (1957). Groundwater was not formally the job of the State Engineer's Office until 1947, after nearly sixty years of surface water management. The 1957 statute enacted a more complete code governing groundwater; see MacKinnon, "The Prospects for Management," 5–7.

37. "Wyecross Ranch," in Johnson, *Trails, Rails and Travails*, 119–20; Curt Meier, personal communication with author, April 27, 2018.

38. Hinckley Consulting and AMEC, *Horse Creek Groundwater/Surface Water*, 3–8–9, 3–14–15 (a chronology of regulatory issues raised on Horse Creek); BOC Petitions Granted Files for Horse Creek Conservation District, I-78–210 (1979); BOC, *Order Record Book 28*, 359–69; Judge Alan B. Johnson, First Judicial District of Wyoming, Memorandum on Petition for Review, John Meier and Son, Inc. v. Horse Creek Conservation District and State Board of Control for the State of Wyoming, December 17, 1980.

39. Hinckley Consulting and AMEC, 1–3, 2–12, 2–17, 3–14–15, also figures 3–4; the report notes that the state engineer authorized the report in response to Horse Creek Conservation District requests in 2009 that groundwater wells be regulated in priority along with surface water rights (which would mean that groundwater wells, all of relatively late date, could not be pumped until after most surface water rights had been satisfied). In late 2011, the State Engineer's Office (SEO) held a public meeting to review report results in hopes that better information would help both sides reach an accommodation, according to an SEO October 19, 2011, press release. The following state engineer orders provide additional information: "Order of the State Engineer—Horse Creek Basin, July 19, 2013" and "First Amended Order of the State Engineer—Horse Creek Basin, May 31, 2017," at https://sites.google.com/a/wyo.gov/seo/; SEO presentation, Horse Creek Public Hearing presenting 2014–2016 water use data, February 15, 2017.

40. The current owner of the old townsite company ranch and its reservoir (known as Pasture 66 Reservoir) is Curt Meier, elected treasurer of Wyoming in November 2018. Meier's lawyer in the groundwater disputes, Hageman, was an unsuccessful candidate for the Republican nomination for Wyoming governor in 2018. Senator Curt Meier, personal communication with author, April 27, 2018. Goshen County Assessor's office, Account # R00220912, includes some sixty acres of irrigated cropland, owned by Mead Land and Livestock LLC of Cheyenne, annual report in Filing ID 2001–000427453, Wyoming Secretary of State's office, Cheyenne, showing registered agent is Matthew H. Mead (governor of Wyoming, 2011–19). Horse Creek Conservation District, owner of the Hawk Springs Reservoir, reported its entire service area as over 10,500 acres of irrigated land. Wyoming Water Development Commission, *State of Wyoming 2015*, 4

41. The court called for close scrutiny in abandonment cases and saved water rights from abandonment for a variety of reasons: the water had been used somehow, through a stream or another ditch, Van Tassel v. Cheyenne, 54 P.2d 906 (Wyo. 1936); there was no water available, Simmons v. Ramsbottom, 68 P.2d 153 (Wyo. 1937); some "fault or neglect" on the part of a water right owner needs to be shown for abandonment (whether this means the owner had to "intend" to abandon has been hotly debated), Ramsey v. Gottsche, 69 P.2d 535 (1937); intent is not necessarily required, but the abandonment must be voluntary, not forced by circumstances, Scott v. McTiernan, 974 P.2d 966 (Wyo. 1999); those claiming abandonment had to be clearly affected by the fate of the contested water right or they had no standing to bring an abandonment case, Hagie v. Lincoln Land Co. 18 F.Supp. 637 (D.Wyo. 1937), cited with approval by the Wyoming Supreme Court

in Platte County Grazing Association v. Board of Control, 675 p.2d 1279 (Wyo. 1984), Cremer v. State Board of Control, 675 P.2d 250 (Wyo. 1984). Further, the rule of abandonment in Wyoming, as in other states, meant simply that water had to be used only *at some point* in the required period of years. One good wetting of a field amid five years of otherwise non-use can be enough to defeat an abandonment charge. Jacobs, Tyrell, and Brosz, *Wyoming Water Law*, 11.

The discussion is lively on whether there an "intent" to abandon a water right is required. Under *Ramsey*, 69 P.2d 535 (Wyo. 1937), the new owner of a water right would not lose it through abandonment, since he showed no "fault or neglect" in not using the water, when after years of damaging floods he started diversion repairs in a reasonable time. The court in Ward v. Yoder (1960) (water right with grassed-in ditch, abandonment charge succeeds when filed after owner starts to clean ditch but before water is used again) declared no "intent" was necessary to find abandonment. Administrators, however, have read the court's decisions differently, Cooper, *History of Wyoming Water*, 78 (Cooper is a former member of the Board of Control, as superintendent of Water Division III.). As long-time state engineer Floyd Bishop, put it, "*no one* intends to abandon a water right!," because the rights are too valuable; interview with the author, December 13, 2010. Bishop backed amendment of the abandonment statute in 1973, with language (passed by the legislature with little hesitation) that explicitly states a water right would be forfeited if its holder "fails, either intentionally or unintentionally, to use the water . . . during any five (5) successive years"; Wyo. Stat. Ann 41–3-401(a). Wyoming, *Digest of Senate and House Journals*, 645–49. But members of the Board of Control believe the state court essentially demands evidence of intent to abandon. The court pronouncement in Scott v. McTiernan, 974 P.2d 966 (Wyo. 1999) that abandonment must be "voluntary" seems to require inquiring into the state of mind of the water user who failed to use water. In *Scott*, irrigated lands once held by the same owner had been split up. One owner of a resulting portion filled in the ditch that took water to lands now owned by a neighbor, making that water unavailable for five years—while reassuring the neighbor that he would open the ditch back up again. Then this deceptive landowner brought an abandonment charge against the neighbor after the requisite five years had passed. The board found that the neighbor's failure to take the recourse available in civil law to get the ditch opened back up was sufficient "fault or neglect" (under earlier Wyoming Supreme Court cases) to justify a declaration of abandonment. The Wyoming Supreme Court disagreed, based on a somewhat abstract sense of equity that contrasted with the board's practical sense of how water use and water users actually work on the ground. The court's result suggested that the board should weigh the attitude of water users, which to a practical mind means examining their intent. For more on the board's wrestle with Supreme Court abandonment decisions, see Cooper, *History of Wyoming Water*, 89–90.

42.   Wheatland Irrigation District v. Laramie Rivers Co., 659 P.2d 561 (Wyo. 1983) at 565–66. In most of its abandonment decisions, the court did not take the modest approach articulated in *Wheatland Irrigation*. Here, however, the court found abandonment where the Board of Control (attempting to follow the court's earlier decisions) had

found none. The court held that an irrigation company that had spent years to get state financial aid and had finally gotten the funds and put repairs underway on its reservoir nonetheless could lose its water right to abandonment when the abandonment claim was filed before the repairs could be completed and water put to use. The case complemented the holding of Sturgeon v. Brooks, 281 P.2d 675 (Wyo. 1955) that successful repair of a reservoir and the putting to use of its water could defeat an abandonment claim filed two years after the water use recommenced. The state's investment of major funds in Laramie Rivers Co.'s repair of its dam at Lake Hattie may, however, have persuaded the Board of Control that in this case the company had made enough of a revival effort to defeat abandonment.

43. Cooper, *History of Wyoming Water*, 89–90. Administrators have also been aware of how difficult, and therefore unlikely, it can be for water users to bring an abandonment action against a neighbor. For that reason, administrators have long sought a legislative reversal of the original court decision on Horse Creek and Crow Creek in 1917, which said that Wyoming water administrators could not themselves use abandonment as a tool to clean out old unused rights. Administrators felt they had to get authority to bring abandonment actions themselves, since "the people evidently are not interested in the matter" and rarely file abandonment claims, the northwest Wyoming Water Div. III superintendent Lou Blakesley wrote in 1920; in Emerson, *Fifteenth Biennial Report*, 83. Quotations from Bishop, *Thirty-seventh Biennial Report*, 24; Bishop, *Thirty-eighth Biennial Report*, 37. State Engineer Floyd Bishop requested research on the problem of unused rights that resulted in a 1970 *Land and Water Law Review* article detailing the problem in Wyoming; see McIntire, "The Disparity Between State Water Rights." In 2012 State Engineer Pat Tyrrell told the Board of Control that he would like to see a legal requirement that the State Engineer's Office review water rights periodically and abandon those not in use. "There are 200k acres of adjudicated rights in this state that aren't used," Tyrrell said (Board of Control, August 2012, notes in author's files). The Wind-Big Horn River adjudication discussed in chapter 5 did review and cancel unused permits in that river basin, and superintendents in other divisions, as in Division IV, have attempted "clean-ups" of unused rights when the opportunity presents itself in a petition coming before the board that reveals the existence of unused rights. Jade Henderson (former superintendent Div. IV), personal communication with the author, October 8, 2018.

44. In Hughes v. Lincoln Land Co., 27 F. Supp.972 (Wyo. 1939), US District Court Judge Blake Kennedy (who had vivid memories of the early 1900s in Wyoming, when he had been active in Republican politics) ruled that the owner of a territorial water right could use the water to irrigate different lands he owned, since the water right was a property right. (The case involved the same townsite company, Lincoln Land, with different lands and water rights on Horse Creek, as in the case discussed in this chapter.) Kennedy relied on the Little Horse Creek case of 1904 to say that water rights in Wyoming had been held to be property rights and "no statute which the State might subsequently pass could abridge that property right or reduce its value without intrenching upon the constitutional right of the owner." He acknowledged the 1909 statute passed in reaction

to the Little Horse Creek case but said that statute could not be implemented to violate constitutional rights; *Hughes v. Lincoln Land Co.* at 973–74. Kennedy's ruling has been subsequently applied to mean that pre-1909 water rights were not affected by the 1909 no-transfers statute. This is relevant to the transfer of rights from the Ringsby Ranch to the Wheatland Irrigation District in the 1960s, discussed later in this chapter.

45.   Larson, *History of Wyoming*, 414–16; Cassity, *Wyoming Will Be Your New Home*, 251–91.

46.   At issue in water transfer laws is not the question of whether a user can sell his water right to another person who will use the water in the same place in his stead. Water rights can be sold along with the irrigated fields or industrial sites they serve—and irrigation water rights have since early statehood in Wyoming considered to be sold along with the land even if the deed fails to mention them; *Frank v. Hicks*, 35 P. 475 (Wyo. 1892). What has caused struggles over water transfers is not the sale to another person, but the sale to another kind of use, or another place of use—the transfer of water away from the original purpose for which the right to the water was established. For the 1909 exception to the ban on such transfers, see Wyo. Stat. Ann. §41-3-102, 103.

47.   Emerson went on to explain: "Rights acquired under certain conditions might be found worthless by changes afterwards allowed other rights"; *Fifteenth Biennial Report*, 50. Water managers in Wyoming and other states have consistently been motivated by a pair of longstanding concerns: antipathy to speculation in water rights, as a danger to stable communities; and appreciation of the interdependence of water users tied to stream hydrology, where use patterns could be disrupted by unfettered transfer of water rights. See also Mead, *First Biennial Report*, 58–59. For further discussion of western states seeking to avoid "injury to other water users" by transfers, see Trelease and Lee, "Priority and Progress," 21–22; Robinson and MacDonnell, *The Water Transfer Process*, 3–3.

Wyoming was more determined than other states to enshrine those concerns into a ban on transfers. Initially focusing on surface water, Wyoming's cautious approach to water right transfers stems from the simple facts that one person's water use may reduce the volume of water available to others, and every stream has its own peculiar hydrology. Few streams have had their hydrology documented, as on Horse Creek, but irrigators will regularly refer to the unique hydrology they deal with—deep gravel or lack of it, for instance—and how that determines just how they'd like their neighbors to use water and whether they'll call in state administrators to enforce priority. Author interviews with irrigators on Bates Creek, Natrona County, Wyoming, and East Fork of New Fork River, Sublette County, Wyoming, December 2011 through January 2012, in author's files. The combination of water use that reduces flows and the hydrology particular to each stream creates a certain pattern of water availability on a stream. That pattern makes users inter-dependent and can make rules like transfer restrictions appear necessary to state staff and water users alike. See Ostrom, *Understanding Institutional Diversity*, 24–25.

48.   Bishop family lore: Heritage Book Committee, *Pages from Converse County's Past*, 56–59. The original organizers of the LaPrele Ditch and Reservoir Co. in 1905 were Dr. J. M. Wilson (born in Scotland, raised in Ohio, and moved to Douglas in 1886;

manager of the Platte Valley Sheep Co., 1896–1925); B. J. Erwin, a Missouri native arriving in Douglas in 1900 as Congregational minister, who soon switched to ventures in mineral development and cattle ranching; and W. F. Hamilton, a teacher who managed the government hay farm at Ft. Laramie in 1883 and became a townsite developer in Douglas for a Cheyenne bank, as well as a sheep rancher. Eventually, a new company got the project approved in 1923 and turned over to the Douglas Reservoirs Water Users Association, under the old 1893 Carey Act. A small society grew up around LaPrele farms, as one farm girl said of life there in the 1920s: "Pioneering was very difficult as the equipment was very primitive, housing very meager, and Wyoming weather very rugged for many of these people. Many hardships were endured, many defeats and failures encountered but the challenge was great. The efforts were all worthwhile when one saw a new country developed, a community grow and prosper, and friendships made that lasted a lifetime"; Heritage Book Committee, *Pages from Converse County's Past*, 77, 184, 245, 643–44, 665–67.

49.   How helpful it was to a water superintendent to be a good shot is demonstrated in the story that L. C. Bishop's son Loren Emerson Bishop, born 1911, told about his father as superintendent of Water Division I: "As water superintendent, on occasion he had to shut down headgates and stop ranchers from using water they wanted and needed. One occasion when I was with him at Pass Creek near Saratoga, a rancher threatened to shoot the local water commissioner so Dad was called to turn the water off. As we drove up to the ranch we saw a big hawk swoop near his chickens. Dad grabbed the 30–06 and killed the hawk high in the air with his first shot. After that exhibition, the headgate was closed with no trouble," Heritage Book Committee, 57. Federal planners of Seminoe Reservoir sought an early priority date for the dam, but L. C. Bishop, whose Division I covered the North Platte, staunchly objected: the controversy is traced in the General Correspondence files, National Archives, RG 115, Entry 7. The files include North Platte Drainage Basin, Kendrick Project, subgroup 031, letters: District Counselor to Bureau of Reclamation Commissioner Elwood Mead November 8, 1933; Chief Engineer to District Counselor November 17, 1933; L. C. Bishop to Commissioner Mead June 18, 1934; Commissioner Elwood Mead to Interior Secretary Harold Ickes January 22, 1935; Commissioner Mead to Wyoming State Engineer Edwin Burritt February 5, 1935. Seminoe was built with a 1930s priority date, along with Alcova Reservoir to raise water sufficiently to enter a long canal to Casper, in an overall project known as the Kendrick Project, named for former Wyoming US senator John Kendrick; "Kendrick Project," Reclamation (website), Projects & Facilities, https://www.usbr.gov/projects/index.php?id=340.

50.   Works Projects Administration, *Wyoming: A Guide to its History*, 334. Trelease and Lee, "Priority and Progress," 32.

51.   On the need to divert more water than the fields can consume, which was "well understood by early Wyoming lawgivers, but finds less understanding among non-irrigators today," see Cooper, *History of Wyoming Water*, Appendix A, 106.

52.   Board of Control, citing Petition of Town of Greybull, November 14, 1940, *Order Record Book 10*, 223–35; Shell Creek, Permit 430 in BOC, *Tabulation of Adjudicated . . .Div. III, 1999*, 158.

53. Town of Lander in BOC, *Order Record Book 7*, 593; Town of Greybull, BOC, *Order Record Book 10*, 223–35 (1940); Union Pacific Railroad in BOC, *Order Record Book 11*, 56; Trelease and Lee, "Priority and Progress," 33–34.

54. *Casper Star-Tribune*, "1st Lt. Floyd Bishop, Cheyenne" (in series "They Served with Honor: WWII"), July 17, 2011, https://trib.com/honor/wwii/st-lt-floyd-bishop-cheyenne/article_3630aea5–1491–532d81a6–1d099a88f0a3.html.

55. Reisner, *Cadillac Desert*, 225–63; Cooper, *History of Wyoming Water*, 67–69.

56. Trelease and Lee, "Priority and Progress," 64–68; Wyoming Statutes Annotated §41-3-110,111, "right to acquire temporary water rights for highway or railroad roadbed construction" first adopted in 1959.

57. Bishop, *Thirty-seventh Biennial Report*, 24; Bishop, *Thirty-eighth Biennial Report*, 37.

58. Trelease and Lee, "Priority and Progress," 62–64; Barnes, "Dave Johnston Power Plant Water Rights," 4, 6, 7–9; Randall Tullis, Water Division I Superintendent, interview with author, January 11, 1999.

59. The Laramie River was adjudicated by the Board of Control in 1903, and after litigation the final adjudication was made by the District Court of Laramie County at the end of 1912. BOC, *Tabulation of Adjudicated . . .Div. I, 1996*, 70. The earlier board action is noted by the Wyoming Supreme Court in Laramie Rivers v. Levasseur, 202 P.2d 680 (Wyo. 1949) at 682; Teele and Ewing, *The Economic Limits*, 8; Trelease and Lee, "Priority and Progress," 40–43. A Carey partner there was a young Irish nobleman, Horace Plunkett, later famed for work with Irish farmer cooperatives. The Wyoming Development Co. project and the town it called "Wheatland" was not a cooperative, though Plunkett and Carey did once discuss cooperatives with Wheatland farmers. The original project intent was for-profit; investors however often found themselves underwriting the company and scrambling for loans to keep afloat. Woods, *Horace Plunkett in America*, 52–54, 85, 109, 150, 155–58.

60. Trelease and Lee, "Priority and Progress," 43–67. Banner Engineering, "Proposed Methods for Augmenting"; Wyoming, *Opinions of the Office of the Attorney General*, 55–56 (April 1957).

61. After a review of the historic use, diversion, and return flow patterns on the ranch, the Board allowed a transfer of some of the pre-1909 rights in their entirety, while cutting other pre-1909 rights by one-quarter to one-half their original amount of water as a condition of the transfer. Board of Control, "Wheatland Irrigation District," BOC, *Order Record Book 16*, 1–26; Trelease and Lee, "Priority and Progress," 44–46.

62. Trelease quotation on "property aspects" is from 1960: Trelease, *Severance of Water Rights*, 39–40. For economic arguments in favor of water markets and transfers, see Trelease in 1960 citing J. W. Milliman, "Water Law and Private Decision Making" 54. In 1966, Trelease published his seminal law review article on transfers, expanding upon his 1960 report to the legislature, with Trelease and Lee, "Priority and Progress," see especially 70–73, 75.

63. Trelease and Lee, "Priority and Progress," see especially 70–75; quotation at 70.

64. Some legislators read the 1966 Trelease and Lee article closely: the copy owned by

Willard Rhoads of Cody was heavily marked up (in author's possession). But a large majority in the Legislature, including Rhoads, endorsed the bill embodying the Board's view of transfers rather than Trelease's. Wyoming, *Digest of Senate and House Journals*, 919–21; 1973 Wyo. Sess. Laws 170 §1; temporary change statute is 1959 Wyo. Sess Laws 148 §1.

65.   1974 Wyo. Sess. Laws 23 §1. The county economic loss issue was added in 1974 in response to opposition to the transfer of the hay ranch rights to Wheatland. Trelease and Lee, "Priority and Progress," 44. Cases demonstrating application of the 1973 statute include Basin Electric Power Coop. v. State Board Of Control, 578 P.2d 557 (Wyo. 1978); Green River Development Co. v. FMC Corp., 660 P.2d 339 (Wyo. 1983); Garber v. Wagonhound Ranch & Livestock Co., 279 P.3d 525 (2013).

66.   For examples of the scrutiny applied when a water right is moved to a new use, see these board reviews of irrigation rights being moved to a municipality: BOC Petitions Granted Files, City of Cody, III-2008–1-3 and City of Riverton, III-2009–2-10. The board often ends up reviewing moves that have already been made, sometimes years earlier, of water under an old water right to a new spot. Such moves can include moving an irrigation right to different lands for better crop yield or for an improved method of irrigation like pivot sprinklers: BOC Petitions Granted Files (all): James and Melisa Crouch, III-2010–2-4; MRDN Corp., IV-2011–3-4; Gordon Drum, II-2017–4-6; for examples of board scrutiny of proposed moves, looking for use either in the lands being moved from or being moved to, see BOC Petitions Granted Files (all): Overland Trail Cattle Co., I-2007–2-6; Dale E. Irthum, I-2018–2-6; D&T Ranches and S&J Farms, I-2011–3-10; Casper Alcova Irrigation District, I-2011–4-1; Walton F. and Rita Cherni Smith, II-2011–2-2. In the case of irrigation districts, where the board has focused on total diversion and return flow for the entire district, the board has in the past sometimes not required evidence from individual fields. Irrigation districts' individual fields have increasingly been examined for evidence of use in transfer proposals; contrast BOC Petitions Granted Files (both) R.E., M., L. and S. Kautz, I-2000–1-3 with Larry and Pat Goyen and Goshen Irrigation District, I-2011–4-2. Discussion of the Goyen proposal revealed that the large Goshen district on the North Platte typically ran a "lottery" market system to move unused rights to lands that were being watered without a water right. For board recommendation of revival, see for example BOC Petitions Granted Files, James and Deborah Housel, III-2010–4-7. The board is particularly likely to make a revival recommendation if the rights involved have an old date, useful in the board's view for keeping that water in use in Wyoming rather than flowing to downstream states.

67.   A requirement for five years consistent water use to support a transfer is recorded in the latest version of the board's written rules; State Engineer's Office, *Regulations and Instructions, Part IV*, chap. 5, secs. 15(c)6, and 16(i). Even where return flows from past irrigation went into a lake in a closed basin, unused by other water right holders, the board appropriately interpreted the 1973 transfers statute to disallow transfer of more than the amount of water consumed by the irrigated crops, the Wyoming Supreme Court ruled in Basin Electric Power Cooperative v. State Board of Control, 578 P.2d 557 (1978). The court specifically noted the relevance of abandonment concerns, saying "issues of nonuse and misuse are

inextricably interwoven with the issues of change of use and change in the place of use . . . even without the formal initiation of abandonment proceedings under the statutes. If an appropriator, either by misuse or failure to use, has effectively abandoned either all or part of his water right through noncompliance with the beneficial-use requirements imposed by law, he could not effect a change of use or place of use for that amount of his appropriation which had been abandoned"; *Basin Electric*, 564. Watering occasionally to avoid an abandonment action is not enough to establish sufficient historic use to move the water right in question to new lands, the board said firmly in BOC Petitions Granted Files, Brad Reese, I-2011-4-3 and Bard Ranch, I-2011-3-12.

68.   For complaints about board transfer reviews, see "Meeting Minutes," from Wyoming Legislature's Joint Agriculture, State and Public Lands and Water Resources Committee, September 27, 2018, 5–6, https://wyoleg.gov/InterimCommittee/2018/05–20180927MeetingMinutes.pdf.

69.   Anderson and Hill, "The Evolution of Property Rights," 137–38; Squillace, "A Critical Look," 307, 338, 340–41; Gould, "Water Rights Transfers," 35–36; Squillace, 338, 340–41; Squillace, "Water Marketing," 6–9.

70.   Anderson, *Tapping Water Markets*, 5–16, 42–46, 51–54, 65–74, 148–60.

71.   Mead, "The Growth of Property," 12.

72.   Lasky, "From Prior Appropriation to Economic Distribution," Pts. 1, 2, and 3, (April 1929) 162, 216; (June 1929): 270; (November 1929): 35, 45–46, 58.

73.   Anderson and Hill, "The Evolution of Property Rights," 137–38; Squillace, "A Critical Look," 307, 338, 340–41; Gould, "Water Rights Transfers," 35–36; Squillace, "Water Marketing," 6–9; Anderson, *Tapping Water Markets*, 72–73; Howe and Goemans, "Water Transfers and Their Impacts," 1055–65; *Colorado's Water Plan*, chap. 6, sec. 6.4.

74.   *Green River Development Co.* 660 P.2d 339 (Wyo.1983). In 1977, an old irrigation venture that had largely failed, holding unused water right permits dating from as early as 1908, made a deal with Pacific Power and Light Co. to sell some of its unused permits to the power company to serve a coal-fired power plant. The plan called for changing the original permits so that the water could be used 134 miles downstream from the original proposed location and be used for industry rather than irrigation. The irrigation company, known as Green River Development Co., asked for the state engineer's blessing. Four years later, the state engineer approved the plan. Ranchers and mining companies along the river were outraged and came to the Board of Control to protest the ghost of an old, large, and unused water permit suddenly coming to life with priority dates that would predate and disrupt some of their long-established water uses. The case led to a painful division between the state engineer and his superintendents, whose vote against allowing such a move was upheld by the Wyoming Supreme Court.

Chapter 5

1.   Pomeroy, *Treatise on the Law*, §100. Full quote (punctuation is as printed): "But the principal defect of the system, the one capable of working the greatest injustice, is

inherent in the very theory itself, in its fundamental conception. This defect is the total absence of any limit to the extent of a prior appropriation,—to the amount of water which may be taken,—except the needs of the purposed for which it is made. The prior appropriator, in order to carry out a purpose regarded by the law as beneficial, of great magnitude,—such, for example, as an extensive system of hydraulic mining, or the irrigation of a large tract of farming lands, or, doubtless, the supply of a municipality,—may divert and consume, without returning to its natural channel, the entire water of a public stream, no matter what may be its size or length, or the natural wants of the country through which it flows."

2.   As Wyoming state engineer Floyd Bishop commented in 1972, in non-drought years, "as is usually the case, most of our water administration problems involve personality differences and minor squabbles between water users which often are not directly a result of water shortages, and often originate with a quarrel over some other item"; Bishop, *Forty-first Biennial Report*, 4. John Teichert, superintendent of Div. IV from the 1960s–80s, said he early on found places that were "a no man's land. No one knew which ditch was which and if they did they weren't talking"; and elsewhere locals named as the "4th of July Ditch" a ditch that got filled up when the water commissioner was away on a bender over the 4th of July; Teichert, "Reflections of a Water Administrator," 10, 12.

3.   Two anonymous water users to author, in author's files, September 2020; Water Division I Superintendent Randall Tullis, to author, ca. May 2005.

4.   Tyrrell, "Instream Flow Overview," slide 6, "General Water Administration," states, "No active priority administration without a formal written call—'Free River' concept governs; Excess diversions for irrigation allowable if not under regulation and no waste," State Engineer PowerPoint Presentation, January 11, 2006, in author's files.

5.   *The River is Free* is a pamphlet prepared by the League of Women Voters of Wyoming and a graduate student in civil and environmental engineering from the University of Wyoming, n.d., but distribution presumably preceded a league-sponsored conference on the subject of wild and scenic rivers, held in Casper in early 1972. A copy can be found in the Stroock Papers, box 97, file folder "Water, 1967–85." The pamphlet text, discussing the antecedents and passage of the federal Wild and Scenic Rivers Act of 1968, makes it clear that the league sought to spark serious consideration in Wyoming of wild and scenic rivers–type legislation, inspired by the federal act. The first sentences of the pamphlet read, under the title "The River is Free": "America's recognition of the changing social values and of the diminishing unpolluted, free-flowing streams, prompted the 1963 study by the Departments of the Interior and Agriculture which led to passage by Congress of the 1968 Wild and Scenic Rivers Act, Public Law 90–542. America's waterways have always played a prominent part in its history and growth; but until recently, the alternative of preserving our nation's water resource as a free-flowing stream was neither recognized nor explored. The increased need for power, the affluent desires for the home, and our expanding industry have decreased the number of rivers in their natural state that are free from pollution, while the recreational need for them grows."

6.   Pomeroy, *Treatise on the Law*, §100.

7. Since the nineteenth and into the twenty-first century, ranchers and fishermen alike continue to be dismayed by fish ending up in irrigation ditches. A collaboration between the Wyoming Game and Fish Department and local ranchers to put in a new ditch headgate structure with special screens to keep fish from entering the ditch in the foothills by the Big Horn basin was heralded in 2018; see DiRienzo, "New Path Keeps Fish Healthy," 9. Marc Reisner describes the flurry of post–World War II federal dam building that affected Wyoming in *Cadillac Desert*, 145–50, 194–202. Dates and data on individual Bureau of Reclamation dams can be found at https://www.usbr.gov/projects/, last updated April 25, 2017.

8. Knight et al., *Mountain and Plains*, 8, 83–85, 109–10.

9. Federal Writer's Project, *Wyoming: A Guide*, 307–15, 389–91, 348, 245–47; Skaggs, "Creation of Grand Teton." Jackson, Wyoming, ski history is summarized at "Snow King Mountain Resort History," Snow King Mountain (website), https://snowkingmountain. com/snow-king-mountain-resort-history/. Wyoming Department of Administration, *Wyoming Data Handbook*, 21; "Wyoming County Profiles: Teton County," Wyoming Department of Administration and Information, Economic Analysis Division (website), http://eadiv.state.wy.us/Wy_facts/Teton2017.htm.

10. Federal Writer's Project, *Wyoming: A Guide*, 348. See, Finis, *Wind River Trails*, 6–10, containing a brief description of Mitchell's mountain fish-stocking and guest-guiding starting in 1930.

11. Marjane Ambler, "Bombardier Conservationist: Tom Bell and *High Country News*," Encyclopedia. Wyoming State Historical Society (website), WyoHistory.org, published November 16, 2016, http://www.wyohistory.org/encyclopedia/bombardier-conservationist-tom-bell-and-high-country-news. Ambler, who knew Bell well and interviewed him at length, describes his life and conservation passions in detail. See also Lillian Schrock, "Famed Wyoming Conservationist Tom Bell Dies in Lander," *Casper Star-Tribune*, August 31, 2016, quoting a Bell nephew.

12. Discussion of Wyoming's coal reserves and potential to host major power plants—some suggesting the state could be a "National Sacrifice Area"—produced a 1971 government-industry report predicting more than forty power plants would be built on the Northern Plains of Wyoming and Montana; Ambler, "Bombardier Conservationist." The nuclear project is detailed in Ann Chambers Noble, "The Wagon Wheel Project," Encyclopedia, Wyoming State Historical Society (website), WyoHistory.org, published November 8, 2014, https://www.wyohistory.org/encyclopedia/wagon-wheel-project. In 1958 the El Paso Natural Gas Co. was frustrated in attempts to produce from known reserves of natural gas on the edge of the Red Desert near Pinedale, south of the Wind River range. The company proposed to the Atomic Energy Commission in 1958 that the area be included in a program for using nuclear explosions underground to liberate natural gas as part of its search for peacetime uses of nuclear devices. In 1968, the company signed a contract with the agency to study a possible pilot explosion in the area; three pilot tests of such explosions were conducted in New Mexico and Colorado from 1967–73, but the Pinedale explosion never took place, largely as a result of local public

opposition. Bell's *High Country News* reported on the proposed Powder River basin coal-fired power plants and the aqueduct to serve industrial development in January 2 and May 26, 1972, editions. Archives of *High Country News* are available at http://www.hcn. org/issues?b_start:int=1120, and links to the 1972 articles are on p. 54 of the archives list.

13.   League of Women Voters, "A Free River," 8–10, Stroock Papers, box 97, file folder "Water, 1967–85."

14.   "Brief," dated May 22, 1968, accompanying the water right application filed January 13, 1969 in the State Engineer's Office (SEO) and signed by Tom Bell, as executive director of the Wyoming Outdoor Coordinating Council. The twenty-two-page brief was signed by L. W. Isaacs, C. L. Noble, Elmer George, Alvin B. Pearson Jr., Carroll R. Noble (all of the Cora-Pinedale area), and by Casper people representing the Wyoming Wildlife Federation, the Wyoming Outdoor Coordinating Council, and the Isaak Walton League. The brief and the permit application originals (along with responses from Floyd Bishop, the state engineer, and the special attorney general, Jack Gage, assigned to his office), are filed in the "Rejected Applications" files, "Thomas Bell, Temporary Filing No. 20, 1/173," SEO (hereafter cited as "Temporary Filing No. 20"). Economic studies are cited in the brief at 7–15. The conclusion about shrinking water supplies for a growing recreation industry, due to the subsidized competition from other users, is from a 1962 study funded by Resources for the Future, Inc. and published as Wollman et al., *The Value of Water*, 90–91.

15.   Brief, 2, "Temporary Filing No. 20."

16.   Application text, "Temporary Filing No. 20."

17.   T. Paul Stauffer, interview with the author, May 11, 2017; Stauffer, "Did I Fish Too Much?," 238–44.

18.   For reservation size since 1950 and today, see Eastern Shoshone and Northern Arapaho Tribes, *Draft Agricultural Resource Management Plan*, 15; O'Gara, *What You See in Clear Water*, 50–53, 165–67. For the Shoshone and Arapaho tribes' push for federal creation in 1934 of the roadless area of wilderness in the Wind River Mountains that still exists today, see Aragon, "The Wind River Indian Reservation," 15–16.

19.   O'Gara, 32–33, 49–53; US BOR, *Riverton Project History: 1976*, 15; 1977, 6–17.

20.   Quotations from the Treaty of 1868 between Shoshone and Bannock Tribes and the United States, Article II and IV, cited in Roncalio, Report Concerning Reserved Water Right, 59. Riverton's interest in groundwater and tribal response are cited in Roncalio, 7, from testimony from the Riverton city manager.

21.   Susan Williams, remarks, to 2008 conference "The Winters Centennial: Will the Commitment to Justice Endure?," reprinted in Cosens and Royster, *The Future of Indian and Federal Reserved Water Rights*, 170–72.

22.   Wilson, "Farming and Ranching," 325–26, 242–45, 328–29, 148–55; Shoshone Tribe v. U.S., 299 U. S. 476 (1937) (the tribe won compensation for the permanent settlement of the Northern Arapaho on the reservation). Act of July 27, 1939, 53 Stat. 1128–30 was enacted by Congress to follow the US Supreme Court's directions in the Shoshone suit. It also restored to the reservation the unclaimed ceded acreage. Reservation

unemployment was 24 percent in 1960, but with only 37 percent of employable adults working full time, underemployment was also high; Wilson, 151, citing Bureau of Indian Affairs, 1960 Population and Income Census for the reservation. For the 1968 Senate report, see US Congress, *Indian Education: A National Tragedy—A National Challenge*, 13–14, 21; for the effort to create tribal-run schools at Wind River, see Hipp, *Sovereigh Schools*, especially 22–64 on conditions prompting that effort; the quote on what tribal schools provide is from Wyoming Indian Schools (webpage), https://www.wyomingindianschools.com/32479.

23.   In 1988, state agencies estimated reservation unemployment at 79 percent, according to a brief filed with the US Supreme Court by the tribes in the water case; cited in Rusinek, "A Preview of Coming Attractions," 382; Champagne, "Organizational Change and Conflict," 14–15; and Rusinek, 404, citing Susan Williams's argument to the US Supreme Court. See Cosens, "The Legacy of Winters v. United States," in Cosens and Royster, *The Future of Indian and Federal*, 5–14; Roncalio, *Report of Special Master*, 13–16; Robison, "Wyoming's Big Horn General Stream," 267–72; O'Gara, *What You See In Clear Water*, 172–76. In 1952, Congress waived US sovereign immunity from suit in state courts if the suit was for "general adjudication" of an entire stream system; Robison (264–66) details the law on this point. As trustee for the tribes, the United States would be the defendant in the case regarding Indian water rights created by treaty (the tribes entered the case on their own behalf, as defendants, in 1978). The state of Wyoming much preferred a Wyoming state court to decide the question of tribal water rights. But to achieve that, the state had to undertake adjudication of all rights in the quarter of the state traversed by the Wind-Big Horn River (one river, changing its name on entering the Big Horn basin). The result was that, as the special master put it, the case included "several thousand defendants" claiming privately held Wyoming water rights (Roncalio, 2). One result, after the case ended nearly forty years later, was a massive cleanup of the unused portion of old, never-adjudicated permits in that part of the state.

24.   Christopulos, *Annual Report of the State Engineer, 1977*, 16 (State Engineer George Christopulos had long been deputy to Bishop). Christopulos wrote that through the lawsuit, he expected an end to uncertainty about federal claims in the basin—not only claims for the tribes (the United States acting in its trust capacity) but for federal lands such as national forests. Richard Baldes, personal communication with author, April 19, 2017.

25.   For coal-fired power plants, see US Bureau of Reclamation, *Report of Phase 1*. The projected spate of coal-fired power plants was never built. Most Wyoming coal was shipped out of state to power plants elsewhere, as railroads competed fiercely for the business. But a few new power plants were built, including one in Wheatland (see chapter 6 for discussion of environmental concerns raised by construction of the Wheatland plant). For the plant's boost to Wheatland's economy, see a discussion in 2000 by the 1980 town planner, Steve Achter, in Wyoming Rural Development Council, *Wheatland Community Assessment*, 9. Wyoming's one-term (1969–1971) US congressman John Wold, geologist, invested in federal coal leases in the Powder River basin east of the Big Horns that paid off well as big mines were built there in the 1970s. Newcomer Casper (website) Obituary,

http://www.newcomercasper.com/Obituary/131341/John-Wold/Casper-Wyoming; Bishop, *Forty-first Biennial Report*, 17–19.

26. For federal development on the Colorado, see a brief version in US Bureau of Reclamation, Upper Colorado Region (website), "Colorado River Storage Project," https://www.usbr.gov/uc/rm/crsp/index.html, last updated November 13, 2018. For detail on the 1948 compact, see www.ucrcommission.com). Riesner's *Cadillac Desert* provides a richly detailed version. For Wyoming's disappointing reservoir project on the Green River, see Linenberger, "The Seedskadee Project," 1997, and Wyoming Water Development Office (website), "Feb. 2011 Technical Memorandum: Use of Wyoming's Contract Storage Water in Fontenelle Reservoir," 2, http://waterplan.state.wy.us/plan/green/2010/finalrept/fontenelle.html.

27. Bishop, *Thirty-eighth Biennial Report*, 20–21; Bishop, *Forty-first Biennial Report*, 1–2.

28. Teichert, "Reflections of a Water Administrator," 3; Bishop, *Thirty-eighth Biennial Report*, 6–7.

29. Bishop, *Thirty-seventh Biennial Report*, 9–10.

30. Bishop, *Thirty-eighth Biennial Report*, 48. The superintendent commented: "Almost without exception whenever I have discussed the potential use of water for industrial purposes with local water users, I have observed a negative attitude on the part of most irrigation water users, which would appear to be the results of a lack of knowledge relative to laws and procedures which must be complied with before an irrigation right can be changed to an industrial use."

31. "The mortmain grip," in Trelease and Lee, "Priority and Progress," 70; Trelease and Lee, 73.

32. Bishop, *Fortieth Biennial Report*, 30–31; Bishop, *Forty-first Biennial Report*, 21–22.

33. State engineer-initiated abandonment actions were officially authorized by statute in 1973 Wyo. Sess. Laws chap. 176, sec. 1 W. S. 41-3-402(a). But as an assistant attorney general who represented the State Engineer's Office recalls, water users deeply resented seeing a Wyoming water rights administrator (not just a neighbor) trying to prove that a user had abandoned a water right. No state engineer has tried to use that tool again, according to Lawrence Wolfe; personal communication with author, February 2010. The statute lies on the books unused. Meanwhile, unused adjudicated rights and unused permits, with the shadows of uncertainty they cast on active rights, still abound on state records today. Wholesale review of unused rights has occurred only in the extraordinary case of the Wind-Big Horn and Shoshone Rivers, brought on by the state of Wyoming's decision to challenge tribal water rights on the Wind River. More commonly, unused rights or permits are dealt with one by one, if at all. Unused permits, when encountered, are considered expired after twenty years and can be "reinstated" only with affidavits that the water was in fact used; from Loren Smith, personal communication with author, May 24, 2018. Cases involving unused adjudicated water rights come regularly before twenty-first century water administrators. Often, a superintendent of one of the state's four water divisions will seek to clean up the record books so distributing water properly

will be easier, especially in dry years. A superintendent can sometimes manage to convince water users formally to give up some old unused water rights. See, for instance, in BOC Petitions Granted Files, Gaspar Wright, I-U-2007–1-4 and I-U-2006–2-8 through 14. Or in other examples, new owners of land with water rights attached decide to put in a residential subdivision, or sometimes a pivot irrigation system, and they may give up some old water rights that simply don't fit the new picture. See BOC Petitions Granted Files: Austin, Michael and Teresa, IV-2007–2-4 (subdivision); Double L Ranch #4, IV-2009–3-9 (subdivision); Mark Lyman Revocable Trust, III-2011–1-11/12 (pivot). What the board terms "voluntary abandonment"—abandonment declaration sought and acknowledged by the water user in order to accomplish some other goal—has therefore become far more prevalent than traditional "abandonment." In a recent example of the Board of Control's interest, however, in preserving high-priority rights if possible; see the May 2018 petition before the Board of Control, Lonesome Star Ranch LLC et al., II-2018–2-3. The landowners installing pivot irrigation asked for a change in place of use and movement of points of diversion and means of conveyance, plus voluntary abandonment of a portion of water rights, from creeks northwest of Sheridan. Division II superintendent Dave Schroeder saw the proposal as a "good cleanup," but there were a variety of issues to be resolved before it could occur. In an August 2018 discussion of the proposal, Schroeder suggested revival of part of a territorial right that appeared unused. State Engineer Pat Tyrrell explained to the landowner's consulting engineer: "You could reestablish that use on existing lands, and then move them. If they've lapsed in use, we want to see it used. We don't want to get rid of an 1881 or 1884 water right either, believe me"; August 2018 discussion notes on file with the author.

34.   On putting standard practices into statute, Henry E. Smith, a legal theorist, has made clear how that is encouraged when the audience affected by local practice begins to widen, as happened for Wyoming when national energy development demands suddenly focused on state coal and water resources. Smith notes that the question of "information cost" affecting a new audience attempting to follow local rules can be a major driver towards rule standardization, formalization and codification of longstanding practice in property law. That pressure grows as the potential audience grows wider and wider. See Merrill and Smith, "Optimal Standardization in the Law of Property"; and Smith, "Community and Custom in Property," and "The Language of Property." In addition to the transfers statute of 1973, in the mid-1960s, soon after Bishop came into office, the Wyoming legislature accepted his request to codify other state water management practices. See, for instance, codification of the administrators' long-time practice of requiring water users to seek administrative approval of changes in point of diversion and/or means of conveyance of water: Bishop, *Thirty-seventh Biennial Report*, 21–22; 1965 Wyo. Sess. Laws 138.

35.   Wyoming water users' belief that Wyoming still has a prohibition on transfers, even into the twenty-first century, have been expressed to author in numerous conversations in various parts of the state.

36.   Chris Propst, "Rock Springs, Wyoming," Encyclopedia, Wyoming State Historical

Society (website), WyoHistory.org, published November 8, 2014, http://www.wyohistory.org/encyclopedia/rock-springs-wyoming; Wyoming Industrial Siting Act, 1975: Wyo. Stats. Ann. 35–12–101 through 35–12–119.

37.   Sarah Gorin, "Taxing Wyoming's Minerals: Severance Taxes and Permanent Funds," Encyclopedia, Wyoming State Historical Society (website), WyoHistory.org, published May 1, 2015, https://www.wyohistory.org/encyclopedia/wyoming-minerals-severance-taxes-and-permanent-funds; Wyoming Taxpayers Association (website), Permanent Wyoming Mineral Trust Fund FAQ, http://www.wyotax.org/PMTF.aspx.

38.   Green River Development Co. v. FMC, 660 P 2d 339 (1983) (State of Wyoming brief, author's files, specifically identifies Jim Bridger Power Plant as the proposed recipient of the water). The State Engineer's Office presented its case for new rules to eliminate temporary filings for water permits at a public hearing in January 1985: "Adoption of Rules and Regulations to establish a procedure for the state engineer to reject or advance to permit status all Temporary Filings"; January 17, 1985, presentation by Deputy State Engineer Jeff Fassett, author's files. The "Problem Description" in the presentation, p. 1, states, "A great majority of these applications, filed primarily as a result of the energy development booms in Wyoming, have been in a 'hold' status because the project sponsors have been unable to find the necessary project financing and the demand for the water to warrant the construction of the facilities."

39.   Reisner, *Cadillac Desert*, 317–43.

40.   Larson, *History of Wyoming*, 566–68.

41.   Wyoming Statutes Annotated 41–2–112 through 124, first enacted in 1975 with additions running into the early 2000s. Funding ratios: Wyoming Water Development Commission, "Operating Criteria," chap. 4, secs. D, E. See also, Warren Wilson, "Herschler's Water Package Put Together Without Needs Study," *Casper Star-Tribune*, February 25, 1982. Total water development spending 1908–2015: Wyoming Legislative Service Office, *Wyoming Water Development Commission*, 6–7. The "informal political alliance" that kept the federal Bureau of Reclamation amply funded for decades has been described as an "iron triangle." It allows congressional committee members to "bring home the bacon," the agency to expand its budgets and staff, and interest groups to get federal funds. The costs of projects are thus dispersed to taxpayers nationwide, but project benefits are concentrated on certain groups; McCool, *Command of the Waters*, 5–12.

42.   Wyoming Water Development Commission (WWDC), "Operating Criteria," chap. 3, prov. A. Wyoming Legislative Service Office, *Wyoming Water Development Commission*, contains examples of WWDC spending in grants and low-interest loans for agricultural projects (rehabilitation projects were funded largely from oil and gas tax revenues; new reservoirs were funded largely from coal tax revenues): Wheatland Irrigation District on the Laramie River received about $1 million for rehabilitation projects 1994–2010 (107–9); Horse Creek Conservation District received $8.5 million to rehabilitate Hawk Springs Reservoir in 1989 (49); LaPrele Irrigation District received nearly $1.5 million for repairs to its tunnel and canals in 1985 (59); the Greybull Valley Irrigation District (including landowners from both the Bench and Farmer's canals, among others), received

$40 million to build Roach Gulch Reservoir in 2005 (47); Midvale Irrigation District on the Wind River received $6.4 million for a variety of rehabilitation projects completed 1999–2013 (67–68); Heart Mountain Irrigation District and Shoshone Irrigation District, on the Shoshone, received $2.3 million and $10.7 million respectively for rehabilitation projects completed 2001–2015 (50, 89–92); Goshen Irrigation District on the North Platte received $5.1 million for rehabilitation projects completed 1991–2013 (44–45); the state shouldered the entire $31.5 million cost to complete the High Savery reservoir for southern Wyoming irrigators in Carbon County, completed in 2010 (52). Wyoming's contribution of nearly half the cost of the expansion of the Buffalo Bill Dam and Reservoir came largely from coal tax funds; WWDC (website), "1996 Legislative Report: Completed Projects, Project 9," http://wwdc.state.wy.us/legreport/1996/comproj.html. The $40 million Shoshone Municipal Pipeline, serving towns along the Shoshone River, was completed in 1992 largely with coal tax funds, and rehabilitated in 2012 largely with oil and gas tax revenues; WWDC, *2017 Legislative Report*, 90–91.

43. Criteria required for water development projects, with the emphasis on putting to use previously unappropriated water, and on new storage capacity, are listed in Wyoming Statutes Annotated, 41–2–121 (a)(ii)(A) and (B). University critiques of Wyoming water development included Jacobs and Taylor, "Wyoming's Water Development Policy," 261–67.

44. For the proposed "Deer Creek Dam" authorized by the legislature in 1985, see Wyoming Water Development Commission (website), "1996 Legislative Report: Completed Projects, Project Reports, 23," http://wwdc.state.wy.us/legreport/1996/proj23.html. For the Laramie River Power Project in Wheatland, see discussion in chapter 6.

45. The 1985 agreement between the US Bureau of Reclamation and Wyoming for modifications of Buffalo Bill Dam: of the total expected cost of $106.7 million, Wyoming agreed to contribute $47 million; "A Partnership for the Future," US Bureau of Reclamation Agreement #5–07–60–WR175, March 29, 1985, in box 97, folder "Water," Thomas Stroock Papers. For Governor Milward Simpson and his work on North Platte River pollution, see Larson, *Water Quality of the North Platte*, 1; Mixer, "Brief History of the North Platte River." "What I remember most about the river in those days was the smell. In the spring of the year, when the water was released from the dams for irrigation, it picked up the human sewage and refinery waste that had accumulated over the winter and carried them downstream. That slug of stuff eliminated not only all of the fish along the way, but the bottom food as well. You could smell it for 3, 4 miles away"; Larry Peterson, District Fisheries Manager, Wyoming Game and Fish Department, quoted in Gannon, "A Sick River is Returned to Nature," 35–37, 83–85. For discussion of the Bureau's review of its North Platte dam operations under Endangered Species Act requirements, see chapter 6.

46. The proposed Sandstone Dam was expected to yield thirty-two thousand acre-feet of water downstream. Jacobs and Taylor, "Wyoming's Water Development Policy," 261–67 found that dam benefits were far outweighed by costs (including foregone interest on funds that could have been invested), and public investment in the projects should be limited to the public benefits projects could legitimately be shown to provide— usually quite a limited amount. Summary of this research was circulated to the Wyoming

Legislature in February 1988. The University of Wyoming's president's disavowal is marked in a letter to colleagues; "Dear Colleagues," from February 17, 1988, UW Pres. Terry Roark and College of Agriculture Dean Lee Bulla Jr. in author's files. Representative Pat O'Toole was interviewed by the author, as news reporter, on March 10, 1988.

47.   For US Army Corps of Engineer's 1992 denial of a Clean Water Act sec. 404 permit for Sandstone due to lack of identified industrial buyers (a problem also noted by the University of Wyoming economists), and the revised proposal for "High Savery" dam, yielding twelve thousand acre feet, plus irrigator inability to pay a 25 percent share of the smaller dam's costs, see Wyoming Water Development Commission (WWDC), *Sandstone Dam: Project Summary*, 1–6; and WWDC, "2004 Legislative Report, #69 Project: High Savery" (website), Active Projects, http://wwdc.state.wy.us/legreport/2004/chap4.html#a. For the subsequent West Fork Dam proposal in the same area; see Angus M. Thuermer Jr., "Fight For $80M Dam Boils as House Strips Funds," WyoFile (website), February 27, 2018, https://www.wyofile.com/fight-80m-dam-boils-house-strips-funds/; Thuermer, "West Fork Dam Kept Alive With $4.7 Million," WyoFile (website), March 13, 2018, https://www.wyofile.com/west-fork-dam-kept-alive-4-7-million/; Andrew Graham, "Legislature Ends Session With Education Cuts, Again," WyoFile (website), March 30, 2018, https://www.wyofile.com/legislature-ends-session-education-cuts/. For former legislator Pat O'Toole, see Brian Almer, "Family Farm Alliance Op-Ed by President Patrick O'Toole—CSU's Water in the West Symposium," Barn Onair & Online (website), April 26, 2018, https://brianallmerradionetwork.wordpress.com/2018/04/26/04–26–18-family-farm-alliance-op-ed-by-president-patrick-otoole-csus-water-in-the-west-symposium/.

48.   From 1980 through 2015, nearly $647 million from Wyoming water development funds went to planning and building municipal projects, with about $120 million of that for rehabilitation; some $276 million went to purely agricultural projects, with about $143 million of that for rehabilitation; Wyoming Legislative Service Office, *Wyoming Water Development Commission*, 7; WWDC, *2017 Legislative Report*, 1–8.

49.   "Subject: Application for Permit to Appropriate Water on the Green River from Green River Lake to Warren Bridge," memorandum, January 28, 1969, Jack R. Gage, special assistant attorney general to Floyd A. Bishop, on file with author; Mike Purcell, personal communication with author, July 11, 2018.

50.   Floyd A. Bishop, state engineer, to Thomas A. Bell, executive director, Wyoming Outdoor Coordinating Council, Inc., July 17, 1969, copy on file with author.

51.   The State Engineer's Office posts on its website under its "Home" tab a list of beneficial uses, some tallied in statute and additional uses recognized by the state engineer. One recent legislative attempt to require that water uses under consideration as "beneficial" must be officially recognized and listed as such by the legislature was in 2012, proposed SF 76, which died in committee: https://www.wyoleg.gov/Legislation/2012/SF0076.

52.   League of Women Voters, *The River is Free*, Thomas Stroock Papers (full citation in note 4); Rauchfuss, *Final Report of the Stream*, see "History and Committee Formation," 2.

53.   T. Paul Stauffer (Stream Preservation Feasibility Study Committee Chairman) in discussion with the author, May 4, 2017.

54.   Stauffer, discussion; and Stauffer, "Did I Fish Too Much?," vol. 1, 246.

55.   In Rauchfuss, *Final Report of the Stream*, see: "History and Committee Formation," 2; secs. 41-(3) to (5), (9) of proposed bill "Creation of Wyoming River Protection System," pp. 23–30, 34 of Report; secs. 41–1.24 thru 1.25 of proposed bill "Protection of Stream Channels," pp. 11–13 of Report.

56.   Rauchfuss, *Final Report of the Stream*, 2.

57.   T. Paul Stauffer (committee chairman) to Governor Stanley K. Hathaway, October 1, 1974, included in *Final Report*.

58.   For the arguments by agricultural interests in legislative debate more than ten years after the Stream Preservation Committee Final Report, see multiple news reports, MacKinnon, Anne, *Casper Star-Tribune*, February 13, 1985. In 2006, State Engineer Pat Tyrrell explained to the Wyoming Water Development Commission that instream flow rights could "somewhat reduce the amount of time a stream is in 'free river' condition" if the state Game and Fish Commission sought regulation to ensure the flows covered by a right and the State Engineer's Office considered the regulation justified to avoid potential damage to the fishery; State Engineer PowerPoint Presentation, January 11, 2006, slides 15 and 13, in author's files. Among the legislative proponents of legislation allowing recognition of a water right to protect instream flows was Tom Stroock, an independent oilman in Casper and avid fly fisherman, who described opponents' arguments as summoning up chimeras and "things that go bump in the night"; MacKinnon, "Instream Flow Bill Historic Marker," *Casper Star-Tribune*, March 17, 1986.

59.   Stauffer, "Did I Fish Too Much?," 245–48. The Wyoming Supreme Court decision on river floating, see Day v. Armstrong, 362 P.2d 137 (1961). Written comments to the committee are preserved in *Letters, Testimony, and Minutes of the Stream Preservation Feasibility Study*, October 1, 1974, bound, unpaged volume at Wyoming State Library. Wyoming Outdoor Council comments, July 8, 1974, are in section titled "Letters and Testimonies to the Stream Preservation Committee from the Public at Large." Wyoming Farm Bureau and Ruth Rudolph letters are in section titled "Testimonies from the Public Hearing Held on Nov. 19, 1973."

60.   State Engineer Bishop's comments to the League of Women Voters Seminar on Wild, Scenic, and Recreational Rivers, held in Casper April 30, 1972, are in the volume cited above, enclosed with his June 19, 1973 letter to the committee, in section titled "Letters Submitted to the Stream Preservation Feasibility Committee from Committee Members and State Agencies." Bishop letter to Governor Stan Hathaway, February 12, 1969, box 208, folder 2, McGee Papers.

61.   On make-up of the Wyoming Legislature, mid-1980s, see Miller, *State Government: Politics in Wyoming*, 84. For initiative process and requirements, Larson, *History of Wyoming*, 322–23, 562–63; and Wyoming Secretary of State, "Initiative and Referendum Summary" (with erroneous date of final instream flow bill passage: it was 1986, not 1985).

62.   For competing forces and legislative maneuvers in 1986, see Anne MacKinnon, "Instream Flow Bill Historic Marker," *Casper Star-Tribune*, March 17, 1986.

63.   Wyoming Laws 1986, chap. 76, sec. 1; Wyoming Statutes Annotated, 41-3-1001 through 1014: Instream Flows; Anderson, *Tapping Water Markets*, 47.

64.   Wyoming Game and Fish Department, "Instream Flow Legislation—1986," 4–6, internal departmental memo, copy in author's files; Tom Annear speaker profile, Instream Flow Council "Flow 2018" (website), https://www.instreamflowcouncil.org/flow-2018-main/flow-2018-speaker-profiles/#Annear; Burkhardt, "An Implementation Case Study: Wyoming's Instream Flow Law," app. B, contains a list of Wyoming instream flow permits approved as of the end of October 1992. A comment in the University of Wyoming's law review in 1986, by Matthew Reynolds, was titled "Wyoming's New Instream Flow Act: An Administrative Quagmire."

65.   Stauffer, "Did I Fish Too Much?," 246, 248, 251–61. Record of House votes on final passage of instream flow bill in author's files. Jade Henderson (retired Division IV superintendent), personal communication with the author, October 8, 2018.

66.   US Geological Survey Gage Data, "U.S.GS 0622800 Wind River at Riverton, 1912–79" (website), https://waterdata.usgs.gov/wy/nwis/dv?cb_00060=on&format=gif_default&site_no=06228000&referred_module=sw&period=&begin_date=1912–04–01&end_date=1979–09–30; US Bureau of Reclamation, "Pick-Sloan Missouri Basin Program, Riverton Unit," 1–3 (1980) and 1–5 (1984). In 1928, after the diversion dam to serve Midvale's long canal had been completed, the project hosted only sixteen farmers. More settlers came in during the Depression as the Dust Bowl drove them off Midwestern farms, and more again after World War II. In the 1960s, however, the federal government had to buy back from farmers some lands that turned out to have bad soils; see Autobee, *Riverton Unit, Pick-Sloan*, 23–29, quotation 32. Wyoming US senator Gale McGee praised Midvale in support of a bill adding it to the Pick-Sloan program for the entire Missouri basin, which meant program hydropower revenues could help pay Midvale's Reclamation debt, see March 20, 1970, testimony, p. 5, box 237, folder 6, McGee Papers. Subsidies went to Bureau of Reclamation projects across the West, through foregone interest payments on production costs, and shifting of costs to non-irrigators benefiting from projects; see Wahl, *Markets for Federal Water*, 27–46; and US Government Accountability Office, *Bureau of Reclamation: Availability of Information*, app. 3, 41–47, with summary figures for the Riverton Unit (covering Midvale), Shoshone, and North Platte Bureau of Reclamation projects, showing that of the three federal irrigation projects in Wyoming the Riverton Unit has received the largest proportional share of subsidies.

67.   US Bureau of Reclamation, "Pick-Sloan Missouri Basin Program, Riverton Unit," 1–3 (1980), 1–5 (1984).

68.   Blomberg, "Tribal Fishery Restoration"; Bergersen and Cook, "Impacts of Water Level Manipulations," 1; Bergersen, Cook, and Baldes, "Winter Movements of Burbot," 141–45; US Bureau of Reclamation, 1–3 (1980), 1–5 (1984); Autobee, *Riverton Unit, Pick-Sloan*, 23–29. The "first dibs" on Wind River flows under the 1906 water right were provided to the LeClair and Riverton Valley Irrigation Districts when the State Board of

Land Commissioners gave those districts portions of the 1906 right in 1916–17; Hoopen-garner, "To Make the Desert Bloom," 165–66, 176. The "Tri-Partite Agreement" between Midvale, LeClair and Riverton Valley districts, modified over time, details exactly how the sharing of the 1906 right works, and is described in Wyoming Water Development Commission, *LeClair Irrigation District*, 3–4, 61–63. The first statute generally authorizing exchanges, with approval of the state engineer and no injury to other water right holders, was passed in 1947; Wyo. Stat. Ann. § 41-3-106.

69.   Bergstedt, "Fishery and Macroinvertebrate Response," iii–iv, 2–5, 25, 32–33, 37–39, 56–57, 67–71: at 2, "Because these sediments impede the flow of water into the Wyoming Canal (the Midvale canal served by Diversion Dam), canal operators routinely flush them from above the dam by opening the dam gates and sluicing the material downstream. Irrigation company records indicate that this was done twenty-five times during the 1988 irrigation season and thirty-two times in 1989. Suspended solid levels during these events have increased from as low as 2 mg/L to over 14,000 mg/L in minutes."

70.   For tribal oil revenues, see O'Gara, *What You See in Clear Water*, 133–39. The voluminous filings and decisions in what became known as the Big Horn Adjudication are all compiled in a comprehensive database, the Big Horn River Adjudication Case (BHRAC), housed at Wyoming State Engineer's Office. Key court decisions related to the tribes' reserved rights include *General Adjudication of All Rights to Use Water in the Big Horn River System, District Court of Fifth Judicial District, Wyoming*, May 10, 1983, BHRAC index JoffeDeco5101983.pdf; *General Adjudication of All Rights to Use Water in the Big Horn River System, District Court of First Judicial District, Wyoming*, June 8, 1984, and May 24, 1985, BHRAC index N5M6UK0000.pdf and 11112t.pdf, respectively; General Adjudication of All Rights to Use Water in the Big Horn River System, 753 P.2d 76 (Wyo. 1988) (Big Horn River System), affirmed by an evenly divided US Supreme Court, Wyoming v. United States, 492 U.S. 406 (1989)—the divided court issued no written opinion. One western justice, Sandra Day O'Connor, had to recuse herself because of a conflict of interest, as water claims of other tribes affected her family's Arizona ranch. A draft opinion she wrote for a proposed majority before realizing that she had to recuse herself was likely to have considerably reduced the tribes' water right, in consideration of conflict with existing non-Indian water use on the river and an assessment of whether new irrigation projects were likely to be built on a reservation. O'Connor's recusal left the court divided 4:4 on the case, so the Wyoming Supreme Court decision was left standing as the final decision of Big Horn River System; 753 P.2d 76 (Wyo. 1988). Presentation by Susan Williams (attorney representing Shoshone tribes at the US Supreme Court) cited in Cosens and Royster, *The Future of Indian and Federal Reserved Water Rights*, 10–11, 172.

71.   Cosens, "The Legacy of Winters v. United States," in Cosens and Royster, 5–9; Shurts, *Indian Reserved Water Rights*.

72.   In periodic snowmelt forecasts, the Bureau of Reclamation from 2002–18 has estimated the thirty-year average annual flow of the Wind River as 540,000–570,000 acre-feet. See for example, Basin 2002–18, at US Bureau of Reclamation, Wyoming Area

Office, 2008 (website), "Bighorn River Basin Snowmelt Runoff Forecast," published May 2008, https://www.usbr.gov/newsroom/newsrelease/detail.cfm?RecordID=21821.

73.   O'Gara, *What You See in Clear Water*, 80–81, 88–93. Review of the claims of non-Indians for treaty-date rights took years and three more decisions: General Adjudication of All Rights to Use Water in the Big Horn River System, 803 P.2d 61 (Wyo. 1990) (Big Horn II), General Adjudication . . . in the Big Horn River System, 899 P.2d 848 (Wyo. 1995) (Big Horn IV), and General Adjudication . . . in the Big Horn River System, 48 P.3d 1040 (Wyo. 2002) (Big Horn V); Cosens and Royster, *The Future of Indian and Federal*, 172–75, featuring presentations by attorney Susan Williams (former attorney representing the tribes before the US Supreme Court) and Gordon "Jeff" Fassett, former Wyoming state engineer.

74.   Memorandum Opinion, Judge Alan B. Johnson, April 29, 2010, at 2–3, in James E. Large v. Fremont County, 709 F. Supp. 2d 1176 (D. Wyo. 2010) (No. 05-CV-0270), at 1–15. This was a federal Voting Rights Act violation case, brought by the Northern Arapaho tribe, which successfully forced Fremont County to drop its at-large districting for county commission elections. The court found such a system tended to prevent tribal citizens from electing a commissioner who might represent their interests.

75.   Cosens and Royster, *The Future of Indian and Federal*, 172, with Susan Williams (former attorney representing the tribes before the US Supreme Court) and affidavit of State Engineer Jeff Fassett; October 3, 1990, BHRAC index 10–8-14t.PDF (full citation in note 68); O'Gara, *What You See in Clear Water*, 226; Rusinek, citing news reports and briefs filed in the appeal of *Big Horn River System*, 753 P.2d 76 to the US Supreme Court, in "A Preview of Coming Attractions?," 392–93.

76.   Special Master Teno Roncalio, the initial reviewer of evidence and arguments who filed recommendations with the state district court, found that congressional intent in the 1868 treaty was "to provide a permanent homeland for the Indians," and therefore water had been reserved for many purposes beyond agriculture, including fish and wildlife. Roncalio argued that to say that water for the reservation was reserved only for an agricultural way of life (as the district courts and the Wyoming Supreme Court subsequently did) "is to unreasonably limit the terms of the Treaty entered into by a Congress and a nation whose own history surpassed its narrow agricultural beginnings"; *Report of Special Master*, 274. Both the state district court and the Wyoming Supreme Court rejected this argument and ruled that the treaty terms clearly established the reservation for use by the tribes for agricultural purposes, and that purpose therefore governed how much water the tribes' reserved water rights would cover; *Big Horn River System*, 753 P.2d at 94–97, 99. For the Arizona Supreme Court's endorsement of interpreting a reservation as "homeland" for tribes on the Gila River in that state, see General Adjudication of All Rights to Use Water in the Gila River System and Source, 35 P.3d 68, 76 (Ariz. 2001).

77.   Cosens and Royster, *The Future of Indian and Federal*, 10; Rusinek, "A Preview of Coming Attractions," 372; Franks, "The Use of the Practicably Irrigable Acreage," 549, 578; Shay, "Promises to a Viable Homeland," 578–79.

78.   Arizona v. California, 439 U.S. 419 (1979), at 421–22, said tribal water awards

quantified under the "practicably irrigable acreage" (PIA) standard did not afterwards have to be used for irrigation. *Big Horn River System*, 753 P.2d at 100 suggested the "futures" award measured by the PIA standard could be used for non-irrigation purposes. A final version of the Wind River Water Code, after some time as an interim policy, was adopted by the Shoshone and Arapaho Joint Business Council in 1991.

79. *Big Horn River System*, 753 P.2d at 99; Parshall, *Twelfth Biennial Report*, 133–34 (gauging station reports); Cosens and Royster, eds., *The Future of Indian and Federal Reserved Water Rights*, (Williams' presentation); O'Gara, *What You See in Clear Water*, 193, 228–29. The lower court, in First District Court June 8, 1984 decision, said, "The reserved water right quantified by [the lower court] does not deny the Tribes the ability to regulate instream flows in order to maintain what may be considered necessary water for optimum fish habitat, nor does the opinion limit any such power that may exist on the part of the Tribes. The Tribes may seek to dedicate their stream flows for fish habitat by using water reserved to them by the decision"; 10–11. The Wyoming Supreme Court was silent on this issue in *Big Horn River System*, 753 P.2d 76—leaving room for the instream flow dedication and litigation of 1990.

80. Tribes' Motion for Order to show cause why further relief should not be granted, July 30, 1990, with attached affidavit of Tribal Water Engineer Catherine Vandemoer, and Exhibits A-C, located in the BHRAC database, index 10–8–14r.PDF, at State Engineer's Office (full citation in note 68). Tribes' Memorandum in support of the Shoshone and Arapaho Tribes' Motion for Order to show cause why further relief should not be granted, July 30, 1990, 3–5 (includes schematic map of river and gauging points, referred to in Vandemoer's affidavit) in the BHRAC database, index 10–8–14s.PDF.

81. States' Exhibit 1, affidavit of State Engineer Jeff Fassett, August 3, 1990, attached to State's Response to Tribes Motion for Order to show cause why further relief should not be granted, August 3, 1990, in BHRAC database, index 10–8–14t.PDF, pp. 1–3; State's Response, August 3, 1990, at 10–8–14t.PDF, pp. 4–5, 8–9. For later descriptions of the issues and actions of the summer of 1990, see General Adjudication of All Rights to Use Water in the Big Horn River System (Big Horn III), 835 P. 2d 273 (Wyo. 1992) at 275–76; Cosens and Royster, *The Future of Indian and Federal*, 173–75 (presentations by Williams and Fassett, 2008); Robison, "Wyoming's Big Horn General Stream," 288–90.

82. Tribes' Motion for Order to show cause, July 30, 1990, in BHRAC index 10–8–14r. PDF. The district court ruling in the case that became Big Horn III, 835 P. 2d 273 in the state supreme court was issued March 11, 1991, in BHRAC index GQVXO70000.pdf (full citation in note 69), with key language at pages 12–14, 18–19, and the quotation in text at 14. The district judge commented, at 13: "The fact that a Tribe changes the use of its reserved water from agriculture to something else does not mean, ipso facto, that the reserved water loses its special status, and that the Tribe is suddenly subject to the jurisdiction of the state."

83. *Big Horn III*, 835 P. 2d at 278–85, 290. Under Wyoming's 1986 instream flow law, an "existing right" transferred to instream flow use had to go through all the standard scrutiny under Wyoming's transfer law, including historic use of the right, and the tribes'

"futures" rights had not been used; and, of course, the right would have to be handed over to the state of Wyoming, see Wyoming Statutes Annotated § 41-3-1007.

84. Katharine Collins, "Fear of Supreme Court Leads Tribes to Accept an Adverse Decision," October 19, 1992, *High Country News*; O'Gara, *What You See in Clear Water*, 243–44; Cosens and Royster, *The Future of Indian and Federal*, 174 (Fassett presentation, 2008).

85. Rusinek, "A Preview of Coming Attractions?," 398–404. He cited Official Transcript of Proceedings before the Supreme Court of the United States; *Wyoming v. United States*, and a *Casper Star-Tribune* newspaper report of the oral argument, April 26, 1989, both of which were available to the Wyoming Supreme Court justices in Cheyenne.

86. *Big Horn III*, 835 P. 2d at 278–80.

87. *Id.* at 295, referring to the landmark US Supreme Court school desegregation case, Brown v. Board of Education of Topeka, 347 U.S. 483 (1954).

88. *Big Horn III*, 835 P. 2d at 294.

89. *Id.*

90. *Id.* at 296.

91. *Id.* at 297, 298–300, 303.

92. *Id.* 303–4.

Chapter 6

1. Ginger Paige, (associate professor of water resources, University of Wyoming Department of Ecosystem Science and Management) interview with the author, March 1, 2018.

2. Colorado River Research Group (CRRG), "When is a Drought not a Drought?," 2018. CRRG is a "self-directed" group of ten veteran Colorado River scholars in water resource management, river science, water law and public policy. https://www.coloradoriverresearchgroup.org/.

3. Eric and April Barnes, interview with author, April 16, 2017. For origins of "Fontenelle" name, see "Wyoming Places," Wyoming State Library (website), http://places.wyo.gov/. The near-failure of Fontenelle Dam in 1965 is described with references on the website of the Association of State Dam Safety Officials, http://damfailures.org/case-study/fontenelle-dam/; Baker, "The Fontenelle Dam Incident," 15. On the non-use of Fontenelle for irrigation, a summary report for the Wyoming Water Development Office puts it this way: "Because of the relative aridity of the central Green River Basin, irrigation first began along the tributaries leading from the various mountain ranges that fringe the basin. As happened early on in much of Wyoming, tributaries were more quickly developed than the larger watercourses they fed. Today, the development of irrigation works in the basin still is defined by these early efforts. The bulk of irrigation in the basin occurs along tributaries. The largest reservoir in the interior of the basin, Fontenelle Reservoir, is downstream of virtually all of the upper Green River irrigated areas, unavailable to other sub-basins, and therefore is virtually unused for irrigation," in "Technical Memoranda, Green River Basin

Plan" Wyoming State Water Plan (website), http://waterplan.state.wy.us/plan/green/2001/
techmemos/aguse.html. For details on the refuge, see "Seedskadee National Wildlife Refuge:
About the Refuge," US Fish and Wildlife Service (website), updated June 24, 2015, https://
www.fws.gov/refuge/Seedskadee/about.aspx.

4.   The Civil War veteran and lawyer who settled on Fontenelle Creek was Charles
Holden, a delegate to Wyoming's constitutional convention who argued there for wom-
en's suffrage; the New York farm boy who went off to California in the gold rush, came to
settle on Fontenelle creek, and married a young Illinois woman teaching school in nearby
Green River was Daniel Rathbun, whose ditches now water the Barneses' ranch. *Progressive Men of Wyoming*, 235–36, 154–55; Eric and April Barnes, interview with the author,
April 16, 2017.

5.   Eric and April Barnes, April 16, 2017.

6.   Dave Rosgen, professional hydrologist and geomorphologist, spent twenty years
with the US Forest Service and in 1985 created the consulting group Wildland Hydrology to design and implement river restoration and to train professionals in watershed
assessment and management, river restoration, and monitoring; in "About Us," www.
wildlandhydrology.com/about. For fish habitat restoration structures and discussion,
see Rosgen, *Applied River Morphology*, 8–20 to 8–43. His is one of the major approaches
to understanding and working with rivers that have been adopted by federal agencies,
including the Forest Service. He described the motivation for his lifework at the Applied
Fluvial Geomorphology Short Course, in Pinedale, Wyoming, July 15, 2002, attended by
the author.

7.   "Building Wyoming's Tax Structure for the 21st Century," Report of the State of
Wyoming Tax Reform 2000 Committee (website), https://www.wyoleg.gov/1999inte/
t2000/final1.htm. Charts depicting state revenues overall and from the minerals sector,
1981 to projected 2010, found in Consensus Revenue Estimating Group, *Wyoming State
Government Revenue*, 2005.

8.   Dustin Bleizeffer, "Coalbed Methane: Boom, Bust and Hard Lessons," Encyclopedia, Wyoming State Historical Society, WyoHistory.org, published March 29, 2015,
https://www.wyohistory.org/encyclopedia/coalbed-methane-boom-bust-and-hard-lessons.

9.   State Engineer's Office, "Guidance: CBM/Ground Water Permits," issued in
2004, describes the rationale for the declaration that production of water from coalbed
methane wells would be considered a beneficial use of Wyoming water, as well as consequent permitting requirements.

10.   Ruckleshaus Institute, *Water Production from Coalbed Methane*, v, 10–32. The
report shows that in 2003, coalbed methane wells, most of them in the Powder River
basin, produced just under seventy-five thousand acre-feet of water. For the obstacles to
centralized treatment, transportation, and use of CBM gas, see Kimball et al., "Technical
Memorandum," to the 2006–2007 Wyoming Coalbed Methane Task Force, summariz-
ing a Wyoming State Geological Survey report on the feasibility of a desalination plant
for coalbed methane water in the Powder River Basin, and Bleizeffer, "Coalbed Methane:
Boom, Bust" (full citation in note 8). The state of Montana sued Wyoming claiming that

pumping of groundwater in CBM production in the Tongue River basin in the early 2000s had depleted water flow in the Tongue River that Montana should have received, but the special master appointed by the US Supreme Court to hear that case ruled that CBM production in Wyoming had not depleted water available to Montana. WY State Engineer's Office (SEO), *State of Wyoming Water Year 2015*, 106. For summary of SEO action by 2006 and the legal basis for SEO and Wyoming Department of Water Quality regulation of coalbed methane water to avoid the problems it created, see MacKinnon and Fox, "Demanding Beneficial Use," 369–99. For landowner complaints of CBM water releases in ephemeral streams creating water quantity and quality problems, as well as lack of adequate monitoring and enforcement under the Clean Water Act by state officials in the Wyoming Department of Environmental Quality, see Swartz v. Beach, 229 F. supp. 2d 1239 (D.Wyo. 2002), at 1248–49 (eventually settled out of court after the federal judge in Wyoming dismissed most of the state and company defendants' motion to dismiss the landowner's complaint). In December 2007, the SEO office issued a press release stating, "This week, State Engineer Pat Tyrrell asked for an explanation of how natural gas operations are putting water to 'beneficial use' when little to no gas is being collected from some of their wells. The wells of interest are at least five years old, and are located in the Clear Creek and Crazy Woman Creek drainages (in the western portion) of the Powder River basin. A total of 296 wells are involved." For cancellation of groundwater permits for excess water production, see Tyrrell, *State of Wyoming 2008*, 43, stating, "Forty seven (47) Coal Bed Methane (CBM) Use permits were suspended due to the long-term production of water and the failure to produce gas."

11.   For Wyoming CBM production data, see "Wyoming Natural Gas Gross Withdrawals from Coalbed Wells," US Energy and Information Administration (website), https://www.eia.gov/dnav/ng/hist/ngm_epg0_fgc_swy_mmcfa.htm; Bleizeffer, "Coalbed Methane: Boom, Bust" (full citation in note 8); MacKinnon and Fox, "Demanding Beneficial Use," 373–78.

12.   Bureau of Reclamation (BOR), *Platte River Recovery Implementation Program*, 24 (hereafter cited as *PRRIP FEIS*); Farrar, "Platte River Instream Flows," 39. On BOR subsidies to irrigated agriculture, see Wahl, *Markets for Federal Water*, 3–126. For unrepaid costs for the Goshen Irrigation District, Wyoming's share of the North Platte Project served by Wyoming reservoirs including Pathfinder, see Wahl's chart, 35. Wahl explains that the ongoing federal subsidies for federal irrigation projects derive from such factors as the federal reclamation office not charging interest on construction costs and shifting costs to other users, like hydropower consumers, 27–39. Recent Bureau of Reclamation figures (2016) show that of the three federal irrigation projects in Wyoming, only the irrigation districts on the Shoshone have paid over half of project construction costs—for the North Platte, the irrigation districts have paid around a third of construction costs and on the Wind River, the Midvale Irrigation District has paid less than 5 percent of construction costs, according to BOR Great Plains Region, "Statements of Project Construction Cost and Repayment, as of 9/30/2016, for the Shoshone Project, the North Platte Project, and the Riverton Project"; report created on author's request.

13.   Flows of Wyoming's major rivers are tabulated in WWDC, *Wyoming Framework Water Plan*, vol. 1, 4–2. US Army Corps of Engineers summary of existing storage in the Platte basin as of the mid-1970s is quoted by the federal district court in Nebraska, in *Nebraska v. REA*, 12 Env't Rep. Cas (BNA) 1166 (D. Neb 1978), appeal vacated and dismissed, 594 F.2d 870 (8th Circ. 1979), at 1178.

14.   Currier, Lingle, and VanDerwalker, *Migratory Bird Habitat on the Platte*, 13–18.

15.   The 30 percent of annual flows reaching the habitat area in 1970 reflects flows there versus total river flows measured in the 1890s—when thousands of irrigation canals tapping into the river had already been built. *PRRIP FEIS*, 26–31; Farrar, "Platte River Instream Flows," 38–42; Currier, Lingle, and VanDerwalker, 14 (flows and channel width), 18: "Little remains today of the open-channel, nearly treeless floodplain of pre-settlement times. During the past 100 years, reductions in peak and average flow in the Platte have resulted in a decrease in scouring and shifting of the alluvial streambed and allowed extensive forest development on the floodplain" (citations omitted). The pallid sturgeon fish is also one of the species of concern. Found in the Missouri River and the lower (mainstem) Platte, the fish is not well adapted to the channelized river conditions created in the last one hundred years: *PRRIP FEIS*, 33.

16.   *Nebraska v. REA*, (D. Neb 1978), *appeal vacated and dismissed*, 594 F.2d 870 (8th Circ. 1979), 1179–90. Nebraska, and other intervenors, successfully argued that the Rural Electric Association had failed to meet the requirement of §7 of the Endangered Species Act (ESA), in approving a loan guarantee for the project based only on its own determination (not a biological assessment by the Fish and Wildlife Service) that the habitat would not be adversely affected. Similarly, the court agreed with the challengers to the project that the Corps of Engineers had violated the ESA's §7 requirement by issuing a permit for the project under the Clean Water Act, before the Fish and Wildlife Service completed its biological assessment of the habitat impact. The ESA § 7 requires all federal agencies to consult with the Secretary of the Interior before authorizing, funding, or carrying out an action which may jeopardize an endangered or threatened species; 16 U.S.C. §1536 (a)(2). One commentator describes Nebraska's lawsuit against the power plant as an "end-run" to protect not so much the endangered species, but downstream Nebraska irrigators: "It was easier to get water for this purpose [the endangered species] than to reopen a 1945 equitable apportionment" of the river made by the US Supreme Court in *Nebraska v. Wyoming* (325 U.S. 589, 1949); Tarlock, "The Endangered Species Act," 20–21. After Wyoming launched plans for a $45 million dam on a North Platte tributary in 1986, however, Nebraska did launch a new *Nebraska v. Wyoming* lawsuit to reopen the original US Supreme Court decree allocating North Platte water between the two states (and also raised claims of impact on habitat in that suit). See further discussion in text.

17.   *PRRIP FEIS*, ii–5; Fassett, "Endangered Species Management," 19.

18.   Fassett, 19; Zallen, "Integrating New Values with Old Uses," 1–4; Pinchot, "The Long Struggle for Effective Federal Water Power Regulation," 9; "About the Program," Platte River Recovery Implementation Program (website), https://platteriverprogram.org/about/program-details; *PRRIP FEIS*, Summary, 1–2.

19. Fassett, "Endangered Species Management," 19–20; WWDC, "PRRIP versus Individual Consultation (Program vs. No Program)," presentation regarding the proposed Platte River Recovery Implementation Program (PRRIP), at a meeting hosted by the WWDC, January 20, 2004, in Casper, and public discussion (notes in author's files); Zallen, "Integrating New Values with Old Uses,"18.

20. For the Deer Creek Dam, a $45 million project authorized by the Wyoming Legislature in 1985, see chapter 5 and WWDC, *1985 Legislative Report*, 38–41, and WWDC, *1986 Legislative Report*, 34–36; for the *Nebraska v. Wyoming* lawsuit, see Tyrrell, *State of Wyoming 2002 Annual Report*, 65–73 for a summary of the original filings, the progress of the litigation, and the final settlement—giving a good sense of the time, effort and multiple hearings and negotiations entailed in the *Nebraska v. Wyoming* lawsuit, filed in 1986.

21. Compare irrigated acreage maps, Wyoming State Water Plan, Platte River Basin Water Atlas (2006); above Pathfinder, http://waterplan.state.wy.us/plan/platte/2006/atlas/above/above_agricultural_irrigated_acreage.htm; and near the Nebraska state line, http://waterplan.state.wy.us/plan/platte/2006/atlas/guernsey/guernsey_agricultural_irrigated_acreage.htm. The plan notes that there are no irrigation districts above Pathfinder but thirteen on the Wyoming stretch of the river near the Nebraska line.

22. Randall Tullis (Wyoming Division I superintendent), "Wyoming Compliance with Key Issues Related to Modified North Platte Decree Implementation," presentation to CLE International Conference on Wyoming Water Law, April 25, 2008, in author's files. *Nebraska v. Wyoming* was settled in 2001, after agreement on key principles were settled almost literally on the courthouse steps in 2000, and all the parties hammered out the details in ensuing months. Tyrrell, *State of Wyoming 2002 Annual Report*, 71. For the settlement on the courthouse steps, see Brodie Farquhar, "Wyo, Nebraska Approve North Platte Deal," *Casper Star-Tribune*, March 14, 2001, quoting Wyoming Assistant Attorney General Tom Davidson. The settlement dealt with allocation issues and included Wyoming's commitment not to pursue the Deer Creek Dam. It also reflected the determination of both Wyoming state engineer Fassett and water development director Purcell that the endangered species habitat issues would not be addressed in the settlement of the lawsuit over decree violations; rather, they were kept for the Platte River Recovery Implementation Program which achieved final adoption by all three states and the federal government in 2006.

23. For a description of the governance committee, its origins and operations, see "Governance Committee," Platte River Recovery Implementation Program (website), https://platteriverprogram.org/group/governance-committee. For the matter of water under the first increment of the program not amounting to what US FWS wanted, see *PRRIP FEIS*, 48–50.

24. Tyrrell, *State of Wyoming 2002 Annual Report*, 71–73; "Modified North Platte Decree," Wyoming State Engineer's Office (website), http://seo.wyo.gov/interstate-streams/know-your-basin/platte-river-basin; documents on "North Platte River Settlement," Nebraska Department of Natural Resources (website), https://dnr.nebraska.gov/water-planning/north-platte-river-settlement; and see citations in footnote 22 above.

25.   *PRRIP FEIS*, 45; WWDC, *2007 Legislative Report*, 87–89 (Pathfinder Modification) and 91–93 (PRRIP). Legislative approvals 2006 information at *PRRIP FEIS*, 2006 Session Laws chap. 99, Section 6 and Pathfinder Modification, 2006 Session Laws chap. 105, 99-3-105 (b). See also Theesfeld and MacKinnon, "Giving Birds a Starting Date," 110–19. The idea of replacing storage space lost to sedimentation, as a means to expand an existing dam, was applied to Pathfinder from the experience of the state of Wyoming and the Bureau of Reclamation (i.e., John Lawson) in expanding Buffalo Bill Dam on the Shoshone River. Mike Purcell, personal communication with the author, October 29, 2018.

26.   The largest irrigation district served by the Bureau of Reclamation dams on the North Platte is the Pathfinder Irrigation District in Nebraska. In 2005 the president of the district commented, "Dealing with endangered species issues is very difficult, and for most water users they find it hard to understand why they are even affected. Bottom line the Endangered Species Act is not going to go away. The PRRIP (Platte River Recovery Implementation Program) provides a reasonable basin-wide approach to addressing endangered species, which as water users we find much more palatable than what might result from individual consultations with FWS (US Fish and Wildlife Service). For this reason the Pathfinder Irrigation District supports the implementation of the PRRIP"; Pathfinder Irrigation District letter to Director, December 9, 2005, located on file at the Wyoming Water Development Office, Cheyenne. A letter in support of the related Pathfinder Modification project, from a member of the Pathfinder district, had also commented: "Fish and wildlife people need to remember that 110 years ago before the Pathfinder and other dams were built for irrigation there was *no water* in the river for fish and wildlife after the spring run-off," (emphasis in the original). Kelley Brothers letter to Director, on file at Wyoming Water Development Office, Cheyenne; WWDC, *2007 Legislative Report*, 87–89 (Pathfinder Modification).

The Bureau of Reclamation's Kendrick Project to irrigate lands near Casper envisioned sixty-six thousand irrigated acres but as of the early 2000s irrigated only twenty-four thousand acres. Anderson Consulting Engineers (ACE), *Executive Summary for Casper Alcova Irrigation*, 1–2. State aid to the Casper-Alcova Irrigation District included over $1.3 million from state water development accounts for water delivery system rehabilitation and digital mapping. State of Wyoming, Session Laws 2004, chap. 118; 2005 chap. 147; 2006, chaps. 99 and 105. Bureau of Reclamation correspondence files from 1933–35 show the pressure to get the project going, for jobs and more agriculture around Casper. Mead, heading the bureau, determined from Orville Beath of the University of Wyoming that at Kendrick, soils containing selenium (a necessary nutrient but dangerous in concentration) were unlikely to cause problems under irrigation. Letters in General Correspondence, Water Appropriations, Entry 7, RG 115, Subgroup 031, National Archives, Denver: November 8, 1933, District Counselor to Bureau of Reclamation Commissioner Elwood Mead; November 17, 1933, Chief Engineer to District Counselor; June 18, 1934 L. C. Bishop to Commissioner Mead; January 22, 1935, Commissioner Elwood Mead to Interior Secretary Harold Ickes; February 5, 1935, Commissioner Mead to Wyoming State Engineer Edwin Burritt. In RG 115, Entry 7, Correspondence, Subgroup 400.02, Land

Classification and Soil Surveys: December 5, 1934 Commissioner Mead in discussion with Orville Beath of selenium research on project lands; January 3, 1935 Commissioner Mead to Orville Beath; January 11 and January 14, 1935 Orville Beath-Commissioner Mead; and February 1, 1935 Construction Engineer Bashore to Commissioner Mead.

Fifty years later, studies in 1986–96 documented that irrigation of soils on the Kendrick lands near Casper had led to selenium content in irrigation drainage waters high enough to kill or deform the embryos of birds using the area, and impair reproduction of fish; other studies showed potential public health risk from consumption of too many ducks and geese on the project, or fish from the North Platte near Casper. Peterson, Jones, and Morton, *Reconnaissance Investigation of Water Quality*, 39–41; Klasing, *Agricultural Drainage Water Contamination*, i–ii; Ramirez and Dickerson, *Monitoring of Selenium Concentration*, 1. Selenium contamination from irrigation drainage water gained national attention in the 1980s at the Kesterson National Wildlife Refuge in California; the Department of Interior later identified the Kendrick project near Casper as one of nine top areas of selenium concern in the western United States. Remediation after 2000 included state of Wyoming funds for project irrigators to switch from flood irrigation to pivots and line dirt canals to reduce water use—thus addressing both selenium issues and the likely water supply shortages to the project due to the Pathfinder Modification plan. Harris, *Death in the Marsh*, 1–13, 190–213; Natrona County Conservation District, Casper-Alcova Irrigation District, and Kendrick Watershed Steering Committee, Technical Bulletin #1, January 2003, in author's files; ACE, *Executive Summary for Casper Alcova*, 1; and "Cody Shale, Selenium & Water Quality," Natrona County Conservation District (website), http://natronacountyconservationdistrict.com/whatisselenium.html discusses ongoing monitoring and selenium management with landowners.

27.   WWDC, *2007 Legislative Report*, 87–89 (Pathfinder Modification); Stipulation and Settlement Agreement before the Wyoming State Board of Control, Docket No. I-2008-1-7, filed October 16, 2008; and US Bureau of Reclamation, amended petition, Docket No. I-2008-1-7; Board of Control hearing, Docket No. I-2008-1-7, October 20, 2008.

28.   "Proposed First Increment Extension Environmental Assessment," Platte River Recovery Implementation Program (website), https://www.usbr.gov/gp/nepa/platte_river/index.html. Platte River Recovery Implementation Program, February 9, 2018, PowerPoint presentation by Harry LaBonde, director, Wyoming Water Development Office, in author's files. Audio recordings of public meetings for the Platte River Recovery Implementation Program (PRRIP), held in several locations, are in the author's files and available from Wyoming Water Development Commission (WWDC). They include February 6, 2018 in Torrington, WY, just west of the Nebraska line; February 7 in Saratoga, Wyoming, in the upper Platte basin; and February 8 in Casper, Wyoming, just east of Pathfinder Reservoir.

29.   Audio recordings of public meetings on PRRIP, held February 7, 2018, in Saratoga, Wyoming, available from WWDC, and in author's files.

30.   Wohl, Ellen E. *Virtual Rivers: Lessons from the Mountain Rivers*, 1–37, esp. 25: Wohl

provides an excellent description for the lay person of river geomorphology. Sheridan, joint City of Sheridan–US Army Corps of Engineers project, "Sheridan Ecosystem Restoration Project," US Army Corps of Engineers (website), http://www.nwo.usace.army.mil/Missions/ Civil-Works/Planning/Planning-Projects/Sheridan-WY/. For City of Casper (website), see "Platte River Revival," http://www.casperwy.gov/residents/environment_and_waste/ platte_river_revival. The history of the restoration effort in Casper, starting with the launch of a riverside walking trail in the 1980s and moving toward the 2006 public-private creation of the Platte River Revival to clean river banks, eliminate non-native trees, and restore fish habitat through the city, is outlined in "North Platte River Restoration," a downloadable 2006 presentation at https://wgfd.wyo.gov/WGFD/media/content/PDF/ Habitat/Aquatic%20Habitat/WWA_RiverRestoration_PlatteRiver.pdf. City of Laramie's brochure, "Laramie River Greenbelt Trail" https://cityoflaramie.org/DocumentCenter/ View/2709. For the City of Evanston (website), see Bear River Greenway "History," http://www.evanstonparksandrec.org/153/History. Cheyenne Chamber of Commerce's (website), "Crow Creek," http://www.cheyennechamber.org/crow-creek. For Pinedale, see Annear and Bulger, "Progress at Pine Creek," 34–39.

31.   The author participated in the state engineer's ad hoc group in the 1990s. Feck and Nibbelink, in two stream studies, *Watershed Analysis of Steam Flow . . . Clear Creek Watershed*, 15, and *Watershed Analysis of Steam Flow . . . Popo Agie Watershed*, 14. Examples of failed temporary instream flow bills: SF 72, 2003 (died in Senate committee), SF 106, 2005 (died in Senate committee), SF 51, 2007 (passed Senate but died in House), SF 71, 2009 (died in Senate committee). In fall 2004, the Wyoming Water Development Commission (WWDC) director successfully opposed funding for the City of Evanston for its project to "rehabilitate" the Bear River flowing through the city. In 2015, the commission turned down a City of Casper proposal for restoring the North Platte through the city on a stretch where the work would protect the city's drinking water well fields. WWDC project applications and recommendations, 2004 and 2015, in author's files. In 2004, a lead staffer at the WWDC noted in an email that in rejecting past proposals to fund stream restoration projects, "the reasoning was that there is a huge volume of stream channels that need work and this tends to divert resources away from water development"; John Jackson, September 24, 2004, in author's files. For Board of Control adjudications of instream flow water rights, see board agendas and minutes from 2011 and 2012.

32.   For a description of Bureau of Reclamation water releases on the North Platte above Casper to clean spawning gravels for trout, see Dallman, "Flushing Flows to Enhance Trout Habitat," 1. And for an example of how those "flushing flows" have become standard, see USBR, Wyoming Area Office, "Fluctuation of North Platte River flow Downstream of Gray Reef Dam," press release February 26, 2004, https://www.usbr.gov/newsroom/// newsrelease/detail.cfm?RecordID=701. For other Bureau activity to improve fish habitat on the Platte, see Dallman, "Pathfinder Dam Celebrates Restored Flow," 13; Copeland, "Birth of a Fishing Town," 12–19. For flows from the enlarged Buffalo Bill Dam, see John H. Lawson, "Buffalo Bill Reservoir Enlargement Winter Release Operation Agreement," PowerPoint presentation June 18, 2004 (in author's files); Annear, "Securing the Shoshone,"

37–43; and letter, August 18, 2004, Heart Mountain Irrigation District to Wyoming Water Development Commission, Cheyenne office files (Heart Mountain is one of the biggest single irrigation districts served by the Buffalo Bill Dam and other facilities in the Bureau of Reclamation's Shoshone River Project). Though Wyoming law requires that reservoirs can fill only once each year—allocating to the reservoir owners only enough water from a stream to fill the reservoir to capacity once—the high flows of the Shoshone had meant that with the original dam, "the Irrigation Districts enjoyed multiple fills on their storage account in most runoff periods," the Heart Mountain district said. Adding new storage space to the dam meant new accounts in the dam, new contenders for Shoshone River water, including fish that needed winter flows, and irrigators being limited to "one fill" of their accounts. Most irrigators had not contemplated that when they supported state investment in Buffalo Bill expansion. "We do not believe that the Senators and Representatives of the State of Wyoming, at the time of the dam modification project was undertaken [*sic*], intended to spend in excess of $50 million to mainly benefit the Shoshone fishery," the Heart Mountain district complained in its letter. For Popo Agie River, see "Healthy Rivers Initiative," at Popo Agie Conservation District (website), http://www.popoagie.org/.

33. The "Just Add Water" series in the Game and Fish Department's *Wyoming Wildlife* magazine is focused "on how some of Wyoming's once languishing rivers have been developed into prolific fisheries for anglers and precious resources for communities. Though it may seem it's as easy as just adding water, far more goes into the work of shaping world-class fisheries." See for instance Annear and Bulger, "Progress at Pine Creek," 34, which begins: "Rivers give us a lot, but maybe their best lesson is teaching us about hope. No matter how bad things get for a river, it never quits working. A river never loses sight of its duty to provide life, even while holding out hope for better times. Sometimes it takes human intervention to steer back to those better times." Others in the series include Annear, "Securing the Shoshone," 37–43. For Game and Fish Department work with Wheatland Irrigation District to improve water levels for fish in one of the district's reservoirs, see Bulger, "Polishing a Gem," 31–37. For Paul Hagenstein, the first irrigator to donate an irrigation water right to become an instream flow right, see *Sweetwater Now* "Obituaries," https://www.sweetwaternow.com/paul-carl-hagenstein-jr-august-03-1927-may-26-2018/; Annear and Bulger, 34; and Anne MacKinnon, "From Hay Fields to Fish Flows: Pinedale Irrigator First in Wyoming to Convert Water Right for Fish," WyoFile (website), published November 15, 2011, https://www.wyofile.com/?s=Hagenstein. For water markets aiding instream flow in other states, see Anderson, *Tapping Water Markets*, 87–106 ("Buy that Fish a Drink" chapter).

34. Johnson, "Reclamation and Water Conservation," 1–2; John Lawson, personal communication to author, April 9, 2020; Natrona County Conservation District, "Cody Shale, Selenium, and Water Quality," http://www.natronacountyconservationdistrict.com/whatisselenium.html. WWDC, *2019 Legislative Report*, pp. 4–83 through 4–85; www.gillettewy.gov/city-government/departments/utilities/water/water-conservation; www.cheyennecity.org/1456/Water-Conservation.

35. Peck et al., "Irrigation-Dependent Wetlands Versus Instream Flow," 842–55;

Blevins, "Valuing the Non-Agricultural Benefits of Flood Irrigation"; Blevins et al., "The True Value of Flood Irrigation"; Intermountain West Joint Venture, 2020 "Digging Deeper into Flood Irrigation," https://iwjv.org/wp-content/uploads/2020/01/IWJV_9927_Intermountain-Insights_Irrigation_v4.pdf. One proposed conservation incentive is allowing users to sell the rights to any water they conserve. Several western states authorize that. Wyoming maintains the policy that unused water left in a stream becomes available to the next user in priority. Anderson, Tapping Water Markets, 63–70. For potential complexity in changes to sprinkler irrigation, see BOC Petitions Granted Files, Bates Creek Cattle Co., I-2018-3-1 and I-U-2018-3-2.

36.   O'Toole is the Wyoming 2014 Leopold Conservation Award Recipient, Ladder Ranch, Savery, WY, https://sandcountyfoundation.org/our-work/leopold-conservation-award-program/otoole-family-ladder-ranch; and board member of the Intermountain West Joint Venture, created by the US Fish and Wildlife Service with private partners to spur migratory bird habitat conservation, https://iwjv.org/management-board/. "Little Snake River Small Dams and Reservoirs," WWDC, *2000 Legislative Report*, 4–54 to 4–56. Ron Vore, former head of Small Water Projects Program, WWDC, personal communication with the author, November 17, 2018.

37.   Sunrise Engineering, Inc., *Kirby Creek Water Watershed Plan*, 1–8; Post, Buckley, Schuh & Jernigan, *Kirby Creek Watershed Level 1 Study*, Executive Summary, 1–3; Wyoming Department of Environmental Quality, *Wyoming's Draft 2016/18 Integrated*, 38–39; Wyoming Association of Conservation Districts, *Wyoming Watersheds Progress 2009*, 4, 16–17. For a description of the small water projects program, see "Small Water Projects," WWDC (website), http://wwdc.state.wy.us/small_water_projects/small_water_project.html. Watershed degradation through erosion on western rangelands generally was highlighted starting with the 1936 report to Congress, in US Department of Agriculture, Forest Service, *The Western Range: A Report on the Western Range*.

38.   In 1980, US senator Malcolm Wallop, R-Wyoming, successfully amended the introductory portion of the Clean Water Act to include section 101(g): "It is the policy of Congress that the authority of each State to allocate quantities of water within its jurisdiction shall not be superseded, abrogated or otherwise impaired by this Act. It is the further policy of Congress that nothing in this Act shall be construed to supersede or abrogate rights to quantities of water which have been established by any State. Federal agencies shall co-operate with State and local agencies to develop comprehensive solutions to prevent, reduce and eliminate pollution in concert with programs for managing water resources." Adler and Cameron, "Virtually Nonexistent Poison Runoff Controls," 171–98, in *The Clean Water Act Twenty Years Later*; Eddy-Miller and Gerhard, *Results of Nitrate Sampling in the Torrington, Wyoming*, 1999; "Reverse Osmosis Tackles Nitrates in Wyoming Wells," WaterWorld (website), published May 1, 2004, https://www.waterworld.com/articles/print/volume-20/issue-5/awwa-exhibitors/reverse-osmosis-tackles-nitrates-in-wyoming-wells.html. Denise Lucero, district manager for the North Platte, Lingle-Fort Laramie, and South Goshen Conservation Districts, personal communication with the author, December 13, 2018.

39.   "About Us," Wyoming Association of Conservation Districts (website), http://conservewy.com/ABOUT.html; Wyoming Association of Conservation Districts, *Wyoming Watersheds Progress 2009*, 4; Wyoming Department of Environmental Quality (DEQ), *Wyoming's Draft 2016/2018 Integrated*, 38–39 (Kirby Creek), 74–80 (Little Snake River); Anderson Consulting Engineers, Inc., *Final Report for Little Snake River*, 3–11/4, 3–108/125 (All three watershed studies cited are accessible at http://library.wrds.uwyo.edu, under Wyoming Water Development Commission Watershed Studies). Two "Section 319 Nonpoint Source Pollution Stories" written for EPA by the Wyoming DEQ regarding Muddy Creek, are https://www.epa.gov/sites/production/files/2015-10/documents/wy_lowermuddy.pdf, and https://www.epa.gov/sites/production/files/2015-10/documents/wy_muddymckinney.pdf. Little Snake River Conservation District Director Larry Hicks, as a state senator, successfully sponsored legislation in 2015 requiring the state water quality agency to create new water quality standards to accommodate streams that could not meet extant standards, due to irrigation withdrawals (Wyo. Stats. 35–11–302(c) (sponsorship and legislative history available at http://www.wyoleg.gov/Legislation/2015/SF0126). The agency, noting it had no power under the Clean Water Act to regulate water quality problems created by the exercise of water rights, avoided time and expense by simply taking the few streams once noted as locations where irrigation withdrawals affected stream water quality potential and moving them into categories whose standards the streams could meet; Lindsay Patterson, (Wyoming DEQ), personal communication with the author, May 24, 2018. After the 2016 fish kill on the Shoshone River when Willwood Dam spewed sediment down the river, Wyoming DEQ created three working groups to restore aquatic life and habitat and "reduce and/or eliminate" future need to release damaging amounts of sediment from the dam, while "Willwood Irrigation District's right to divert water under state water laws shall not be impacted by the efforts of this initiative." Opening paragraph of "Willwood Dam Advisory Committee and Working Groups"; in Wyoming DEQ, *Willwood Working Group 1 Final Report*. US Geological Survey, 2017, "U.S.GS Real-Time Monitoring of suspended-Sediment Concentrations" in Willwood Dam Operating Recommendations Summary of November 7, 2017, and Willwood Dam Operating Recommendations. The Willwood Dam Work Groups document series and activities are accessible for navigation and download, as of December 26, 2018, at http://deq.wyoming.gov/wqd/willwood-dam-and-shoshone-river/.

40.   Anderson, *Tapping Water Markets*, 8; WY State Engineer's Office, *2010 Annual Report*, 40 and *2018 Annual Report*, 24; Angus Thuermer, "Why a Wrinkle in Wyo Water Law is Worth Millions," WyoFile (website), published September 4, 2018, https://www.wyofile.com/water-sale-state-property-private-profits-and-zero-taxes/; Bob Davis, "Municipal Temporary Transfers," presentation at Wyoming Water Association Annual Meeting, October 17, 2019. Hydraulic fracturing was associated with groundwater contamination in Fremont County (western Wyoming) but state government resisted a finding that drinking water wells were affected, so no regulatory action affected the drilling company, DiGiulio, "Reconciling Oil and Gas Development and Groundwater Protection," and Andrew Graham, "Missing Science, Disagreement Surrounds DEQ's Final Pavillion

Report," WyoFile (website), published December 6, 2016, https://www.wyofile.com/missing-science-disagreement-surrounds-deqs-final-pavillion-report/.

41.   Wyoming Wildlife and Natural Resource Trust (WWNRT), *Status Report 2018*, 3-5.10. 20. For trends in land use and ways to preserve open space in Wyoming, see Ruckelshaus Institute, "Wyoming's State of the Space," 2009.WWNRT, 3–4, 9, 11, 16–19. For one example of work involving the Wildlife Trust, Trout Unlimited, and other partners including the US Natural Resource Conservation Service, the US Fish and Wildlife Service, the Bureau of Land Management, the National Fish and Wildlife Foundation, the Wyoming Landscape Conservation Initiative and the Rocky Mountain Elk Foundation, see "Lonetree Ranch," in *Wyoming Wildlife*, 28.

42.   While Wyoming coal burned in out-of-state power plants made Wyoming a major US energy supplier by the mid-1980s and for the next two decades, by 2021, renewables like solar and wind were expected to surpass coal in electric power generation nationwide; "EIA expects US electricity generation from renewables to soon surpass nuclear and coal," US Energy Information Administration (website), published January 30, 2020, https://www.eia.gov/todayinenergy/detail.php?id=42655. Wyoming coal production and mineral revenue figures: Wyoming State Geological Survey, https://www.wsgs.wyo.gov/energy/coal-production-mining; Consensus Revenue Estimating Group, "Wyoming State Government Revenue," tables 4 and 7; Katie Klingsporn, "Powering Down: Examining Coal's Shaky Ground from Jim Bridger," WyoFile (website), published September 3, 2019, https://www.wyofile.com/powering-down-examining-coals-shaky-ground-from-jim-bridger/; Michael Madden, "No Easy Answers," WyoFile (website), published November 26, 2019, https://www.wyofile.com/no-easy-answers-wyo-wrestles-with-a-tangle-of-fiscal-worries/; WWDC, *2019 Legislative Report*, pp. 2–8 to 2–10; Nick Reynolds, "307 politics: Voters show continued distaste for tax increases as budget questions continue to mount," published August 22, 2020, https://trib.com/news/state-and-regional/govt-and-politics/307-politics-voters-show-continued-distaste-for-tax-increases-as-budget-questions-continue-to-mount/article_0c36a505-328-59a7-ad9e-19aa6dd12714.html; Wyoming governor Mark Gordon, "Governor calls first 10 percent state budget cuts devastating but necessary," August 26, 2020, https://governor.wyo.gov/media/news-releases/2020-news-releases/governor-calls-first-10-state-budget-cuts-devastating-but-necessary; Camille Erickson, "Proposals to raise sales and use taxes dismissed by Wyoming lawmakers," published August 25, 2020, https://trib.com/news/state-and-regional/proposals-to-raise-sales-and-use-taxes-dismissed-by-wyoming-lawmakers/article_400aae83-f5ca-5c2a-b345-d9a90f208295.html.

43.   Third National Climate Assessment vol. 2 (website), Great Plains 2014, https://nca2014.globalchange.gov/report/regions/great-plains#statement-16856. National Science Foundation grants to the University of Wyoming fostered improved computer modeling and then, with a five-year award through 2018, detailed work in environmental hydrology and geophysics (establishing a center with that focus) to understand and model how water moves through Wyoming's complex geology and landscape. For examples, see UW News 2011 (website), "Wyoming, Utah Researchers Join Forces to Understand

Complex Water Problems [. . .]," published November 3, 2011, http://www.uwyo.edu/uw/news/2011/11/wyoming,-utah-researchers-join-forces-to-understand-complex-water-problems-facing-western-states.html; and the Wyoming Center For Environmental Hydrology and Geophysics (WYCEHG) and Wyoming EPSCOR websites, http://www.uwyo.edu/epscor/wycehg/ and http://www.uwyo.edu/epscor/about/. Ginger Paige, personal communication with the author, March 1, 2018.

44.   Ginger Paige, personal communication with the author, March 1, 2018.

45.   Wyoming State Climate Office (website), "Percentage of Wyoming in each Drought Category by Week," http://www.wrds.uwyo.edu/drought/droughttimeline.html. A summary of findings from a 2006 workshop is Steve Gray's "Water, Drought and Wyoming's Climate: Final Report." Gray moved on to become director of the US Geological Survey's Alaska Climate Science Center, and his predecessor Jan Curtis joined the National Water and Climate Center of the US Department of Agriculture's Natural Resource Conservation Service. WWDC's statewide water plan, issued in 2007 (known as the *Wyoming Framework Water Plan*), avoided any use of the words "climate change," see especially secs. 3.1.3, 7.1.4, 7.1.5, 7.1.6.

46.   WWDC, "Middle Big Horn River Watershed, Level I Study, Consultant Contract for Services, Attachment A" 1–30, is part of larger PDF document generated from the July 15, 2018, WWDC and Select Water Committee meeting. Available at http://wwdc.state.wy.us/commission/eNotebook/201806-eNotebookMeeting.html; 14 of 30 (289 in the PDF) states, "In effort to provide information to the conservation districts' future water use and management efforts, the Consultant shall provide a section in the final report that summarizes current climate science. The intention is not to debate whether climate change is real, perceived, or human caused. Rather, the Consultant will present a state-of-the-art science summary so that the Sponsors can use the information for planning purposes, allowing for better planning and preparation for climate variability and associated extreme weather events." Barry Lawrence (WWDC deputy director-planning), personal communication with the author, June 11, 2018. Presentation to WWDC by commission staff, "River Basin Planning: What Does the Future Hold?," August 23, 2017, PowerPoint slides in author's files. Jason Mead (WWDC deputy director-dams and reservoirs) personal communication with the author, June 13, 2018. See also WWDC staff PowerPoint update on modernization, "What Do We Have up Our Sleeves," published March 10, 2020, http://seo.wyo.gov/interstate-streams/water-forum/presentations. The engineering community has been engaged for a decade in debates on how to plan water management infrastructure now that engineers can no longer rely on the idea of "stationarity"(stationarity is defined as "the idea that natural systems fluctuate within an unchanging envelope of variability"); Milly et al., "Stationarity Is Dead: Whither Water Management?"; Galloway, "If Stationarity Is Dead, What Do We Do Now?"; and Milly et al., "On Critiques of 'Stationarity Is Dead: Whither Water Management?'"

47.   Ginger Paige, personal communication with the author, March 1, 2018; WYCEHG website, http://www.uwyo.edu/epscor/wycehg/; Gordon, "Return Flow in Northeastern Wyoming," 11; Blevins et al., "The True Value of Flood Irrigation"; John

Fialka, "Drought: Cataclysms Were Predictable for 1,000 years. That's Changing," ClimateWire (website), published June 8, 2018, https://www.eenews.net/climatewire/stories/1060083879.

48.   DeVisser and Fountain, "A Century of Glacier Change," 103–16; VanLooy et al., "Spatially Variable Surface Elevation," 98–113; Hall et al., "Snow Cover, Snowmelt Timing," 87–93; Cheesbrough et al., "Estimated Wind River Range," 818–28.

49.   Gary Collins (tribal water engineer), "Wind River Reserved Water Rights," Wind River Reservation Tour Booklet, August 24, 2000, in author's files; Eastern Shoshone and Northern Arapaho Tribes, "Looking to the Future of the Wind River Indian Reservation," in *Wind River Indian Needs Determination Survey*; Aragon, "The Wind River Indian Tribes," 17; Mitch Cottenoir (tribal water engineer), personal communication with the author, October 25, 2018.

50.   John Fialka, "Drought: A Slow Disaster Scorched Wyo. No One Saw It Come or Go," ClimateWire (website) June 6, 2018, https://www.eenews.net/climatewire/stories/1060083629. In DeVisser and Fountain, "A Century of Glacier Change," tables 2, 3, and 8, pp. 109–12: the authors estimate contributions to snowmelt for Dinwoody Creek, in turn tapped by Dinwoody Canal for the reservation system, and Bull Lake Creek, which runs to the Bull Lake Reservoir serving Midvale irrigation district. Cody Knutson, "Project Proposal: The Wind River Indian Reservation's Vulnerability to the Impacts of Drought and the Development of Decision Tools to Support Drought Preparedness" (estimated project duration: 6/2015–5/2017), in author's files; McNeeley, "Sustainable Climate Change Adaptation"; Cohn et al., "Seems Like I Hardly See Them around Anymore," 405–29; Jonathan Friedman, S. McNeeley, M. Cottenoir, A. C'Bearing, J. Wellman, "The History of Water on Wind River Reservation to Inform Climate Adaptation," PowerPoint presentation to Wind River Water Resources Control Board, n.d., notes in author's files; Jennifer Wellman, "Wyoming EPSCOR Program at Wind River," presentation October 30, 2017, http://www.uwyo.edu/epscor/_files/documents/wig/wig%202017%20wellman.pdf.

51.   Wyoming Session Laws, 1989, chap. 145; Wyoming Session Laws, 2003, chap. 78; Wyoming Water Development Commission, Annual Legislative Reports, 1992–2014 (on file at the Wyoming Water Development Office, Cheyenne, and 1996–2014 accessible online at http://wwdc.state.wy.us/legreport/legreports.html); Mitch Cottenoir (tribal water engineer), personal communication with the author, October 25, 2018; Cottenoir and Baptiste Weed (deputy tribal water engineer), personal communication with the author, March 15, 2018.

52.   Eastern Shoshone and Northern Arapaho Tribes, *Agricultural Resource Management Plan*, 30; McNeeley, "Sustainable Climate Change Adaptation"; Hanna, "Native Communities and Climate Change: Executive Summary"; Cottenoir and Weed, personal communication with the author, March 15, 2018.

53.   Preston and Engle (1928) "Report of Advisors on Irrigation on Indian Reservations," from US Congress Committee on Indian Affairs Hearings. This and later reports through 1981 are cited in McCool, *Command of the Waters: Iron Triangles, Federal*

Development, and Indian Water, 154; Shay, "Promises to a Viable Homeland," 557–58; John Anevski (Chief, Division of Water and Power, Office of Trust Services, Bureau of Indian Affairs), Statement before the Senate Committee on Indian Affairs, Field Hearing on the Wind River Irrigation Project, April 20, 2011, https://www.bia.gov/sites/bia.gov/files/assets/as-ia/pdf/idc013562.pdf; Anne-Marie Fennell (Director, Natural Resources and Environment, General Accounting Office), "Indian Irrigation Projects: Deferred Maintenance and Financial Sustainability Issues Remain Unresolved," testimony on March 4, 2015 to the Senate Committee on Indian Affairs, reprinted in Flores, *Indian Irrigation Projects*, 1–16.

54.   "Testimony of Mitchel T. Cottenoir, Tribal Water Engineer, Shoshone and Arapaho Tribes of Wind River Reservation. Legislative Hearing on S. 438," on March 4, 2015, reprinted in Flores, *Indian Irrigation Projects*, 31–37; "Barrasso Secures Key Wyoming Project in Bipartisan Water Infrastructure Bill," October 10, 2018, https://www.barrasso.senate.gov/public/index.cfm/news-releases?ID=A3C64752-3229-4C11-9793-37961CE394D7; "President Trump Signs Barrasso's Water Infrastructure Legislation into Law," October 23, 2018, https://www.barrasso.senate.gov/public/index.cfm/news-releases?ID=D1E986DB-466E-4CF2-9253-9C4734605076. Mitch Cottenoir and Baptiste Weed, personal communication with the author, March 15, 2018; Wind River Water Resources Control Board, discussion notes in author's files, August 16, 2017, and June 20, 2018; WWDC, *2019 Legislative Report*, 3–60.

55.   Cottenoir and Weed, personal communication with the author, March 15, 2018; Eastern Shoshone and Northern Arapaho Tribes, *Agricultural Resource Management Plan*, 8–9; Tetra Tech, *Executive Summary, Little Wind River Drainage Level II*; Wind River Water Resources Control Board, discussion notes in author's files, August 16, 2017, June 20, 2018 and August 1, 2019; WWDC, *2019 Legislative Report*, 2–8 to 2–10.

56.   WWDC, *2019 Legislative Report*, 3–8.

57.   Tetra Tech, *Big Wind River Drainage Level II*, 1–4.

58.   For WWDC plans for improvements at Midvale, Leclair, and Riverton Valley irrigation districts, see listings of plans under those headings at the Water Resources Data System library at the University of Wyoming, http://library.wrds.uwyo.edu/wwdcrept/wwdcrept.html.

59.   McNeeley, "Sustainable Climate Change Adaptation," 400–401. Elsewhere in the United States, tribes and their non-tribal neighbors have similarly fought divisive legal battles over natural resource use, for many years. Yet where there has been imperative need for a resource, and it becomes impossible to access it without cooperation, there have been examples of progress—slow but steady—in cooperating tribal-state management of natural resources. Singleton, *Constructing Cooperation: The Evolution of Institutions*, 66–80, 143–45 (discussing an example of work on salmon fishing allocation and habitat undertaken by Puget Sound area tribes and the state of Washington after bitter litigation on salmon issues.)

60.   WY State Engineer's Office, *Wyoming and the Colorado River: A Report*, 3, 45.

61.   WWDC, "Technical Memorandum," 2011.

62.   US Bureau of Reclamation, *Executive Summary: Colorado River Basin*, 1–10; Water Education Foundation, Colorado River Project River Report, 1–4; US Bureau of Reclamation, "Another dry year in the Colorado River Basin increases the need for additional state and federal actions," press release May 9, 2018, https://www.usbr.gov/newsroom/newsrelease/detail.cfm?RecordID=62170.

63.   "The law and politics of the Colorado River have long been driven primarily by the efforts of all the basin states, except California, to prevent California's actual use from ripening into a permanent right," water law commentator Dan Tarlock noted in 2001, Tarlock, "The Future of Prior Appropriation," 784. State Engineer's Office (SEO), *Wyoming and the Colorado River*, 11–32, 49–51 provides succinct descriptions of key actions related to the Colorado River compacts in recent years. To guide the state of California in cutting its use of the Colorado River back to its actual compact allocation of 4.4 million acre-feet per year, all seven Colorado River states and the Department of Interior negotiated for years to produce the Interim Surplus Guidelines issued in 2001, and California succeeded in adopting a plan for its major water users to accomplish the cutback in 2003, but the drought forced California to implement that cutback much more quickly than originally planned; see Water Education Foundation (website), "Colorado River Water Use 4.4 Plan," https://www.watereducation.org/aquapedia/colorado-river-water-use-44-plan. For the 2007 Interim Shortage Guidelines, see US Bureau of Reclamation (website), "Colorado River Interim Shortage Guidelines and Coordinated Operations for Lake Powell and Lake Mead," updated June 5, 2015, https://usbr.gov/lc/region/programs/strategies.html; and SEO (cited above), 29–30. Pat Mulroy, former longtime manager of the Southern Nevada Water Authority, June 6, 2019, conference presentation describes how the states achieved the 2007 shortage guidelines during the drought: available on video at https://www.getches-wilkinsoncenter.cu.law/2019/06/20/40th-annual-gwc-summer-conference/ (Day 1 Video). For a general discussion of why the many deals and compromises involved in politics are more likely to lead to successful water solutions than centralized management, see Tarlock (2001), and Schlager and Blomquist, *Embracing Watershed Politics*.

64.   State Engineer's Office (SEO), *Wyoming and the Colorado River*, 60–61; Bidtah Becker, "Water & Tribes Initiative," conference presentation, June 7, 2019. Work addressing the water rights, uses, and needs of key tribes on the Colorado River (all outside Wyoming) is discussed in the December 2018 Colorado River Basin Ten Tribes Partnership Tribal Water Study (website), updated December 13, 2018, https://www.usbr.gov/lc/region/programs/crbstudy/tribalwaterstudy.html. The Colorado River Salinity Control Program was enacted by Congress at the behest of basin states in 1974 (see http://coloradoriversalinity.org/). The Upper Colorado River Endangered Fish Recovery program was started in 1988 with a cooperative agreement among the governors of Colorado, Utah, and Wyoming, the secretary of the interior and the administrator of the Western Area Power Administrations (producing power from major federal dams on the river), http://www.coloradoriverrecovery.org/general-information/about.html. The program website includes considerable historic information on the fish involved (humpback chub, bonytail, Colorado pikeminnow, and razorback sucker). A succinct analysis of what

caused those native fish to become endangered is provided in Valdez and Muth, "Ecology and Conservation of Native Fish," 157–78. For experimental releases for the Grand Canyon, see Melis et al., "Three Experimental High-Flow Releases." Information on the subsequent 2016 and 2018 releases, see Glen Canyon High Flow Experimental Release (website), ethttps://www.usbr.gov/uc/rm/gcdHFE/index.html. For releases to the Colorado River Delta, see International Boundary and Water Commission, *Minute 319 Colorado River*. For pulse flows into Mexico, see Postel, *Replenish: The Virtuous Cycle*, 19–42. For overviews of these recent issues on the river, see also SEO, 51–61.

65.   The Colorado River Research Group (CRRG) published in 2014 a paper titled "Charting a New Course for the Colorado River: A Summary of Guiding Principles." While praising notable policy reforms on the river since 2000, the paper argues that "few have addressed the underlying problems, but rather have 'bought time' and planted the seeds for more lasting and comprehensive solutions to emerge." Among the group's recommended guiding principles are, "The solutions that are most cost-effective, reliable, equitable and quickly implemented are those focused on conservation, reallocation, and voluntary shortage sharing"; and, "Preferred policy options are those that are flexible and iterative, use science and economics, and that feature a sound collaborative structure that allows constant reassessment and adjustment over time (within well-defined rules and process)," 3–4. Brad Udall of Colorado State University warned in September 2017 of the inadequacy of Colorado River basin response to climate change; Water Education Foundation, Colorado River Project River Report, 7; Brad Udall, personal communication with the author, October 2, 2018. Udall and a colleague reported in early 2017 that the impact of climate change on the Colorado River is likely to be much more serious than had been assumed by that time; Udall and Overpeck, "The 21st Century Hot Drought," 2404–18. In 2018 he and other colleagues reported on research into the causes of low flows on the river, Xiao, Udall, and Lettenmaier, "On the Causes of Declining Colorado River," 6739–56.

66.   CRRG, "When is a Drought Not a Drought?," 2018.

67.   Purcell Consulting, "Colorado River Compact Administration Project Prepared for Wyoming Attorney General's Office and the State Engineer," 2005, in author's files; State Engineer's Office, *Colorado River Compact Administration Program*, 2008, in author's files. The 2005 report said that "given the competition for water in the Colorado River Basin, Wyoming should be the leading authority on its water use in the Green River Basin. It is not good business to rely on information developed by others in the event of a controversy or a valid call for curtailment of use under the Upper Colorado River Basin," 14. The 2008 report adopts that mission (cover letter, January 2, 2008).

68.   The 2016 State Engineer's Office report, *Wyoming and the Colorado River*, 32–33, 49, 61, contains clear descriptions of what could prompt a Colorado River curtailment affecting Wyoming and the work being done to prepare for such an event including compiling better data on the Wyoming's consumptive use of water in the Green and Little Snake basins.

69.   A State Engineer's Office PowerPoint presents data on the river and its reservoirs

and projections for the future, as of fall 2019; at http://www.uwyo.edu/uwe/wy-dm-ucrb/meeting-documents/november-2019-wy-dm-public-meetings.final.pdf. The website contains materials from the SEO outreach efforts on the feasibility of a demand management program through 2020.

70.   Quotes from Wyoming state engineer Pat Tyrrell and figures and descriptions of the provisions negotiated come from the public presentation Tyrrell held in Baggs, Wyoming (in the basin of the Little Snake River) on October 9, 2018; background information at http://www.uwyo.edu/uwe/wy-dm-ucrb/index.html. Members of the Wyoming Legislature started paying close attention to Colorado River matters, holding several committee meetings including an educational session on Colorado River issues and possible conservation approaches: see materials and minutes of a June 14, 2018, meeting of the Joint Committee on Agriculture, State and Public Lands and Water Resources, and the Legislature's Select Water Committee (overseeing the work of the Wyoming Water Development Commission), available at http://www.wyoleg.gov/Committees/2018/J05. See especially a presentation by Larry MacDonnell, "Shepherding 'Compact Security Water' Under Wyoming Law: What are the issues?," https://www.wyoleg.gov/InterimCommittee/2018/05–20180613 ShepherdingCompactSecurityWaterUnderWyomingLaw.pdf.

71.   Colorado River District, "West Slope Risk Study: Review and Status Report," presented to Wyoming legislative committees June 14, 2018 (see detail regarding the committees in previous footnote), https://www.wyoleg.gov/InterimCommittee/2018/05–20180613shepardingupdate.pdf. US Bureau of Reclamation, "Agreement Among the United States of America, through the Department of the Interior," 2014. The State Engineer's Office announced the opportunity for using the pilot funds on its website, with links to the Pilot System Water Conservation Program RFP notices for 2016 and 2017, as well as application forms.

72.   Upper Colorado River Commission, *Final Report: Colorado River System Conservation*, 14–15. Eric and April Barnes, interview with the author, April 16, 2017.

73.   Blevins et al., "The True Value of Flood Irrigation." Kendy, "Impacts of Changing Land Use," 12, 25–27, 32.

74.   State Engineer's Office, "Wyoming Demand Management Feasibility for the Upper Colorado River Basin," public presentation, November 2019, http://www.uwyo.edu/uwe/wy-dm-ucrb/meeting-documents/november-2019-wy-dm-public-meetings.final.pdf.

75.   State Engineer's Office public presentation, November 2019; state representative Albert Sommers, "Upper Green River Water Eyed by Down-river States," *Pinedale Roundup*, July 18, 2019, https://pinedaleroundup.com/article/upper-green-river-water-eyed-by-down-river-states.

76.   In 1929, Lasky (November, part 3, p. 56), argued that increasingly complex interstate relations and the negotiations role of a state engineer would have this effect.

77.   For an initial report on state government's latest attempt to build toward a less-minerals-dependent future, see https://www.endowyo.biz/.

BIBLIOGRAPHY

Primary Sources

ARCHIVAL MATERIAL

Baxter, George. Biographical file. American Heritage Center, University of Wyoming, Laramie.

Beck, George T. Papers. "Autobiography" (unpublished manuscript). Collection 59, box 7, folder 4. American Heritage Center, University of Wyoming, Laramie.

Brock, J. Elmer. Papers. Collection 00102. American Heritage Center, University of Wyoming,

"Buffalo, WY." Clipping files. American Heritage Center, University of Wyoming, Laramie.

"Buffalo, WY." Clipping files. Johnson County Public Library, Buffalo, WY.

Carey Family Papers. Collection 01212. American Heritage Center, University of Wyoming, Laramie.

Correspondence Files. State Engineer's General Correspondence 1913–1920. RG 0037, AS 1687. Wyoming State Archives, Cheyenne.

General Correspondence. Box 64, file 801, Federal Court Matters, misc., file 1. Records of the Bureau of Indian Affairs, Wind River Agency. RG 75, National Archives, Denver, CO.

General Correspondence, Water Appropriations. Entry 7. Kendrick Project, Bureau of Reclamation, Rocky Mountain Region. Record Group 115, National Archives, Denver, CO.

Gordon, John H. Biographical File. American Heritage Center, University of Wyoming, Laramie.

Johnston, Clarence T. Biographical File. American Heritage Center, University of Wyoming, Laramie.

Kennedy, T. Blake. "Memoirs." Unpublished manuscript. Box 1, T. Blake Kennedy Papers. Collection 00405, American Heritage Center, University of Wyoming, Laramie.

McGee, Gale W. Papers. Collection 9800. American Heritage Center, University of Wyoming, Laramie.

Mead, Elwood. Papers. Collection 05258. American Heritage Center, University of Wyoming, Laramie.

Mead, Elwood. Papers. WRCA 055. Water Resources Collections and Archives. Special Collections & University Archives, University of California, Riverside.

Mead-Van Orsdel Correspondence File. Box 2, 1901–1904, K-M, RR 517, Wyoming State Archives, Cheyenne.

Mondell, Frank. Papers. Collection 01050. American Heritage Center, University of Wyoming, Laramie.

Morton Family Papers. Chicago History Museum

Oral History Files. Homesteader Museum, Powell, WY.

Smith, J. R. Papers. Historical subject files, no. 18. Johnson County Library, Buffalo, WY.

State Engineer Elwood Mead Records 1886–1892. Elwood Mead—Federal Government Officials 1886, 1888–1890. Wyoming State Archives, Cheyenne.

Stroock, Thomas F. Papers. Collection 10356. American Heritage Center, University of Wyoming, Laramie.

Warren, Francis E. Papers. Collection 00013. Digital Collections, American Heritage Center, University of Wyoming, Laramie. http://digitalcollections.uwyo.edu:8180/luna/servlet/ view/all/what/Letterpress%20copy%20books/Wyoming-History-1890-1918?sort=RID,Description,Title,Date_Original.

GOVERNMENT DOCUMENTS

Anderson Consulting Engineers, Inc (ACE). *Executive Summary for Casper Alcova Irrigation District Rehabilitation Needs Analysis*. Cheyenne: Wyoming Water Development Commission, 2003. http://library.wrds.uwyo.edu/wwdcrept/Casper/Casper_Alcova_Irrigation_District-Rehabilitation_Needs_Analysis-Executive_Summary-2003.pdf)

———. *Final Report for Little Snake River/Vermillion Creek Watershed Study, Level 1*. Cheyenne: Wyoming Water Development Commission, 2013.

Annual Report of the Commissioner of the General Land Office for the Year 1885. Washington, DC: Government Printing Office, 1885. https://babel.hathitrust.org/cgi/pt?id=mdp.39015067316656;view=1up;seq=7

Autobee, Robert. *Riverton Unit, Pick-Sloan Missouri Basin Program*. Washington DC: US Bureau of Reclamation, 1996. https://www.usbr.gov/projects/pdf.php?id=173.

Banner Engineering. "Proposed Methods for Augmenting Laramie River Basin Water Supplies." Report to Wyoming Natural Resources Board, Cheyenne, 1955.

Bishop, Floyd A. *Fortieth Biennial Report of the State Engineer to the Governor of Wyoming, 1969–1970*. https://babel.hathitrust.org/cgi/pt?id=uc1.$b791162;view=1up;seq=5.

———. *Forty-first Biennial Report of the State Engineer to the Governor of Wyoming, 1970–1972*. https://babel.hathitrust.org/cgi/pt?id=uc1.$b791163;view=1up;seq=8.

———. *Thirty-Eighth Biennial Report of the State Engineer to the Governor of Wyoming, 1965–1966*. Casper, WY: Prairie, [1967?]. https://babel.hathitrust.org/cgi/pt?id=uc1.$b791160;view=1up;seq=5.

———. *Thirty-Seventh Biennial Report of the State Engineer to the Governor of Wyoming, 1963–1964*. Casper, WY: Prairie, [1965?]. https://babel.hathitrust.org/cgi/pt?id=uc1.$ b791159;view=1up;seq=5.

Blake, John W., Willis Van Devanter, and Isaac P. Caldwell, eds. *Revised Statutes of Wyoming*. Cheyenne, WY: Daily Sun Steam, 1887.

Board of Control (BOC). *Order Record Books 1–28*. Records of the Board of Control, Wyoming State Engineer's Office, Cheyenne.

———. Petitions Granted Files. Records of the Board of Control, Wyoming State Engineer's Office, Cheyenne.

———. *Tabulation of Adjudicated Surface Water Rights of the State of Wyoming, Division I, 1996*. Records of the Board of Control, Wyoming State Engineer's Office, Cheyenne, 1996

———. *Tabulation of Adjudicated Surface Water Rights of the State of Wyoming, Division II, 1999*. Records of the Board of Control, Wyoming State Engineer's Office, Cheyenne, 1999.

———. *Tabulation of Adjudicated Surface Water Rights of the State of Wyoming, Division III, 1999*. Records of the Board of Control, Wyoming State Engineer's Office, Cheyenne, 1999.

———. "Testimony in Proof #3321, April 1898, Boxelder Creek." Records of the Board of Control, Wyoming State Engineer's Office, Cheyenne.

———. "Testimony in Proof #3322, April 1898, Boxelder Creek." Records of the Board of Control, Wyoming State Engineer's Office, Cheyenne.

Bond, Fred. *Fifth Biennial Report of the State Engineer to the Governor of Wyoming, for 1889 and 1900*. Cheyenne, WY: S. A. Bristol, 1901. https://babel.hathitrust.org/cgi/pt?id =mdp.39015011420398;view=1up;seq=9.

———. *Sixth Biennial Report of the State Engineer to the Governor of Wyoming, for 1901 and 1902*. Laramie, WY: Chaplin, Spafford & Mathison, 1902. https://babel.hathitrust. org/cgi/pt?id=mdp.39015011420406;view=1up;seq=1;size=125.

Burritt, Edwin. *Report on Water Rights of Shoshone Irrigation District*. Wyoming State Engineer's Office, Cheyenne, WY, 1935.

Cassity, Michael. *Wyoming Will Be Your New Home . . . : Ranching, Farming and Homesteading in Wyoming, 1860–1960*. Cheyenne: Wyoming State Historic Preservation Office, 2011. http://wyoshpo.state.wy.us/homestead/pdf/ historic_context_study_011311.pdf

Christopulos, George. *Annual Report of the State Engineer, 1977*. Board of Control. Wyoming State Engineer's Office, Cheyenne, WY, 1977.

———. *Annual Report of the State Engineer, 1982*. Board of Control. Wyoming State Engineer's Office, Cheyenne, WY, 1982.

Colorado Water Conservation Board. *Colorado's Water Plan*. Denver, 2015. https://www. colorado.gov/pacific/cowaterplan/plan.

Consensus Revenue Estimating Group. *Wyoming State Government Revenue Forecast Fiscal Year 2005–Fiscal Year 2010*. Cheyenne: State of Wyoming, 2005. http://wyoleg.gov/ budget/CREG/Reports/GreenCREG_Jan05.pdf.

Curtis, Jan, and Kate Grimes. *Wyoming Climate Atlas*. Cheyenne, WY: State Climatologist Office, 2004.

Dallman, Jay. "Flushing Flows to Enhance Trout Habitat," *Plains Talk, News of the Bureau of Reclamation Great Plains Region*, November 1995.

———. "Pathfinder Dam Celebrates Restored Flow." Plains Talk, US Bureau of Reclamation Great Plains Region, October 2002.

Economic Research Service. "Wyoming Agricultural Commodity Figures for 2017." US Department of Agriculture (website), updated November 30, 2018, https://data.ers. usda.gov/reports.aspx?StateFIPS=56&StateName=Wyoming&ID=17854#Pb896c07 8c63b42739e7a45625a3cce88_2_586iT21Roxo.

Economically Needed Diversity Options for Wyoming (ENDOW). *Socioeconomic Assessment of Wyoming, Appendix*. Cheyenne, WY: ENDOW Executive Council, 2017. https://www.dropbox.com/sh/26n3prcsombbuap/ AADdvKgrQL4S48nW16m3vdjLa?dl=0.

Eastern Shoshone and Northern Arapaho Tribes. *Agricultural Resource Management Plan*, Draft. Wind River Indian Reservation, January 2018.

———. *Wind River Indian Needs Determination Survey* (WINDS 2). Fort Washakie, WY: Joint Business Council, 1999.

Eddy-Miller, C. A., and Gary Gerhard. *Results of Nitrate Sampling in the Torrington, Wyoming, Wellhead Protection Area 1994–98*. US Geological Survey Water Resources Investigations Report 99–4164. Cheyenne, WY: USGS, 1999.

Emerson, Frank C. *Fifteenth Biennial Report of the State Engineer to the Governor of Wyoming, for 1919 and 1920*. Laramie, WY: Laramie Printing 1921. https://babel. hathitrust.org/cgi/pt?id=mdp.39015011420547;view=1up;seq=7.

———. *Sixteenth Biennial Report of the State Engineer to the Governor of Wyoming, for 1921 and 1922*. Casper, WY: Mills, [1923?]. https://babel.hathitrust.org/cgi/pt?id=mdp.39 015011420554;view=1up;seq=5

Government Accounting Office. *Bureau of Reclamation: Availability of Information on Repayment of Water Project Construction Costs Could Be Better Promoted*. GAO report 14–764. U. S. Government Accountability Office, 2014. https://www.gao. gov/products/GAO-14-764.

Hall, William Ham. *Irrigation Development: History, Customs, Laws, and Administrative Systems Relating to Irrigation, Water-courses, and Waters in France, Italy, and Spain*. Office of the State Engineer, California. Sacramento: State Printing Office, 1866.

Hinckley Consulting and AMEC. Horse Creek Groundwater/Surface Water Connection Investigation. For the Wyoming State Engineer's Office, Groundwater Division. Laramie, WY, 2011.

International Boundary and Water Commission, United States and Mexico. *Minute 319 Colorado River Delta Environmental Flows Monitoring*, December 2014. https:// www.ibwc.gov/EMD/Min319Monitoring.pdf.

Johnston, Clarence T. *Eighth Biennial Report of the State Engineer to the Governor of Wyoming, for 1905 and 1906*. Laramie, WY: Laramie Republican, [1907?]. https://babel. hathitrust.org/cgi/pt?id=mdp.39015011420414;view=1up;seq=3.

———. *Ninth Biennial Report of the State Engineer to the Governor of Wyoming, for 1907 and 1908*. Cheyenne, WY: S. A. Bristol, 1908. https://babel.hathitrust.org/cgi/pt?id =mdp.39015011420489;view=1up;seq=3.

———. *Seventh Biennial Report of the State Engineer to the Governor of Wyoming, for 1903 and 1904*. Laramie, WY: Laramie Republican, 1905. https://babel.hathitrust.org/cgi/pt?id=mdp.39015011420422;view=1up;seq=1.

———. *Tenth Biennial Report of the State Engineer to the Governor of Wyoming, for 1909 and 1910*. Cheyenne, WY: S. A. Bristol, 1910. https://babel.hathitrust.org/cgi/pt?id= mdp.39015011420497;view=1up;seq=7.

Journals and Debates of the Constitutional Convention of the State of Wyoming, begun at the City of Cheyenne on September 2, 1889, and Concluded September 30, 1889. Cheyenne: The Daily Sun, Book and Job Printing, 1893.

Kimball, Bob, Randy Huffsmith, Mike Smith, and Tom Charles. "Technical Memorandum." CDM Consulting Review of Wyoming State Geological Survey, Powder River Basin Desalination Project Feasibility. Moose, WY, June 29, 2006. https://web.archive.org/web/20070709160512/http://cbm.moose.wy.gov/documents/CDMReviewofWYSt.GeologicalSurvey.pdf.

Klasing, Susan A. *Agricultural Drainage Water Contamination in the Kendrick Reclamation Project Area, Wyoming: A Public Health Perspective for Selenium and Selected Elements*. Irrigation Drainage Program, National Irrigation Water Quality Program, US Department of Interior, 1993.

Lampen, Dorothy. *A Report of an Economic Investigation of Home Conditions on Federal Reclamation Projects*. Washington, DC: US Bureau of Reclamation, 1929.

Larson, L. R. *Water Quality of the North Platte River, East-Central Wyoming*. US Geological Survey Water Resources Investigation Report 84–4172. Cheyenne, WY: US Department of the Interior, 1985. https://pubs.usgs.gov/wri/1984/4172/report.pdf.

Letters, Testimony, and Minutes of the Stream Preservation Feasibility Study. Bound, unpaged volume. October 1974. Wyoming State Library, Cheyenne.

Linenberger, Toni Rae. "The Seedskadee Project." Bureau of Reclamation History Project. Denver, CO: Department of the Interior, 1997. https://babel.hathitrust.org/cgi/pt?id=uc1.31210024875294;view=1up;seq=1.

Management Audit Committee. *Wyoming Water Development Commission*. Cheyenne: Wyoming Legislative Service Office, 2016. http://wyoleg.gov/progeval/REPORTS/2016/WaterReportBinder1–6-2016.pdf.

Mead, Elwood. *First Biennial Report of the State Engineer to the Governor of Wyoming, for 1891 and 1892*. Cheyenne, WY: S. A. Bristol, 1892. https://babel.hathitrust.org/cgi/pt?id=mdp.39015011420356;view=1up;seq=7.

———. *Fourth Biennial Report of the State Engineer to the Governor of Wyoming, for 1897 and 1898*. Cheyenne, WY: S. A. Bristol, 1898. https://babel.hathitrust.org/cgi/pt?id =mdp.39015011420380;view=1up;seq=9.

———. *Second Annual Report of the Territorial Engineer to the Governor of Wyoming, for the Year 1889*. Cheyenne, WY: Bristol & Knabe, 1890. https://babel.hathitrust.org/cgi/pt?id=uc1.b4525854;view=1up;seq=7.

——. *Second Biennial Report of the State Engineer to the Governor of Wyoming, for 1893 and 1894*. Cheyenne, WY: S. A. Bristol, 1894. https://babel.hathitrust.org/cgi/pt?id=uiug.30112110327639;view=1up;seq=9.

——. *Third Biennial Report of the State Engineer to the Governor of Wyoming, for 1895 and 1896*. Cheyenne, WY: S. A. Bristol, 1897. https://babel.hathitrust.org/cgi/pt?id=mdp.39015011420372;view=1up;seq=3.

——. "Water-right Problems of the Bighorn Mountains." Water Supply Paper 23. Washington, DC: Government Printing Office, 1899. https://doi.org/10.3133/wsp23.

Melis, Theodore S., Paul E. Grams, Theodore A. Kennedy, Barbara E. Ralston, Christopher T. Robinson, John C. Schmidt, Lara M. Schmit, Richard A. Valdez, and Scott A. Wright. "Three Experimental High-Flow Releases from Glen Canyon Dam, Arizona: Effects on the Downstream Colorado River Ecosystem." US Geological Survey Fact Sheet 2011–3012, February 2011. https://pubs.usgs.gov/fs/2011/3012/fs2011–3012.pdf.

"Memorial to President Ulysses S. Grant." In *General Laws, Memorials and Resolutions of the Territory of Wyoming Passed at the First Session [. . .], 1869*. Cheyenne, 1870.

"Memorial to the Honorable Commissioner of Indian Affairs." In *General Laws, Memorials and Resolutions of the Territory of Wyoming Passed at the Second Session [. . .], 1871*. Cheyenne, 1872.

"Minutes of Council Meeting," May 28–29, 1912. Shoshone and Arapaho Business Council Proceedings, box 6, file 215. RG 75, National Archives, Denver, CO.

Mullen, William E., ed. *Wyoming Compiled Statutes, Annotated*. Laramie, WY: Laramie Republican, 1910.

Parshall, A. J. *Eleventh Biennial Report of the State Engineer to the Governor of Wyoming, for 1911 and 1912*. Laramie, WY: Laramie Republican, 1913. https://babel.hathitrust.org/cgi/pt?id=mdp.39015011420505;view=1up;seq=5;size=150.

——. *Twelfth Biennial Report of the State Engineer to the Governor of Wyoming, for 1913 and 1914*. Laramie, WY: Laramie Republican, 1914. https://babel.hathitrust.org/cgi/pt?id=mdp.39015011420513;view=1up;seq=1.

Peterson, David A., William E. Jones, and Anthony G. Morton. *Reconnaissance Investigation of Water Quality, Bottom Sediment and Biota Associated with Irrigation Drainage in the Kendrick Reclamation Project Area, Wyoming, 1986–87*. USGS Water Resources Investigations Report 87–4255. Denver, CO: USGS, Department of Interior, 1987.

Post, Buckley, Schuh & Jernigan. *Kirby Creek Watershed Level 1 Study*. Cheyenne: Wyoming Water Development Commission, 2010.

Powell, J. W. *Report on the Lands of the Arid Regions of the United States, with a More Detailed Account of the Lands of Utah. With Maps*. 2nd edition. Washington: Government Printing Office, 1879.

Ramirez, Pedro, Jr., and Kimberly Dickerson. *Monitoring of Selenium Concentration in Biota from the Kendrick Reclamation Project, Natrona County, Wyoming, 1992–96*. US Fish and Wildlife Service Contaminant Report R6/714C/99. USFWS, 1999.

Rauchfuss, Russell E., comp. *Final Report of the Stream Preservation Feasibility Study Committee, October 1, 1974*. Cheyenne, WY: The Committee, 1974.

Skaggs, Jackie. "Creation of Grand Teton National Park." National Park Service History & Culture. Jackson, WY: Grand Teton National Park, 2000. https://www.nps.gov/grte/learn/historyculture/upload/The-Creation-of-Grand-Teton-National-Park_webiste-2000.pdf.

State Engineer's Office. Permit Files. Records of the State Engineer, Wyoming State Engineer's Office, Cheyenne.

Sunrise Engineering, Inc. *Kirby Creek Water Watershed Plan Level 1 Study*. Cheyenne: Wyoming Water Development Commission, 2005.

Teele, R. P. and Paul A. Ewing. *The Economic Limits of Cost of Water for Irrigation: The Wyoming Development Company, Wyoming*. Washington, DC: US Department of Agriculture, 1925.

"Temporary Filing No. 20, 1/173, Thomas Bell." Rejected Applications Files. Wyoming State Engineer's Office Archives, Cheyenne.

Tetra Tech. *Big Wind River Drainage Level II, Phase 1, Storage Feasibility Study*. Cheyenne: Wyoming Water Development Commission, 2016.

———. *Executive Summary, Little Wind River Drainage Level II, Phase 1, Storage Feasibility Study*. Cheyenne: Wyoming Water Development Commission, 2016. http://library.wrds.uwyo.edu/wwdcrept/Wind_River/Little_Wind_River_Drainage-Level_II_Phase_I_Storage-Executive_Summary-2017.pdf.

Trelease, Frank J. *Severance of Water Rights from Wyoming Lands*. Wyoming Legislative Resource Committee, Report 2. Cheyenne, 1960.

Trenholm, Virginia C., ed. *Wyoming Blue Book*. 3 vols. Cheyenne: Wyoming State Archives and Historical Department, 1974.

True, Jas. B. *Thirteenth Biennial Report of the State Engineer to the Governor of Wyoming, for 1915 and 1916*. Laramie, WY: Laramie Republican, 1916. https://babel.hathitrust.org/cgi/pt?id=mdp.39015011420521;view=1up;seq=9.

Tyrrell, Patrick. "Instream Flow Overview." PowerPoint presentation at the Select Water Committee/Wyoming Water Development Commission Workshop, January 11, 2006.

———. *State of Wyoming 2002 Annual Report of the State Engineer's Office*. State Board of Control, Cheyenne, WY, 2002. http://seo.wyo.gov/documents-data/annual-reports.

———. *State of Wyoming 2008 Annual Report of the State Engineer's Office*. State Board of Control, Cheyenne, WY, 2008. http://seo.wyo.gov/documents-data/annual-reports-strategic-plans.

Upper Colorado River Commission. *Final Report: Colorado River System Conservation Pilot Program in the Upper Colorado River Basin*. 2018 http://www.ucrcommission.com/RepDoc/SCPPDocuments/2018__SCPP_FUBRD.pdf.

US Bureau of Reclamation. "Agreement Among the United States of America, through the Department of the Interior, Bureau of Reclamation, the Central Arizona Water

Conservation District, the Metropolitan Water District of Southern California, Denver Water, and the Southern Nevada Water Authority, [. . .]." Agreement No. 14-XX-30-W0574. https://www.usbr.gov/newsroom/docs/2014–07–30-Executed-Pilot-SCP-Funding-Agreement.pdf.

———. *Colorado River Supply and Demand Study: Executive Summary*. US Department of Interior, Washington, DC. December 2012. http://www.usbr.gov/lc/region/programs/crbstudy/finalreport/Executive%20Summary/CRBS_Executive_Summary_FINAL.pdf.

———. *Executive Summary: Colorado River Basin Water Supply and Demand Study*. US Department of the Interior, Bureau of Reclamation, December 2012.

———. "Pick-Sloan Missouri Basin Program, Riverton Unit. Wyoming: Fremont County." 1980. Revised edition, Washington DC: US Government Printing Office, 1984.

———. *Plan of Study for Evaluation of Operation of Existing Reclamation Projects on the Platte River for the Potential to Affect Threatened or Endangered Species*. US Department of the Interior, North Platte River Projects Office, 1990.

———. *Platte River Recovery Implementation Program: Final Environmental Impact Statement* (PRRIP FEIS). US Department of the Interior, Washington DC, 2006.

———. *Report of Phase 1: Study of Mine-Mouth Thermal Power Plants with Extra-High-Voltage Transmission for Delivery of Power to Load Centers*. North Central Power Study Coordinating Committee. Billings, MT, 1971.

———. *Riverton Project History*. 1918, 1976, and 1977. Records on file at Midvale Irrigation District Office, Pavilion, WY.

US Congress. Senate. Committee on Labor and Public Welfare, Special Subcommittee on Indian Education. *Indian Education: A National Tragedy—A National Challenge*. 91st Cong., 2nd sess., Report No. 91-501. Washington, DC: GPO, 1969.

US Department of Agriculture. *Wyoming Agricultural Statistics 2016*. National Agricultural Statistics Service, Mountain Region, Wyoming Field Office. Cheyenne, WY, 2016. https://www.nass.usda.gov/Statistics_by_State/ Wyoming/Publications/Annual_Statistical_Bulletin/WY_2016_Bulletin.pdf.

US Department of Agriculture, Forest Service. *The Western Range: A Report on the Western Range—A Great but Neglected Natural Resource*. 74th Cong., 2d Sess., 1936. Sen. Doc. No. 199. Washington, DC: US Government Printing Office, 1936. Reprint, New York: Arno Press, 1979.

US Department of Commerce. *Irrigation and Drainage: General Report and Analytical Tables and Reports for States, with Statistics for Counties, Volume VII. Fourteenth Census of the United States*. Bureau of the Census. Washington, DC: Government Printing Office, 1922. http://agcensus.mannlib.cornell.edu/AgCensus/censusParts.do?year=1920

US General Land Office. "Report of Fraudulent Claim or Entry, Report of Special Agent Jas. A. George, May 20, 1888." In *Selected Records of the General Land Office and of the Office of the Secretary of the Interior, Appointments Division, Governors of Wyoming*. University of Wyoming Libraries, Laramie, 1955. Microfilm.

US Government Accountability Office. *Bureau of Reclamation: Availability of Information on Repayment of Water Project Construction Costs Could be Better Promoted.* Document GAO 14–764. Washington DC: Government Accountability Office, 2014.

US Reclamation Service. "Cost of Water Per Acre." Washington, DC: Department of Interior, 1913.

Van Orsdel, J. A. *Biennial Report of the Attorney General to the Governor of Wyoming 1901–02.* Laramie, WY: Laramie Republican, 1903.

———. *Biennial Report of the Attorney General to the Governor of Wyoming 1903–04.* Laramie, WY: Laramie Republican, 1905. https://babel.hathitrust.org/cgi/pt?id=umn.31951002548581o;view=1up;seq=5.

Wadsworth, H. E. "Wind River Indian Reservation Annual Report, Narrative," 1912. General Correspondence, box 1, file 107. Records of the Bureau of Indian Affairs, Wind River Agency. RG 75, National Archives, Denver, CO.

Whitehead, J. R. *The Compiled Laws of Wyoming: Including All the Laws in Force in Said Territory at the Close of the Fourth Session of the Legislative Assembly of Said Territory.* Cheyenne, WY: H. Glafcke, 1876.

Works Projects Administration. *Wyoming: A Guide to its History, Highways, and People.* 1941. Original text reprinted, Lincoln: University of Nebraska Press, 1981.

Wyoming. *Digest of Senate and House Journals of the 43rd Legislature of the State of Wyoming.* Cheyenne, WY: Legislative Service Office, 1973.

———. *Opinions of the Office of the Attorney General of the State of Wyoming, 1957–1960.* Cheyenne: Attorney General's Office.

Wyoming Constitutional Convention, 1889. *Journal and Debates of the Constitutional Convention of the State of Wyoming.* Cheyenne, WY: The Daily Sun, 1893.

Wyoming Department of Administration and Information. *Wyoming Data Handbook: Per Capita Personal Income in Wyoming Counties, 1978–1987.* Division of Research and Statistics. Cheyenne: State of Wyoming, 1989.

Wyoming Department of Environmental Quality. "Willwood Dam Advisory Committee and Working Groups, Terms of Reference, June 27, 2017." Cheyenne: WDEQ, 2017. http://deq.wyoming.gov/media/attachments/Water%20Quality/Watershed%20Protection/Willwood%20Dam%20and%20Shoshone%20River/2017–0627_Willwood-Dam_Advisory-Committee_ToR_FINAL.pdf.

———. Willwood Working Group Final Report. Cheyenne, WY: WDEQ, 2017. http://deq.wyoming.gov/media/attachments/Water%20Quality/Watershed%20Protection/Willwood%20Dam%20and%20Shoshone%20River/2017–1016_WorkGroup-1_Final-Report.pdf.

———. *Wyoming's Draft 2016/18 Integrated 305(b) and 303(d) Report.* Water Quality Division. Cheyenne: WDEQ, 2018.

Wyoming Rural Development Council. *Wheatland Community Assessment, Platte County, Wyoming.* Cheyenne, WY: Wyoming Rural Development Council, 2000. http://pluto.state.wy.us/awweb/pdfopener?md=1&did=11115293.

Wyoming Secretary of State. "Initiative and Referendum Summary." Election Division memo, July 2015. http://soswy.state.wy.us/Elections/Docs/IRSum.pdf.

Wyoming State Engineer. *26th Biennial Report 1941–42*. Cheyenne: Wyoming State Engineer's Office, 1942.

Wyoming State Engineer's Office. "Guidance: CBM/Ground Water Permits," Guidance document issued by Ground Water Division. Cheyenne: State Engineer's Office, 2004.

———. "Regulations and Instructions, Part IV, Board of Control." Cheyenne: State Engineer's Office, 2004.

———. *State of Wyoming 2010 Annual Report of the State Engineer*. Board of Control. Cheyenne: State Engineer's Office, 2010.

———. *State of Wyoming Water Year 2015 Annual Report of the State Engineer*. Board of Control. Cheyenne: State Engineer's Office, 2015.

———. *State of Wyoming Water Year 2018 Annual Report of the State Engineer*. Board of Control. Cheyenne: State Engineer's Office, 2018.

———. *Wyoming and the Colorado River: A Report*. Cheyenne: State Engineer's Office, 2016. https://docs.google.com/a/wyo.gov/viewer?a=v&pid=sites&srcid=d3lvLmdvdnxzZW98Z3g6NWUyNTRhNDhkYWUiYWMiYg.

Wyoming Water Development Commission (WWDC). *1985 Legislative Report: Wyoming Water Development Program*. Wyoming Water Development Office, Cheyenne.

———. *1986 Legislative Report: Wyoming Water Development Program*. Wyoming Water Development Office, Cheyenne.

———. *2000 Legislative Report: Wyoming Water Development Program*. Cheyenne: Wyoming Water Development Office, 2000.

———. *2004 Monitoring Report: High Savery Dam*. Water Resources Data System. University of Wyoming, Laramie, 2004. http://library.wrds.uwyo.edu/wwdcrept/High_Savery/High_Savery_Dam-Monitoring_Report-Final_Report-2004.html.

———. *2007 Legislative Report: Wyoming Water Development Program*. Cheyenne: Wyoming Water Development Office, 2007.

———. *2017 Legislative Report: Wyoming Water Development Program*. Wyoming Water Development Office, Cheyenne. http://wwdc.state.wy.us/legreport/2017/2017Rept.pdf.

———. *2019 Legislative Report*. Wyoming Water Development Program. Wyoming Water Development Office, Cheyenne. http://wwdc.state.wy.us/legreport/2019/2019Rept.pdf.

———. *Final Report for LeClair Irrigation District, Level I System Master Plan Study*. 2016. Wyoming Water Development Office, Cheyenne. http://library.wrds.uwyo.edu/wwdcrept/LeClair/LeClair_Irrigation_District-Level_I_System_Master_Plan-Final_Report-2016.pdf.

———. *Operating Criteria of the Wyoming Water Development Program*. Cheyenne, WY: Legislative Select Water Committee, 2018. http://wwdc.state.wy.us/opcrit/WWDPopCriteria.pdf.

————. "Platte River Basin Plan: Executive Summary, May 2006." Cheyenne: Wyoming Water Development Commission, 2006. http://waterplan.state.wy.us/plan/platte/2006/finalrept/Executive_Summary_lowres.html

————. *Platte River Basin Plan Final Report*. Cheyenne: Wyoming Water Development Commission, 2006. http://waterplan.state.wy.us/plan/platte/2006/report.html

————. "The Wyoming Framework Plan: A Summary." Cheyenne: Wyoming Water Development Commission, 2007. http://waterplan.state.wy.us/plan/statewide/execsummary.pdf

————. *Sandstone Dam: Project Summary*. Laramie, WY: Water Resources Data System, 1993. http://library.wrds.uwyo.edu/wwdcrept/Sandstone/Sandstone_Dam-Project_Summary-1993.pdf.

————. *State of Wyoming 2015 Irrigation System Survey Report, Report 3*. Wyoming Water Development Office, Cheyenne. http://wwdc.state.wy.us/irrsys/2015/usage.html.

————. "Technical Memorandum." Wyoming's Contract Storage Water in Fontenelle Reservoir. Wyoming Water Development Office, Cheyenne, 2011. http://waterplan.state.wy.us/plan/green/2010/finalrept/fontenelle.html

————. *Wyoming Framework Water Plan, Volume 1*. 2007 Wyoming Water Development Office, Cheyenne. http://waterplan.state.wy.us/plan/statewide/Volume_I.pdf.

Wyoming Wildlife and Natural Resource Trust. *Status Report 2016*. WWNRT Board and Staff, 2016.

COURT CASES AND FILES

Arizona v. California, 439 U.S. 419 (1979)

Basin Electric Power Cooperative v. State Board of Control, 578 P.2d 557 (1978)

Big Horn River Adjudication Case files (BHRAC). Computer Database. Wyoming State Engineer's Office, Cheyenne.

Cremer v. State Board of Control, 675 P.2d 250 (Wyo. 1984)

Farm Investment Co. v. Carpenter, 9 Wyo. 110, 61 P.258 (1900).

Frank v. Hicks, 35 P. 475 (Wyo. 1892)

General Adjudication of All Rights to Use Water in the Big Horn River System (Big Horn I), 753 P.2d 76 (Wyo. 1988); affirmed, Wyoming v. United States, 492 U.S. 406 (1989)

General Adjudication of All Rights to Use Water in the Big Horn River System (Big Horn III), 835 P. 2d 273 (Wyo. 1992)

Green River Development Co. v. FMC Corp., 660 P.2d. 339 (Wyo. 1983)

Holt v. City of Cheyenne, 22 Wyo. 212, 137 P. 867 (1914).

Horse Creek Conservation Dist. v. Lincoln Land Co., 54 Wyo. 320, 92 P. 2d 572 (1939)

Hughes v. Lincoln Land Co., 27 F. Supp.972 (Wyo. 1939)

In the Matter of Johnson Ranches, 605 P.2d 367 (Wyo. 1980)

Johnson County Civil Case #234, Crazy Woman Decree (1889). Zezas Petition Files. Records of the Board of Control. Wyoming State Engineer's Office, Cheyenne.

Johnston v. Little Horse Creek Irrigating Co., 79 P. 22 (Wyo. 1904)

Johnston v. Little Horse Creek civil case file (1895), Laramie County District Court, Docket # 6–233, box 2, Wyoming State Archives.

Kansas v. State of Colorado, 206 U.S. 46 (1907)

Lewis v. State Board of Control, 699 P.2d 822 (Wyo. 1985)

Lone Wolf v. Hitchcock, 187 U.S. 553 (1903)

Moyer v. Preston, 6 Wyo. 308 (1896)

Parshall v. Cowper, 143 P. 302 (1914)

Platte County Grazing Association v. Board of Control, 675 p.2d 1279 (Wyo. 1984)

Ramsay v. Gottsche, 51 Wyo. 516, 69 P.2d (1937)

Report of Special Master Roncalio: Concerning Reserved Water Right Claims by and on Behalf of the Tribes of the Wind River Indian Reservation. In re Rights to Use Water in the Big Horn River (Civil No. 4993, Wyo. 5th Dist., December 15, 1982).

Scott v. McTiernan, 974 P.2d 966 (Wyo. 1999)

Simmons v. Ramsbottom, 68 P.2d 153 (Wyo. 1937)

Snake River Land Co. v. State Board of Control, 560 P. 2d 733 (Wyo. 1977)

Sturgeon v. Brooks, 281 P.2d 675 (Wyo. 1955)

Swartz v. Beach, 229 F. Supp. 2d 1239 (D.Wyo. 2002)

The Making of Modern Law: U.S. Supreme Court Records and Briefs, 1832–1978. Online Database. Gale, A Cengage Company, accessed December 13, 2018. https://www. gale.com/c/making-of-modern-law-us-supreme-court-records-and-briefs-1832-1978.

United States v. Hampleman, Case No. 763 (D.Wyo. 1916), Civil Case Files, U.S. District Court for Wyoming, RG 21, Box 120, File 753, folder 2, National Archives, Denver, CO.

United States v. Parkins, 18 F.2d 642 (D.Wyo. 1926)

United States v. Rio Grande Dam and Irrigation Co., 174 U.S. 690 (1899)

Van Tassel Real Estate & Live Stock Co. v. City of Cheyenne, 54 P. 2d 906 (1936)

Ward v. Yoder, 355 P.2d (Wyo. 1960)

Wheatland Irrigation District v. Laramie Rivers Co., 659 P.2d 561 (Wyo. 1983)

Wheatland Irrigation District v. Pioneer Canal Co., 464 P.2d 533 (Wyo. 1970)

Winters v. United States, 207 U.S. 564 (1908)

Wyoming v. Colorado, 259 U.S. 419 (1922)

Zezas Ranch Inc. v. Board of Control, 714 P. 2d 759 (Wyo. 1986)

BOOKS AND ARTICLES

Chatterton, Fenimore. *Yesterday's Wyoming: The Intimate Memoirs of Fenimore C. Chatterton, Territorial Citizen, Governor, Builder.* Laramie, WY: Powder River Publishers, 1957.

Clay, John. *My Life on the Range.* 1924. Reprint of original, Norman: University of Oklahoma Press, 1962.

David, Robert B. *Malcolm Campbell, Sheriff: The Reminiscences of the Greatest Frontier Sheriff in the History of the Platte Valley.* Casper, WY: Wyomingana, 1932.

Downing, C.O., with Sharon R. Smith, ed. "Recollections of a Goshen County Home-steader." *Annals of Wyoming* 43, no. 1 (Spring 1971): 53–72.

Flannery, L. G., ed. *John Hunton's Diary, Vol. 3 1878–1879*. Lingle, WY: Guide-Review, 1960.

Gillette, Edward. *Locating the Iron Trail*. Boston: Christopher Publishing House, 1925.

Mead, Elwood. "Government Aid and Direction in Land Settlement," *American Economic Review*, 8. Supp. (March 1918): 72–98.

———. *Irrigation Institutions: A Discussion of the Economic and Legal Questions Created by Growth of Irrigated Agriculture in the West*. New York: Macmillan, 1903.

———. "The Cody Canal in Wyoming." *Irrigation Age* 9, no. 1 (1896): 12–14. https://www.biodiversitylibrary.org/item/60610#page/27/mode/1up.

———. "The Growth of Property Rights in Water." *The International Quarterly* 6 (September–December 1902): 1–12.

———. "The Ownership of Water: Address by Professor Mead before the Farmers at Fort Collins." Denver, CO: Times Print Works, 1887.

Teichert, John A. "Reflections of a Water Administrator." Unpublished manuscript, n.d. (circa 1994). Paper copy.

Secondary Sources

Adler, Robert W., J. C. Landman, and Diane M. Cameron. *The Clean Water Act Twenty Years Later*. Washington, DC: Island Press, 1993.

Anderson, Terry L., Brandon Scarborough, and Lawrence R. Watson. *Tapping Water Markets*. New York: Routledge, 2012.

Anderson, Terry L., and Peter J. Hill. "The Evolution of Property Rights." In *Property Rights: Cooperation, Conflict, and Law*, edited by Terry L. Anderson and Fred S. McChesney, 118–41. Princeton, NJ: Princeton University Press, 2003.

Annear, Tom. "Securing the Shoshone." *Wyoming Wildlife*, May 2017.

Annear, Tom, and Amy Bulger. "Progress at Pine Creek." *Wyoming Wildlife*, July 2017.

Aragon, Don. "The Wind River Indian Tribes." *International Journal of Wilderness* 13, no. 2 (August 2007): 14–17.

Baker, Mark E. "The Fontenelle Dam Incident: The Investigation, Likely Causes, and Lessons (Eventually) Learned." Paper present to Association of State Dam Safety Officials, Annual Conference, Washington, DC, September 2011. http://damfailures.org/wp-content/uploads/2015/07/094_The-Fontenelle-Dam-Incident.pdf.

Ballowe, James. *A Man of Salt and Trees: The Life of Joy Morton*. DeKalb: Northern Illinois University Press, 2009.

Barnes, John. "Dave Johnston Power Plant Water Rights." Master's thesis, University of Wyoming, Laramie, 1993.

Becker, Bidtah. "Water & Tribes Initiative." Conference presentation, June 7, 2019. https://www.getches-wilkinsoncenter.cu.law/wp-content/uploads/2019/06/Bidtah-Becker.-2019-GWC-Summer-Conference-.pdf.

Berger, Robert G. "Conservation Easements in Wyoming—An Overview." Rural Law Center, University of Wyoming College of Law, Laramie, 2011. https://www.uwyo.edu/law/centers/rural-law-center/conservation-easement-conference/berger%20 6–2-11%20uw%20ce%20presentation.pdf.

Bergersen, Eric P., and Mark F. Cook. "Impacts of Water Level Manipulations on Burbot and Lake Trout in Bull Lake, Wind River Indian Reservation, Wyoming." Colorado Cooperative Fish and Wildlife Research Unit, Colorado State University, Fort Collins, 1987.

Bergersen, Eric P., Mark F. Cook, and R. J. Baldes. "Winter Movements of Burbot (*Lota lota*) during an Extreme Drawdown in Bull Lake, Wyoming, U.S.A." *Ecology of Freshwater Fish* 2 (1993): 141–45.

Bergstedt, Lee C. "Fishery and Macroinvertebrate Response to Water Management Practices in the Wind River on the Wind River Indian Reservation, Wyoming." Master's thesis, Department of Fishery and Wildlife Biology, Colorado State University, Fort Collins, 1994.

Blevins, Spencer. "Valuing the Non-Agricultural Benefits of Flood Irrigation in the Upper Green River Basin." Master's thesis. University of Wyoming, 2015.

Blevins, Spencer, Kristi Hansen, Ginger Paige, and Anne MacKinnon. "The True Value of Flood Irrigation." *Western Confluence*, Summer 2016. http://www.westernconfluence.org/the-true-value-of-flood-irrigation/.

Blomberg, Kelli. "Tribal Fishery Restoration on the Wind River Indian Reservation: Forging a Co-Management Agreement." Master's thesis. Haub School of Environment and Natural Resources, University of Wyoming, Laramie, 2016. http://repository.uwyo.edu/enr_plan_b.

Bonner, Robert E. "Elwood Mead, Buffalo Bill Cody, and the Carey Act in Wyoming." *Montana: The Magazine of Western History* 55, no. 1 (Spring 2005): 36–51.

———. *William F. Cody's Wyoming Empire: The Buffalo Bill Nobody Knows*. Norman: University of Oklahoma Press, 2007.

Bulger, Amy. "Polishing a Gem." *Wyoming Wildlife*, April 2017.

Burkhardt, Nina. "An Implementation Case Study: Wyoming's Instream Flow Law." Unpublished paper. Department of Political Science, Colorado State University, Fort Collins, December 1992.

Carnes, F. T. "The Wyoming Development Company." Master's thesis, University of Wyoming, Laramie, 1953.

Carstensen, Vernon, ed. *The Public Lands: Studies in the History of the Public Domain*. Madison: University of Wisconsin Press, 1968.

Champagne, Duane. "Organizational Change and Conflict: A Case Study of the Bureau of Indian Affairs." *American Indian Culture and Research Journal* 7, no. 3 (1983): 3–28.

Cheesbrough, Kyle, Jake Edmunds, Glenn Tootle, Greg Kerr, and Larry Pochop. "Estimated Wind River Range (Wyoming, U.S.A) Glacier Melt Water Contributions to Agriculture." *Remote Sensing* 4, no. 1 (2009): 818–28.

Churchill, Beryl. *Dams, Ditches and Water: A History of the Shoshone Reclamation Project.* Cody, WY: Rustler Printing, 1979.

———. *People Working Together: A 75th Anniversary Salute to Powell, Wyoming.* Powell, Wyoming 75th Anniversary Commission, 1984.

Cohn, Teresa C., William Wyckoff, Matt Rinella, and Jan Eitel. "Seems Like I Hardly See Them Around Anymore: Historical Geographies of Riparian Change along the Wind River." *Water History* 8, no. 4 (December 2016): 405–29.

Colorado River Research Group. "Charting a New Course for the Colorado River: A Summary of Guiding Principles." Boulder: Colorado River Research Group, December 2018. https://www.coloradoriverresearchgroup.org/uploads/4/2/3/6/42362959/crrg_guiding_principles.pdf.

———. *The First Step in Repairing the Colorado River's Broken Water Budget: Summary Report.* Boulder: Colorado River Research Group, December 2014. http://www.coloradoriverresearchgroup.org/uploads/4/2/3/6/42362959/crrg_summary_report_1_updated.pdf.

———. "When Is a Drought Not a Drought? Drought, Aridification, and the 'New Normal.'" Boulder: Colorado River Research Group, March 2018. https://www.coloradoriverresearchgroup.org/uploads/4/2/3/6/42362959/crrg_aridity_report.pdf.

Conkin, Paul K. "The Vision of Elwood Mead." *Agricultural History* 34, no. 2 (April 1960): 88–97.

Cook, Jeannie. *Wiley's Dream of Empire: The Wiley Irrigation Project.* Cody, WY: Jeannie Cook, 1990.

Cooper, Craig. *History of Water Law, Water Rights and Development in Wyoming.* Riverton, WY: Cooper Consulting, 2004. http://wwdc.state.wy.us/history/Wyoming_Water_Law_History.pdf.

Copeland, Matt. "Birth of a Fishing Town." *Wyoming Wildlife*, July 2014.

Cosens, Barbara, and Judith V. Royster, eds. *The Future of Indian and Federal Reserved Water Rights: The Winters Centennial.* Albuquerque: University of New Mexico Press, 2012.

Currier, Paul J., Gary Lingle, and John VanDerwalker. *Migratory Bird Habitat on the Platte and North Platte Rivers in Nebraska.* Grand Island, NE: Platte River Whooping Crane Critical Habitat Maintenance Trust, 1985.

Davis, John W. *A Vast Amount of Trouble: A History of The Spring Creek Raid.* Norman: University of Oklahoma Press, 1993.

———. *Goodbye, Judge Lynch: The End of a Lawless Era in Wyoming's Big Horn Basin.* Norman: University of Oklahoma Press, 2005.

———. *The Trial of Tom Horn.* Norman: University of Oklahoma Press, 2016.

———. *Wyoming Range War: The Infamous Invasion of Johnson County.* Norman: University of Oklahoma Press, 2010.

Deloria, Vine, Jr., and Clifford M. Lytle. *American Indians, American Justice.* Austin: University of Texas Press, 1983.

Demsetz, Harold. "Toward a Theory of Property Rights." *American Economic Review* 57 (1967): 347–59.

DeVisser, Mark, and Andrew G. Fountain. "A Century of Glacier Change in the Wind River Range, WY," *Geomorphology* 232 (2015):103–16.

Dew, Jay R. "Frustrated Fortunes: Francis E. Warren and the Search for a Grazing Policy, 1890–1929." PhD diss., University of Oklahoma, 2007.

Dietz, Thomas, Nives Dolšak, Elinor Ostrom, and Paul C. Stern. "The Drama of the Commons." In *The Drama of the Commons*, edited by Elinor Ostrom, Thomas Dietz, Nives Dolšak, Paul C. Stern, Susan Stonich, and Elke U. Weber, 3–35. Washington, DC: National Academy Press, 2002.

DiGiulio, Dominic. "Reconciling Oil and Gas Development and Groundwater Protection: Lessons from Pavillion, WY." Presentation at the 2018 Getches-Wilkinson Center Summer Conference, University of Colorado, Boulder. https://www.getches-wilkinsoncenter.cu.law/wp-content/uploads/2018/06/Dominic-DiGiulio.Pavillion-WY.pdf.

DiRienzo, Sara. "New Path Keeps Fish Healthy, Anglers Happy on Nowood River." *Wyoming Wildlife*, May 2018.

Donahue, James Q. "Wyoming's Water: Yesterday, Today, and Tomorrow." Unpublished manuscript. Funded in Part by the Wyoming Council for the Humanities, Cheyenne, WY, January 17, 1985. Loose-leaf mimeograph.

Dunbar, Robert G. *Forging New Rights in Western Waters*. Lincoln: University of Nebraska Press, 1983.

Farrar, Jon. "Platte River Instream Flows—Who Needs It?" *NEBRASKAland Magazine*, December 1992.

Fassett, Gordon W. "Endangered Species Management in the Central Platte River Basin." In *Information Series No. 81, Endangered Species Management: Planning Our Future, Proceedings of the 1995 South Platte Forum, Oct. 25–26, 1995.* Colorado Water Resources Research Institute, Colorado State University, Fort Collins, 1995.

Faulkner, Harold E. *American Economic History*. 4th ed. New York: Harper, 1938.

Feck, Jules, and Nathan Nibbelink. "Watershed Analysis of Steam Flow Diversion Influences on Aquatic Ecosystem Health: Clear Creek and Popo Agie Watersheds." Trout Unlimited and The Nature Conservancy, January 2003.

Federal Writer's Project. *Wyoming: A Guide to Its History, Highways and People.* 1941. Reprinted with an introduction by T. A. Larson. Lincoln: University of Nebraska Press, 1981.

Flores, Christopher, ed. *Indian Irrigation Projects: Maintenance and Sustainability Issues.* New York: Nova Science, 2015.

Folke, Carl, Fikret Berkes, and Johan Colding, "Ecological Practices and Social Mechanisms for Building Resilience and Sustainability." In *Linking Social and Ecological Systems: Management Practices and Social Mechanisms for Building Resilience*, edited by Fikret Berkes and Carl Folke, with the editorial assistance of Johan Colding, 414–36. Cambridge, UK: Cambridge University Press, 1998.

Franks, Martha C. "The Use of the Practicably Irrigable Acreage Standard in the Quantification of Reserved Water Rights." *Natural Resources Law Journal* 31, no. 3 (Summer 1991): 549–85.

Frink, Maurice. *Cow Country Cavalcade: Eighty Years of the Wyoming Stock Growers Association.* Denver, CO: Old West, 1954.

Galloway, Gerald E. "If Stationarity Is Dead, What Do We Do Now?" *Journal of the American Water Resources Association* 47, no. 3 (June 2011): 563–70.

Gannon, Robert. "A Sick River is Returned to Nature." *True*, April 1966.

Ganoe, John T. "The Desert Land Act in Operation, 1877–1891." *Agricultural History* 11, no. 2 (1937): 142–57.

Gates, Paul W. "The Homestead Act in an Incongruous Land System." In *The Public Lands: Studies in the History of the Public Domain*, edited by Vernon Carstensen, 315–48. Madison: University of Wisconsin Press, 1968.

Gatzweiler, Franz, Konrad Hagedorn, and Thomas Siko. "People, Institutions and Agro-ecosystems in Transition." Conference paper, Ninth Biennial Conference of the International Association for the Study of Common Property, June 2002. http://dlc.dlib.indiana.edu/dlc/handle/10535/232.

Glennon, Robert Jerome. *Unquenchable: America's Water Crisis and What to Do about It.* Washington: Island Press, 2009.

Gordon, Bea. "Return Flow in Northeastern Wyoming." *The Keg: WyCEHG Newsletter* 4, no. 1 (December 2015): 11. http://www.uwyo.edu/epscor/wycehg/docs/newsletters/wycehg-newsletter-v4-n1-1215.pdf.

Goshen County History Committee. *Wind Pudding and Rabbit Tracks: A History of Goshen County, Wyoming*, vol. 1. Torrington, WY: Goshen County History Book Committee, 1989.

Gould, George A. "Water Rights Transfers and Third Party Effects." *Land and Water Law Review* 23, no. 1 (1988): 1–41.

Gould, Lewis L. *Wyoming: A Political History, 1869–1896.* New Haven, CT: Yale University Press, 1968.

Gray, Steve. "Water, Drought and Wyoming's Climate: Final Report." Workshop proceedings summary, American Heritage Center, University of Wyoming, Laramie, 2006. http://www.uwyo.edu/haub/_files/_docs/ruckelshaus/pubs/2006-water-drought-wyoming-climate-final.pdf.

Gressley, Gene. "The American Cattle Trust: A Study in Protest." *Pacific Historical Review* 30, no. 1 (February 1961): 61–77.

Gunderson, Lance H., C. S. Holling, and Steven S. Light, eds. *Barriers and Bridges to the Renewal of Ecosystems and Institutions.* New York: Columbia University Press, 1995.

Hall, Dorothy K., James L. Foster, Nicolo E. DiGirolamo, and George A. Riggs. "Snow Cover, Snowmelt Timing and Stream Power in the Wind River Range, Wyoming." *Geomorphology* 137, no. 1 (January 2012): 87–93.

Hanna, Jonathan M. "Native Communities and Climate Change: Protecting Tribal Resources as Part of National Climate Policy, Executive Summary." Natural Resource Law Center, University of Colorado School of Law, 2007.

Hansen, Anne C. "The Congressional Career of Sen. Francis E. Warren from 1890–1902." *Annals of Wyoming* 20, no. 1 (January 1948): 3–49.

Hardin, Garrett. "The Tragedy of the Commons." *Science* 162, no. 3859 (December 1968): 1243–48.

Harris, Tom. *Death in the Marsh*. Washington, DC: Island Press, 1991

Hays, Samuel P. *Conservation and the Gospel of Efficiency: The Progressive Conservation Movement 1890–1920*. Cambridge, MA: Harvard University Press, 1959.

Heritage Book Committee. *Pages from Converse County's Past*. Douglas, WY: Heritage Book Committee-Wyoming Pioneer Association, 1986. https://www.conversecounty.org/DocumentCenter/View/325/Converse-Countys-Past-PDF?bidId=.

Hipp, Martha Louise. *Sovereign Schools: How Shoshones and Arapahos Created a High School on the Wind River Reservation*. Lincoln, NE: University of Nebraska Press, 2019.

Holling, C. S., Lance H. Gunderson, and Donal Ludwig. "In Quest of a Theory of Adaptive Change." In *Panarchy: Understanding Transformations in Human and Natural Systems*, edited by Lance H. Gunderson and C. S. Holling, 3–24. Washington, DC: Island Press, 2002.

Holling, C. S., Lance H. Gunderson, and G. D. Peterson. "Sustainability and Panarchies." In *Panarchy: Understanding Transformations in Human and Natural Systems*, edited by Lance H. Gunderson and C. S. Holling, 63–102. Washington, DC: Island Press, 2002.

Hoopengarner, Molly. "To Make the Desert Bloom: How Irrigation Came to the Ceded Portion of the Wind River Indian Reservation." Unpublished manuscript, 1991.

Horwitz, Morton J. *The Transformation of American Law, 1780–1860*. Cambridge, MA: Harvard University Press, 1977.

Howe, Charles, and Christopher Goemans. "Water Transfers and Their Impacts: Lessons from Three Colorado Water Markets." *Journal of the American Water Resources Association* 39, no. 5 (October 2003): 1055–65.

Hoxie, Frederick E. *A Final Promise: The Campaign to Assimilate the Indians, 1880–1920*. Lincoln: University of Nebraska Press, 1984.

Jackson, W. Turrentine. "The Administration of Thomas Moonlight." *Annals of Wyoming* 18, no. 2 (July 1946): 139–62.

———. "The Wyoming Stock Growers' Association: Its Years of Temporary Decline, 1886–1890." *Agricultural History* 22, no. 4 (October 1948): 260–70.

———. "Wyoming Stock Growers' Association: Political Power in Wyoming Territory, 1873–1890." *Mississippi Valley Historical Review* 33, no. 5 (March 1947): 571–94.

Jacobs, James J., and Donald Brosz. *Wyoming's Water Resources*. Pub. B-969R. Laramie, WY: University of Wyoming Cooperative Extension Service, 2000. http://www.wyoextension.org/agpubs/pubs/B-969R.pdf.

Jacobs James J., and Dave T. Taylor. "Wyoming's Water Development Policy: What Are the Costs to the State?" *Western Journal of Agricultural Economics* 14, no. 2 (December 1989): 261–67.

Jacobs, James J., Patrick T. Tyrell, and Donald Brosz. *Wyoming Water Law: A Summary*. Pub. B-849R. Laramie, WY: University of Wyoming Agricultural Experiment Station, 2003. http://www.wyoextension.org/agpubs/pubs/B849R.pdf.

Johnson, Brenton. "Reclamation and Water Conservation." *Irrigator's News Line* 5, no. 2 (2000): 1–2.

Johnson, Elizabeth Wilkinson, ed. *Trails, Rails and Travails*. LaGrange, WY: City of LaGrange, 1988.

Kendy, Eloise. "Impacts of Changing Land Use and Irrigation Practices on Western Wetlands." *National Wetlands Newsletter* 28, no. 3 (May/June 2006): 12–32.

Kinney, Clesson S. *A Treatise on the Law of Irrigation, Section 493*. Washington, DC: Lowdermilk, 1894.

———. *Treatise on Irrigation and Water Rights: and the Arid Doctrine of Appropriation of Waters*. 2nd ed. San Francisco: Bender-Moss, 1912. https://archive.org/details/cu31924019999576/page/n7.

Kluger, James R. *Turning on Water with a Shovel: The Career of Elwood Mead*. Albuquerque: University of New Mexico Press, 1992.

Knight, Dennis, George P. Jones, William Reiners, and William Romme. *Mountain and Plains: The Ecology of Wyoming Landscapes*. 2nd ed. New Haven, CT: Yale University Press, 2014.

Koelling, Robert. *First National Bank of Powell: The History of a Bank, a Community and a Family*. Powell, WY: First National Bank of Powell, 1997.

Kruse, Babs, comp. "The Wind River Reservation, 1865–1910: Historical Photographs and Anecdotes." Adapted and abridged. Ethete: Wyoming Indian High School, 1984.

Larson, T. A. *History of Wyoming*. Lincoln: University of Nebraska Press, 1978.

Lasky, Moses. "From Prior Appropriation to Economic Distribution of Water by the State—Via Irrigation Administration." Pts. 1, 2, and 3. *Rocky Mountain Law Review* 1 (April 1929): 161–216; (June 1929): 270; 2 (November 1929): 35–58.

Lilley, William III, and Lewis L. Gould. "The Western Irrigation Movement, 1878–1902: A Reappraisal." In *The American West: A Reorientation*, edited by Gene Gressley, 57–76. Laramie: University of Wyoming Press, 1966.

Lindsay, Charles. "The Big Horn Basin." PhD diss., University of Nebraska, 1930. ETD collection for University of Nebraska, Lincoln. AAIDP14111.

"Lonetree Ranch." *Wyoming Wildlife*, December 2017.

MacDonnell, Lawrence. "The Development of Wyoming Water Law." *Wyoming Law Review* 14, no. 2 (2014) 327–78. http://dx.doi.org/10.2139/ssrn.2691080.

———. *Treatise on Wyoming Water Law*. Westminster, CO: Rocky Mountain Mineral Law Foundation, 2014.

Mackey, Mike. *Protecting Wyoming's Share: Frank Emerson and the Colorado River*. Sheridan, WY: Western History Publications, 2013.

MacKinnon, Anne. "The Prospects for Management of a Fragmented Aquifer by a Divided Farm Community." Paper presented at Workshop on the Workshop 4, Indiana University, Bloomington, June 2009. http://hdl.handle.net/10535/721.

MacKinnon, Anne, and Kate Fox. "Demanding Beneficial Use: Opportunities and Obligations for Wyoming Regulators in Coalbed Methane." *Wyoming Law Review* 6, no. 2 (2006): 369–99. https://repository.uwyo.edu/wlr/vol6/iss2/7.

McCool, Daniel. *Command of the Waters: Iron Triangles, Federal Water Development, and Indian Water.* Tucson: University of Arizona Press, 1994.

McIntire, Michael V. "The Disparity between State Water Rights Records and Actual Water Use Patterns." *Land and Water Law Review* 5 (1970).

McNeeley, Shannon. "Sustainable Climate Change Adaptation in Indian Country." American Meteorological Society 9 (July 2017). https://journals.ametsoc.org/doi/full/10.1175/WCAS-D-16-0121.1.

Meinzen-Dick, Ruth, and Leticia Nkonya. "Understanding Legal Pluralism in Water and Land Rights: Lessons from Africa and Asia." In *Community-Based Water Law and Water Resource Management Reform in Developing Countries*, edited by Barbara Van Koppen, Mark Giordano and John Butterworth, 12–27. Oxfordshire, UK: CABI, 2007.

Mercer, A. S. *The Banditti of the Plains: Or the Cattlemen's Invasion of Wyoming in 1892.* Reprint, with a forward by William H. Kittrell. Norman: University of Oklahoma Press, 1954.

Merrill, Thomas W., and Henry E. Smith. "Optimal Standardization in the Law of Property: The Numerus Clausus Principle." *Yale Law Journal* 110, no. 1 (October 2000): 1–200.

Miller, Tim R. *State Government: Politics in Wyoming*, 2nd, ed. Dubuque, IA: Kendall/Hunt, 1985.

Milliman, J. W. "Water Law and Private Decision Making: A Critique." *Journal of Law and Economics* 2 (October 1959): 41–63.

Milly, P. C. D., Julio Betancourt, Malin Falkenmark, Robert M. Hirsch, Zbigniew W. Kundzewicz, Dennis P. Lettenmaier, and Ronald J. Stouffer. "Stationarity is Dead: Whither Water Management?" *Science* 319, no. 5863 (February 2008): 573–74.

Milly, P. C. D., Julio Betancourt, Malin Falkenmark, Robert M. Hirsch, Zbigniew W. Kundzewicz, Dennis P. Lettenmaier, Ronald J. Stouffer, Michael D. Dettinger, and Valentina Krysanova. "On Critiques of 'Stationarity is Dead: Whither Water Management?'" *Water Resources Research* 51, no. 9 (2015): 7785–89.

Mitchell, Finis. *Wind River Trails: A Hiking and Fishing Guide to the Many Trails and Lakes of the Wind River Range in Wyoming.* Salt Lake City, UT: Wasatch, 1975.

Mixer, William. "Brief History of the North Platte River." Unpublished manuscript, 1999. Unbound print in author's files.

Nania, Julie; Bob Snow, Jennifer Pitt, Herb Becker, and Bob Johnson. "Session III, Part 2: Looking Forward." Video recording of Arizona v. California at 50: The Legacy and Future of Governance, Reserved Rights, and Water Transfers, Martz Summer Conference, University of Colorado Law School, Boulder, CO, August 2013. https://scholar.law.colorado.edu/arizona-v-california-at-50/7.

North, Douglass C. *Institutions, Institutional Change and Economic Performance.* Cambridge, UK: Cambridge University Press, 1990.

O'Gara, Geoffrey. *What You See in Clear Water: Life on the Wind River Reservation.* New York: Knopf, 2000.

Olson, Ted. *Ranch on the Laramie*. Boston: Little, Brown, 1973.

Ostrom, Elinor. *Governing the Commons*. Cambridge, UK: Cambridge University Press, 1990.

———. *Understanding Institutional Diversity*. Princeton, NJ: Princeton University Press, 2005.

Overton, Richard C. *Burlington West: A Colonization History*. Cambridge, MA: Harvard University Press, 1941.

Peck, Dannelle E., Doanold M. McLeod, John P. Hewlett, and James R. Lovvern. "Irrigation-Dependent Wetlands versus Instream Flow Enhancement: Economics of Water Transfers from Agriculture to Wildlife Uses." *Environmental Management* 34 (December 2004): 842–55.

Peters, B. R. "Joseph M. Carey and the Progressive Movement in Wyoming." PhD diss., University of Wyoming, 1971.

Pexton, John, and Catherine Pexton. "Carey-Bixby Ranch." Unpublished manuscript. Pexton personal collection, last modified May 5, 1995.

Pinchot, Gifford. "The Long Struggle for Effective Federal Water Power Regulation." *George Washington Law Review* 14 (1945–46): 9.

Pisani, Donald J. *Water, Land and Law in the West: The Limits of Public Policy, 1850–1920*. Lawrence: University of Kansas Press, 1996.

Pomeroy, John Norton. *Treatise on the Law of Water Rights as the Same is Formulated and Applied in the Pacific States, Including the Doctrine of Appropriation and the Statutes and Decisions Relating to Irrigation*. St. Paul, MN: West, 1893. https://archive.org/details/atreatiseonlawwooblacgoog/page/n6.

Postel, Sandra. *Replenish: The Virtuous Cycle of Water and Prosperity*. Washington, DC: Island Press, 2017.

Progressive Men of the State of Wyoming. Chicago: A. W. Bowen, 1903.

Read, Molly. *Johnny Gordon Had a Dream*. Wheatland, WY: Readworks, 2006.

Reisner, Marc. *Cadillac Desert: The American West and its Disappearing Water*. New York: Viking, 1986.

Reynolds, Matthew. "Wyoming's New Instream Flow Act: An Administrative Quagmire." *Land and Water Law Review* 21, no. 2 (1986): 455–86.

Riley, Gladys F., comp. "A Memorial to the Members of the Constitutional Convention." *Annals of Wyoming* 12, no. 3 (July 1940): 165–88.

Robison, Jason A. "Wyoming's Big Horn General Stream Adjudication." *Wyoming Law Review* 15, no. 2 (2015) 243–312. https://repository.uwyo.edu/wlr/vol15/iss2/1.

Robinson, Robert S., and Lawrence J. MacDonnell. *The Water Transfer Process as a Management Option for Meeting Changing Water Demands, Volume II*. Natural Resources Law Center. Boulder: University of Colorado School of Law, 1990.

Rose, Carol M. *Property and Persuasion: Essays on the History, Theory, and Rhetoric of Ownership*. Boulder, CO: Westview, 1994.

Rosgen, Dave. *Applied River Morphology*. 2nd ed. Fort Collins, CO: Wildland Hydrology, 1996.

Ruckleshaus Institute of Environment and Natural Resources. *Water Production from Coalbed Methane Development in Wyoming: A Summary of Quantity, Quality and Management Options.* University of Wyoming, Laramie, 2005.

———. "Wyoming's State of the Space." In *Wyoming Open Spaces*, May 2009. http://www.uwyo.edu/haub/_files/_docs/ruckelshaus/open-spaces/2009-state-of-the-space.pdf.

Rusinek, Walter. "A Preview of Coming Attractions: Wyoming v. United States and the Reserved Rights Doctrine." *Ecology Law Quarterly* 17, no. 2 (March 1990): 355–412.

Schlager, Edella, and William Blomquist. *Embracing Watershed Politics.* Boulder: University Press of Colorado, 2008.

Schlager, Edella, and Elinor Ostrom. "Property-Rights Regimes and Natural Resources: A Conceptual Analysis." *Land Economics* 68, no. 3 (August 1992): 249–62.

Shay, Monique C. "Promises to a Viable Homeland, Reality of Selective Reclamation: A Study of the Relationship between the Winters Doctrine and Federal Water Development in the United States." *Ecology Law Quarterly* 19, no. 3 (1992): 547–90.

Shurts, John. *Indian Reserved Water Rights: The* Winters *Doctrine in Its Social and Legal Context, 1880s–1930s.* Norman: University of Oklahoma Press, 2000.

Singleton, Sara. *Constructing Cooperation: The Evolution of Institutions of Co-Management.* Ann Arbor: University of Michigan Press, 1998.

Smith, Helena Huntington. *The War on Powder River: The History of an Insurrection.* Lincoln: University of Nebraska Press, 1966.

Smith, Henry E. "Community and Custom in Property." *Theoretical Inquiries in Law* 10, no. 1 (January 2009): 5–41. http://nrs.harvard.edu/urn-3:HUL.InstRepos:4555851.

———. "Governing Water: The Semicommons of Fluid Property Rights," *Arizona Law Review* 50, no. 2 (2008): 445–78.

———. "The Language of Property: Form, Content, and Audience," Stanford Law Review 55, no. 4 (April 2003): 1105–192.

Spaulding, George W. *A Treatise on the Public Land System of the United States: With References to the Land Laws, Rulings of the Departments at Washington, and Decisions of Courts, and an Appendix of Forms in United States Land and Mining Matters.* San Francisco: A. L. Bancroft, 1884.

Spring, Agnes Wright. "Carey Story Is a Wyoming Saga." *Hereford Journal* (July 15, 1938): 10–31.

Squillace, Mark. "A Critical Look at Wyoming Water Law." *Land and Water Law Review* 24, no. 2 (1989): 307–46.

———. "Water Marketing and the Law." Paper presented at Moving the West's Water to New Uses: Winners and Losers, Natural Resources Law Center, University of Colorado School of Law, Boulder, 1990.

Stauffer, Paul. "Did I Fish Too Much?" Vol. 1. Unpublished manuscript. Printed document in author's files.

Tarlock, A. Dan. "How Well Can Water Law Adapt to the Potential Stresses of Global Climate Change." *University of Denver Water Law Review* 14 (Fall 2010): 1–46.

——. "The Endangered Species Act and Western Water Rights." *Land and Water Law Review* 20, no. 1 (1985): 1–30.

——. "The Future of Prior Appropriation in the New West." *Natural Resources Law Journal* 41, no. 4 (Fall 2001): 769–94.

Tarlock, A. Dan, and Jason Anthony Robison. *Law of Water Rights and Resources.* Thomson Reuters, 2018.

Theesfeld, Insa, and Anne MacKinnon. "Giving Birds a Starting Date: The Curious Social Solution to a Water Resource Problem in the U.S. West." *Ecological Economics* 97 (January 2014): 110–19.

Trelease, Frank J. *Cases and Materials on Water Law.* 3rd ed. St. Paul, MN: West, 1979.

Trelease, Frank J., and Dellas W. Lee. "Priority and Progress—Case Studies of the Transfer of Water Rights." *Land and Water Law Review* 1, no. 1 (1966).

True, Jere, and Victoria Tucker Kirby. *Allen Tupper True: An American Artist.* Bozeman, MT: Museum of the Rockies, 2009.

Tyler, Daniel. *Silver Fox of the Rockies: Delphus E. Carpenter and Western Water Compacts.* Norman: University of Oklahoma Press, 2003.

Udall, Bradley, and Jonathan Overpeck. "The 21st Century Hot Drought on the Colorado River and its Implications for the Future." *Water Resources Research* 53 (2017): 2404–18, doi:10.1002/2016WR019638.

Udall, Bradley, Pat Mulroy, Barton "Buzz" Thompson, Tanya Trujillo, and Eric Kuhn. "Session III, Part 1: Looking Forward." Video recording of Arizona v. California at 50: The Legacy and Future of Governance, Reserved Rights, and Water Transfers, Martz Summer Conference, University of Colorado Law School, Boulder, CO, August 2013. https://scholar.law.colorado.edu/arizona-v-california-at-50/6.

Valdez, Richard A., and R. T. Muth. "Ecology and Conservation of Native Fish in the Upper Colorado River Basin." In *American Fisheries Society Symposium* 45, edited by J. N. Rinne, R. M. Huges, and B. Calamusso, 157–204. Bethesda, MD, 2005.

VanLooy, Jeffrey A., Richard R. Forster, David Barta, and James Turrin. "Spatially Variable Surface Elevation Changes and Estimated Melt Water Contribution of Continental Glacier in the Wind River Range, Wyoming, U.S.A: 1966–2011." *Geocarto International* 28, no. 2 (2013): 98–113.

Wahl, Richard W. *Markets for Federal Water: Subsidies, Property Rights, and the Bureau of Reclamation.* Washington, DC: Resources for the Future, 1989.

Water Education Foundation. Colorado River Project. *River Report*, Winter 2017–18. http://www.watereducation.org/sites/main/files/file-attachments/19677_wef_12pg_website.pdf.

West, Trevor. *Horace Plunkett, Cooperation and Politics: An Irish Biography.* Washington, DC: Catholic University Press of America, 1986.

West Group. *West's Encyclopedia of American Law.* 12 vols. Minneapolis, MN: West, 1998.

Western, Samuel. "Evolving Wyoming Tourism," in *Western Confluence* 5 (Winter 2016): 4–8.

Wiel, Samuel C. *Water Rights in the Western States; The Law of Prior Appropriation of Water* [. . .]. 3rd ed. San Francisco: Bancroft-Whitney, 1911.

Wilkinson, Charles F. *Crossing the Next Meridian: Land, Water, and the Future of the West.* Washington, DC: Island Press, 1992.

Wilson, Paul B. "Farming and Ranching on the Wind River Indian Reservation, Wyoming." PhD diss., University of Nebraska, 1972.

Wilson, Randall K. *America's Public Lands: From Yellowstone to Smokey Bear and Beyond.* Lanham, MD: Rowman and Littlefield, 2014.

Wohl, Ellen E. *Virtual Rivers: Lessons from the Mountain Rivers of the Colorado Front Range.* New Haven, CT: Yale University Press, 2001.

Wollman, Nathaniel, Ralph L. Edgel, Marshall E. Farris, H. Ralph Stucky, and Alvin J. Thompson. *The Value of Water in Alternative Uses: With Special Application to Water Use in the San Juan and Rio Grande Basins of New Mexico.* Albuquerque: University of New Mexico Press, 1962.

Woods, Lawrence M. *Horace Plunkett in America: An Irish Aristocrat on the Wyoming Range.* Norman, OK: Arthur H. Clarke, 2010.

———. *John Clay, Jr., Commissionman, Banker and Rancher.* Spokane, WA: Arthur H. Clarke, 2001.

———. *Moreton Frewen's Western Adventures.* Boulder, CO: Roberts Rinehart for the American Heritage Center, University of Wyoming, 1986.

———. *Sometimes the Banks Froze: Wyoming's Economy and its Banks.* Boulder: Colorado Associated University Press, 1985.

———. *Wyoming's Big Horn Basin to 1901: A Late Frontier.* Spokane, WA: Arthur H. Clark, 1997.

Wyoming Association of Conservation Districts. *Wyoming Watersheds Progress 2009.* Cheyenne: WACD, 2009.

Xiao, Mu, Bradley Udall, and Dennis P. Lettenmaier. "On the Causes of Declining Colorado River Streamflows." *Water Resources Research* 54 (2018): 6739–56. doi:10.1029/2018WR023153.

Yale University. *Obituary Record of Graduates of Yale University Deceased During the Year 1993–1936.* Bulletin of Yale University, October 1936. http://mssa.library. yale.edu/ obituary_record/1925_1952/1935–36.pdf.

Zallen, Margot. "Integrating New Values with Old Uses in the Relicensing of Kingsley Dam and Related Facilities: Making Part of the Problem a Part of the Solution." Presentation at Dams: Water and Power in the New West conference, University of Colorado Natural Resources Law Center, Boulder, CO, June 2–4, 1997.

INDEX

Page numbers in italic text indicate maps.